DICTIONARY
OF PLANT NAMES

DICTIONARY
OF PLANT NAMES

ALLEN J COOMBES

HAMLYN

Cover Photograph:
Reed International Picture Library

Line Drawings:
Allium christophii (title page)
Sternbergia lutea (page 5)
all by Charles Stitt

First published in 1985 in Great Britain by Newnes Books
under the title, *The Collingridge Dictionary of Plant Names*

This edition published in 1994 by Hamlyn Books,
an imprint of Reed Consumer Books Limited,
Michelin House, 81 Fulham Road, London SW3 6RB
and Auckland, Melbourne, Singapore and Toronto

Copyright © Reed International Books Limited 1985, 1994

ISBN 0 600 58187 X

A CIP catalogue record for this book is available
from the British Library

Printed in Finland

Contents

Introduction

The aim of this book is to provide a guide to the derivation, meaning and pronunciation of the scientific names of the more commonly grown plants. The term scientific name is preferable here to Latin name as many names derive from languages other than Latin, for example many derive from Greek or personal names. Whatever their origin, all names are treated as Latin. Generic names are treated as nouns and, as in Latin, have a gender (i.e. masculine, feminine or neuter). Names of species and varieties are adjectival and their endings follow the rule of Latin grammar, e.g. the Latin word for white can be rendered as *albus* (masculine), *alba* (feminine) or *album* (neuter) depending on the gender of the generic name. The use of Latin for plant names can certainly be confusing when first encountered, producing many words of unfamiliar form and uncertain pronunciation. It should be remembered, however, that when a scientific attitude was first taken towards the naming of plants in the 16th and 17th centuries, Latin was a common language among the intellectuals of Europe and it was second nature for many to use it. Today, although Latin has evolved from the Latin used in classical and medieval times to meet the needs of botany it forms a method of communication between botanists of all nationalities.

How to use this book

Generic names and common names are listed alphabetically. The name of each genus (plural genera) is given first e.g. **Acer**, followed by the suggested pronunciation, the family in which it is placed (related genera are placed in the same family) and the derivation of the name, whether from Latin (L.), Greek (Gk.) or other sources. Following this is a short statement of the main use in gardens of the plants listed e.g. as herbaceous perennials, shrubs, etc. and a general guide to hardiness. It should be noted that these are general statements about the genus as represented in gardens and some of the plants listed may be exceptions. Usually, unless otherwise stated, plants are at least fairly hardy in many parts of the country, semi-hardy plants require winter protection in most areas and tender plants are generally only suitable for growing under cover or for summer bedding. A common name is given if it is applied to the whole genus. If a country of origin is given here it applies to the plants listed and not necessarily to the whole genus.

Species are listed alphabetically under each genus. Each name is followed by the suggested pronunciation, the meaning of the name, the common name (if any) and the country of origin unless this is given above. If, instead of a country of origin, the abbreviation cult. (cultivated) is given this indicates that this plant is known only in gardens.

Categories below the rank of species are listed under the relevant specific name. These include those recognised botanically (sub-species, varieties and forms) which are given in small letters (e.g. *Pinus nigra maritima*) and those recognised horticulturally (cultivars and groups). A cultivar is a form of a species or hybrid selected either in the wild or in cultivation and maintained in gardens. These are given with capital initials and single inverted commas e.g. *Hedera hibernica* 'Deltoidea'. Plants which form part of a variable group are given a group name which is given with capital initials but without inverted commas e.g. *Fagus sylvatica* Purpurea.

Common names are cross referenced to the correct scientific name and are given in Roman type. They are not listed if they are the same as or very similar to the generic name.

Hybrids arise both in the wild and in cultivation when two species cross or are crossed to give rise to a distinct plant. When the parents are in different genera the plant is known as an intergeneric hybrid and the 'generic' name is preceded by a multiplication sign e.g. × *Laeliocattleya*. When the parents are in the same genus the 'specific' name is preceded by a multiplication sign e.g. *Abutilon* × *suntense*. Names of graft-hybrids are preceded by a plus sign. Sometimes when parentage is complex or unknown, or a scientific name has not been given to the plant, it is more convenient to place the cultivar or group name of a hybrid directly under the generic name e.g. *Begonia* Rex Cultorum Hybrids. Parents of hybrids are given where known. The hybrids listed are found only in cultivation unless otherwise stated.

A synonym is a scientific name that is no longer accepted for the plant in question. It may be that the name has been mis-applied in gardens and that the plant grown under this name is really something else (e.g. *Helichrysum microphyllum*); it may be that a name has been rejected because there is an earlier name available for the same plant (e.g. *Magnolia conspicua*), or it may be that a plant is better placed in a different genus (e.g. *Alyssum maritimum*). Synonyms are cross-referenced unless they occur in the same genus and there are less than ten entries in that genus; they are indicated by an equation sign =. The abbreviation hort. (*hortorum*, of gardens) following a name means that the plant is known under that name in gardens but there may be another plant to which that name correctly applies.

Allen J. Coombes

Pronunciation

Unlike the use of scientific names, their pronunciation is not governed by rules. The majority of people who use scientific names treat them as if they are in their own language. Where pronunciation is ambiguous by this method, it is common to encounter several ways of saying a word. Ideally pronunciation should be based on the origin of the word in question. Thus names derived from Latin or Greek should be given a Latin pronunciation, names derived from personal or place names should follow the original pronunciation of that name with a Latin pronunciation for any endings, changing the position of stress where necessary. The following table illustrates the classical pronunciation of Latin letters and dipthongs which are often said differently when an English pronunciation is given to the word. It should only be applied to words of Latin or Greek origin.

Latin letter or dipthong	*Classical pronunciation*
a	Short *a* is variously given as either in c*a*t or in *a*bove. Long *a* as in r*a*ther, never as in g*a*te.
ae	As *i* in m*i*te.
au	As *ou* in *ou*t.
c	Always hard as in c*a*t.
e	Short *e* as in l*e*t, long *e* as *a* in g*a*te.
ei	As *a* in g*a*te.
eu	As *oy* in b*oy*.
g	Always hard as in g*a*te.
i	Short *i* as in t*i*n, long *i* as *ee* in k*ee*n.
j	As *y* in *y*es.
s	As in thi*s*, not as in tho*s*e.
u	Short *u* as in f*u*ll, long *u* as *oo* in sh*oo*t, never as in t*u*b.
v	Classically pronounced as *w*.
y	Occurs in words of Greek origin and was pronounced as the French *u*.

Each scientific name is followed by a suggested pronunciation. This is often not the same as the English pronunciation of the name but does not necessarily conform to a complete classical pronunciation. Each word is divided by symbols, each symbol representing one syllable. The stressed syllable is given in italics. The pronunciation of each symbol is as in English, with the ambiguous letters being explained in the following table. For a normal English pronunciation, the scientific name can be pronounced as if it were English, for a more classical pronunciation consult the table on p. ix for words of Latin or Greek origin.

Symbol or letter	Pronounce as in
a	*a*part, c*a*nal
ă	c*a*t
e	l*e*t
ew	f*ew*
ewr	p*ure*
g	*g*ate
i	*i*n
ie	k*i*te
j	*j*am
o	h*o*t
ō	n*o*te
oi	usually as *oy* in boy but classically as *o-i*
our	French p*our*
ow	h*ow*
s	thi*s*
th	*th*in
tH	*th*is
u	f*u*ll
ū	t*u*b
zh	French *je*

Glossary

The use of technical terms has been kept to a minimum but some have been used for brevity and accuracy. These are defined below.

acuminate	Tapered to a long point.	inflorescence	The arrangement of individual flowers on a stem.
anther	The portion of the stamen (q.v.) which bears the pollen.	involucre	A whorl of bracts beneath an inflorescence.
areole	In cacti a cluster of spines.	linear	Narrow, with parallel sides
auricle	An ear-like lobe.	lip	In orchids, a modified petal.
awn	A bristle-like tip.	monocarpic	Fruiting once then dying.
bi-pinnate	Twice pinnate, i.e. pinnate (q.v.) with the divisions pinnate.	mucronate	With a short, abrupt point.
bract	A leaf-like structure beneath a flower or group of flowers.	node	A point on the stem bearing one or more leaves.
carpel	Part of the female portion of a flower consisting of an ovary (or part of one) and stigma.	obovate	Egg-shaped in outline, broadest above the middle.
		ovate	Egg-shaped in outline, broadest below the middle.
column	In orchids, the combined style and stigma.	panicle	A raceme (q.v.) with the branches branched.
compound	Consisting of several parts.	pappus	A tuft of hairs or bristles on the fruit.
corolla	The part of the flower made up of the petals.	pedicel	The stalk of a single flower in an inflorescence.
corymb	A type of raceme (q.v.) with the flowers at more or less the same height making a rounded or flat-topped inflorescence.	peltate	With the stalk of a leaf inserted below the blade and not on the margin.
dioecious	Bearing male and female flowers on separate plants.	perianth	The part of the flower comprising the petals and sepals, generally used when they are similar.
disc	An organ at the base of the ovary.	petiole	The leaf stalk.
entire	Not lobed or toothed.	phyllode	A leaf-like structure consisting of a flattened petiole.
epiphytic	A plant growing non-parasitically on another one.	pinna (plural pinnae)	The primary division of a pinnate leaf.
filament	The stalk of the stamen on which the anther is borne.	pinnule	The secondary division of a bi-pinnate leaf.
glabrous	Without hairs.	pistil	The female part of a flower.
glaucus	With a blue-grey bloom.	pollinium (plural pollinia)	A mass of pollen grains transported as a whole during pollination.
graft-hybrid	A plant which originated when one plant was grafted onto a different one. The tissues combine to create a different plant although there is no mixing of the genes.	proliferous	Reproducing vegetatively.
indusium	in ferns, the covering of the sorus.	pseudobulb	The thickened stem of an orchid.

raceme   An inflorescence bearing single, stalked flowers from a central axis, the youngest at the apex.

receptacle   The apex of a flower stem on which the floral parts are borne.

scape   A leafless flower stalk.

sepal   The parts of the perianth outside of the petals.

sessile   Unstalked.

simple   Not divided or lobed.

sorus (plural sori)   A cluster of sporangia (q.v.)

spadix   Unstalked flowers densely arranged on a fleshy stem.

spathe   A large bract surrounding an inflorescence.

sporangium (plural sporangia)   A capsule containing spores.

stamen   The male part of a flower consisting of an anther and a filament.

strobilus   A cone.

style   The female part of a flower which bears the pollen-receiving organ (stigma) at its tip.

type specimen   A specimen from which a plant was originally described and which defines the application of the name given.

umbel   An inflorescence in which the individual flower stalks all arise from one-point at the apex of the stem, sometimes compound and then with the primary stalks bearing umbels.

A

Aaron's Beard see *Hypericum calycinum*
Aaron's Rod see *Verbascum thapsus*
Abele see *Populus alba*

Abelia a-*bel*-ee-a *Caprifoliaceae*. After Dr
Clarke Abel (1780–1826) who introduced
A. chinensis. Deciduous and evergreen
shrubs.
   chinensis chin-*en*-sis. Of China.
   engleriana eng-gla-ree-*ah*-na. After
   Heinrich Engler (1844–1930). China.
   floribunda flō-ri-*bun*-da. Profusely
   flowering. Mexico.
   × *grandiflora* grăn-di-*flō*-ra. *A. chinensis* ×
   A. uniflora. Large-flowered.
   schumannii sho-*mahn*-ee-ee. After K. M.
   Schumann (1851–1904), German
   botanist. C China.
   triflora tri-*flō*-ra. Three-flowered, the
   flowers are in clusters of three. Himalaya.

Abeliophyllum a-bel-ee-ō-*fil*-lum *Oleaceae*.
From *Abelia* q.v. and Gk. *phyllon* (a leaf)
referring to the similar leaves. Deciduous
shrub.
   distichum dis-tik-um. Two-ranked (the
   leaves). Korea.

Abelmoschus ă-bel-*mos*-kus *Malvaceae*. From
Arabic *abu-l-mosk* (father of musk) referring
to the musk-scented seeds. Tender annuals.
   esculentus es-kew-*len*-tus. Edible (the fruit).
   Ladies' Fingers, Okra. Tropical Asia.
   manihot mah-nee-hot. From *manioc* the
   Brazilian name for cassava (*Manihot
   esculenta*). S.E. Asia.

Abies ă-bee-ayz *Pinaceae*. The classical name.
Evergreen conifers. Fir.
   alba ăl-ba. White (the bark of old trees).
   European Silver Fir. C and S Europe.
   amabilis a-*mah*-bi-lis. Beautiful. Red Silver
   Fir. Pacific Silver Fir. NW United States.
   balsamea băl-*săm*-ee-a. Balsam-producing
   (the bark). Balsam Fir. Canada, E United
   States.
   bracteata brăk-tee-*ah*-ta. With bracts (on
   the cone scales) conspicuous in this
   species. Santa Lucia Fir. California.
   cephalonica kef-a-*lon*-i-ka. Of Cephalonia
   (now Kefallinia) a Greek Island. Greek
   Fir. Greece (W & SW; mts.)
   concolor kon-ko-lor. Of the same colour
   (the leaf surfaces). White Fir. SW United
   States.
   delavayi del-a-*vay*-ee. After its discoverer

Abies (continued)
   the Abbé Jean Marie Delavay (1838–95),
   French missionary in China who
   introduced many notable plants to
   cultivation. China.
   forrestii fo-*rest*-ee-ee. After its discoverer
   and introducer George Forrest
   (1873–1932), a Scottish plant collector
   who made several expeditions to
   China. China.
   grandis grăn-dis. Large. Giant Fir. W N
   America.
   homolepis ho-mō-*lep*-is. With similar scales
   (on the cone). Nikko Fir. Japan.
   koreana ko-ree-*ah*-na. Of Korea. Korean
   Fir. S Korea.
   lasiocarpa lă-see-ō-*kar*-pa. With rough
   cones. Alpine Fir. W N America.
   magnifica mahg-*ni*-fi-ka. Magnificent. Red
   Fir. W United States.
   nordmanniana nord-mahn-ee-*ah*-na. After
   its discoverer Alexander von Nordmann
   (1803–66), German botanist. Caucasian
   Fir. Caucasus, W Asia.
   pindrow pin-drō. The native name. W
   Himalaya.
   procera pro-*kay*-ra. Tall. Noble Fir. W
   United States.
   spectabilis spek-*tah*-bi-lis. Spectacular.
   Himalayan Fir. Himalaya.
   squamata skwah-*mah*-ta. Flaking (the bark).
   Flaky Fir. China.

Abronia a-brō-nee-a *Nyctaginaceae*. From Gk.
abros (delicate) referring to the bracts. Annual
herb.
   umbellata um-bel-*lah*-ta. The flowers
   appear to be in umbels. Pink Sand
   Verbena. W N America.

Abutilon a-*bew*-ti-lon *Malvaceae*. From the
Arabic name for a similar plant. Tender and
semi-hardy shrubs.
   × *hybridum* hort. *hib*-rid-um. Hybrid.
   Name used for various hybrids.
   insigne in-*sig*-nee. Remarkable. Colombia,
   Venezuela.
   megapotamicum meg-a-pot-*ăm*-ik-um.
   From near the Rio Grande, Brazil.
   × *milleri* mil-la-ree. *A. megapotamicum* × *A.
   pictum*. After Philip Miller (1691–1771)
   curator of Chelsea Physic Garden.
   pictum pik-tum. (= *A. striatum*). Painted
   (the flowers). Brazil.
   × *suntense* sŭn-*ten*-see. *A. ochsenii* × *A.
   vitifolium*. From Sunte House, Sussex
   where it was raised in 1967.
   vitifolium vee-ti-*fo*-lee-um. *Vitis*-leaved.
   Chile.

Abyssinian Feathertop see *Pennisetum villosum*.

Acacia a-*kay*-see-a or classically a-*kah*-kee-a Leguminosae (Mimosoideae). Gk. name for *A. arabica* from *akis* (a sharp point) referring to the thorns. Tender and semi-hardy evergreen trees and shrubs. Wattle.

baileyana bay-lee-*ah*-na. After F. M. Bailey (1827–1915) Australian botanist. Cootamundra Wattle. New South Wales.

cultriformis cul-tree-*form*-is. Knife-shaped (the phyllodes). Knife Acacia. E Australia.

dealbata dee-al-*bah*-ta. Whitened (the young shoots and leaves). Silver Wattle. SE Australia, Tasmania.

farnesiana far-neez-ee-*ah*-na. From the garden of the Farnese Palace, Rome. Tropical America (uncertain).

longifolia long-gi-*fo*-lee-a. With long leaves (phyllodes). Sydney Golden Wattle. Australia.

melanoxylon mel-a-*noks*-i-lon. With black wood. Blackwood. S Australia, Tasmania.

mucronata mew-kron-*ah*-ta. Mucronate (the phyllodes). Victoria, Tasmania.

podalyriifolia pod-a-li-ree-i-*fo*-lee-a. With leaves (phyllodes) like *Podalyria* (S African shrub). Pearl Acacia. Queensland Silver Wattle. NE Australia.

pravissima prah-*vis*-im-a. Very crooked (the phyllodes). Ovens Wattle. SE Australia.

riceana ries-ee-*ah*-na. After Mr T. Spring-Rice, Chancellor of the Exchequer 1835–9. Rice's Wattle. Tasmania.

verticillata ver-tik-il-*lah*-ta. Whorled (the phyllodes). Star Acacia, Prickly Moses. Victoria, Tasmania.

Acaena a-*see*-na or classically a-*kie*-na Rosaceae. From Gk. *akaina* (a thorn) referring to the spiny fruits. Herbaceous perennials and sub-shrubs. New Zealand Burr, Bidi-Bidi.

adscendens ad-*send*-enz. Ascending (the branches from procumbent stems). New Zealand.

buchananii bew-kan-*an*-ee-ee. After John Buchanan (1819–98), botanist with the Geological Survey in New Zealand. New Zealand, Australia.

inermis in-*er*-mis. Unarmed. New Zealand.

microphylla mik-ro-*fi*-la. Small-leaved. New Zealand.

novae-zelandiae no-vie-zee-*lahn*-dee-ie. Of New Zealand.

Acalypha a-ka-*lee*-fa Euphorbiaceae. Classical name for the nettle from the similar leaves. Tender shrubs.

hispida his-pid-a. Bristly. Chenille Plant, Red-hot Cat's Tail. Malay Archipelago.

wilkesiana wilks-ee-*ah*-na. After Admiral Charles Wilkes (1798–1877), American explorer of the Pacific. Pacific Islands.

Acantholimon a-kanth-o-*lee*-mon Plumbaginaceae. From Gk. *akanthos* (a thorn) and *limonium* (sea lavender). Spiny sub-shrubs and herbs. Prickly Thrift.

glumaceum gloo-mah-*kee*-um. Having glumes, the papery bracts around the flowers. Caucasus, W Asia.

olivieri o-liv-ee-*e*-ree. After G. A. Olivier (1756–1814) French naturalist. Iran.

ulicinum ew-li-*kee*-num. (= *A. androsaceum*). Like *Ulex*. SE Europe, W Asia.

creticum kray-ti-kum. (= *A. creticum*). of Crete. Crete, W Asia.

venustum ven-*us*-tum. Charming. W Asia.

Acanthopanax see *Eleutherococcus*.

Acanthus a-*kanth*-us Acanthaceae. From Gk. *akanthos* (a thorn) referring to the often spiny leaves and flower bracts. Perennial herbs. Bear's Breeches.

balcanicus bal-*kan*-i-kus. (= *A. hungaricus*. *A. longifolius*) Of the Balkan Mts. SE Europe.

mollis mol-lis. Soft, the leaves are not spiny. SW Europe.

Latifolius lah-tee-*fo*-lee-us. Broad-leaved.

spinosus speen-o-sus. Spiny (the leaves). SE Europe.

Spinosissimus speen-o-*sis*-i-mus. Most spiny. Hybrid gp. between *A. mollis* and *A. spinosus*.

Acer ay-ser or classically a-ker Aceraceae. The classical L. name. Deciduous trees. Maple.

campestre kam-*pes*-tree. Of fields. Common Maple, Field Maple. Europe, N Africa, W Asia.

capillipes ka-*pil*-lee-pays. Slender-stalked (the flowers). Japan.

cappadocicum kap-a-*do*-kik-um. Of Cappadocia, W Asia. Caucasus, W Asia.

lobelii lo-*bel*-ee-ee. After M. Lobel (1538–1616) Flemish botanist. S Italy.

carpinifolium kar-peen-i-*fo*-lee-um. *Carpinus*-leaved. Hornbeam Maple. W N America.

Acer (continued)
circinatum ker-kin-*ah*-tum. Rounded (the leaves). Vine Maple. W N America.
crataegifolium kra-tieg-i-*fo*-lee-um. *Crataegus*-leaved. Japan.
davidii dă-*vid*-ee-ee. After Armand David, see *Davidia*.
   grosseri grō-sa-ree. After Grosser. China.
distylum di-*stil*-um. With two styles (in some species they are united). Japan.
griseum gris-ee-um. Grey (the lower leaf surface). Paper-bark Maple. C China.
japonicum ja-*pon*-i-kum. Of Japan.
   'Aconitifolium' ă-kon-ee-ti-*fo*-lee-um. *Aconitum*-leaved.
   'Vitifolium' vee-ti-*fo*-lee-um. *Vitis*-leaved.
macrophyllum măk-rō-*fil*-um. Large-leaved. Oregon Maple. W N America.
maximowiczianum max-im-ō-*vi*-chē-ah-nūm (= *A. nikoense*) For Carl Johann Maximowicz (1827–91) Russian botanist. Nikko Maple. Japan. C China.
monspessulanum mon-spes-ew-*lah*-num. Montpelier. Montpelier Maple. Europe, N Africa, W Asia.
negundo ne-*gun*-do. From the native name of *Vitex negundo* because of a supposed similarity of leaf. Ash-leaved Maple. Box Elder. N America.
palmatum pahl-*mah*-tum. Hand-like (the leaves). Japanese Maple. Japan.
   'Atropurpureum' aht-rō-pur-*pewr*-ree-um. Dark purple (the leaves).
   'Dissectum' dis-*sek*-tum. Finely cut (the leaves).
   Heptalobum hep-ta-*lo*-bum. Seven-lobed (the leaves).
   Linearilobum lin-ee-ah-ri-*lō*-bum. With linear lobes (the leaves).
pensylvanicum pen-sil-*vahn*-i-kum. Of Pennsylvania. Moosewood. E N America.
   'Erythrocladum' e-rith-rō-*kli*-dum. With red shoots.
platanoides plă-ta-*noi*-deez. Like *Platanus*. Norway Maple. Europe.
pseudoplatanus sood-ō-*plă*-ta-nus. False *Platanus*. Sycamore. Europe.
rubrum rub-rum. Red (autumn colour). Red Maple. E N America.
   'Schlesingeri' shlez-*ing*-ga-ree. After Mr Schlesinger in whose grounds it was found.
rufinerve roof-i-*ner*-vee. Red-veined, referring to the reddish hairs on the leaf veins. Japan.
saccharinum săk-a-*ree*-num. Sugary (the sap). Silver Maple. E N America.

Acer (continued)
saccharum sa-*kah*-rum. L. name for sugar cane. Maple syrup and sugar are made from the sap of this tree. Sugar Maple. E N America.
tataricum ta-*tah*-ri-kum. Of Tatary, C Asia. SE Europe, W Asia.
   ginnala gin-*nah*-la. The native name. Amur Maple. NE Asia.
× *zoeschense* zur-*shen*-see. *A. campestre* × *A. cappadocicum* ssp. *lobelii*. Of the Zoeschen nursery, Germany.

Achillea a-*kil*-lee-a or classically ă-kil-*lee*-a Compositae. After Achilles of Gk. mythology who is said to have used it medicinally. Herbaceous perennials.
ageratifolia a-ge-ra-ti-*fo*-lee-ă. *Ageratum*-leaved. SE Europe.
ageratum a-*ge*-ra-tum. From the resemblance to *Ageratum*. Sweet Nancy. Mediterranean Region.
chrysocoma kris-o-*ko*-ma. With golden hair, referring to the yellow flower heads. SE Europe.
clavennae kla-*ven*-ie. After Niccola Chiavera, died 1617, Italian apothecary. Alps, Balkans.
clypeolata kli-pee-ō-*lah*-ta. Shield-shaped (the flower heads). SE Europe.
filipendulina fi-li-pen-dew-*lee*-na. Like *Filipendula*. W Asia, Caucasus.
millefolium meel-lee-*fo*-lee-um. Thousand-leaved, referring to the finely divided leaves. Yarrow. Europe, W Asia.
ptarmica tar-mi-ka. Gk. name for a plant which caused sneezing, referring to its use for snuff. Sneezewort. Europe, Asia.
   'Taygetea' tay-*gee*-tee-a. From the Taygetos mountains, Greece. The true plant (now *A. aegyptiaca*) is rarely grown. A cv. of garden origin.
umbellata um-bel-*lah*-ta. The flowers appear to be in umbels.

Achimenes ă-kim-ee-neez Gesneriaceae. Possibly from Gk. *chemaino*, meaning to suffer from cold, referring to their tender nature, after King Hakhamash of Turkey (Gk. Achaemenes) or from *achaemenis*, Gk. name of a plant. Tender perennials. Hot-water Plant.
erecta e-*rek*-ta (= *A. coccinea*). Erect. C America, Mexico, Jamaica.
grandiflora grăn-di-*flō*-ra. Large-flowered. Mexico, C America.
heterophylla he-te-rō-*fi*-la. Variably leaved (the leaves of a pair are unequal). Mexico, Guatemala.

Achimenes (continued)
longiflora long-gi-*flō*-ra. Long-flowered.
Mexico to Panama.
mexicana meks-i-*kah*-na. Of Mexico.

Acidanthera *bicolor* see *Gladiolus callianthus*

Acinos a-*kee*-nos *Labiatae*. From *akinos*, Gk.
name for an aromatic herb. Perennial herb.
alpinus ăl-*peen*-us. (= *Calamintha alpina*).
Alpine. C and S Europe.

Acokanthera a-kō-kan-*the*-ra *Apocynaceae*.
From Gk. *akoke* (a point) and *anthera* (an
anther) referring to the pointed anthers.
Tender, evergreen shrub. Bushman's poison.
oblongifolia ob-long-gi-*fo*-lee-a. (= *A
spectabilis*). With oblong leaves. SW
Africa.

Aconitum ă-kon-*ee*-tum *Ranunculaceae*. The
L. name. Poisonous perennial herbs.
Monkshood.
  'Bicolor' *bi*-ko-lor. (= *A*. × *cammarum*
  'Bicolor'). *A. napellus* × *A. variegatum*.
  Two-coloured (the flowers).
carmichaelii kar-mie-*keel*-ee-ee. (= *A.
fischeri A. wilsonii*). After J. R. Carmichael
(1838–70). E Asia.
lycoctonum ly-kok-*tō*-num. From Gk. *lykos*,
a wolf; the Gk. name for Wolfsbane.
Once used to poison wolves. Europe, N
Africa.
    vulparia vul-*pah*-ree-a. Of Wolves.
    Wolf's Bane, Badger's Bane. Europe.
napellus na-*pel*-us. A small turnip, referring
to the roots. Common Monkshood. W
and C Europe.
paniculatum pa-nik-ew-*lah*-tum. With
flowers in panicles. C and E Europe.
variegatum vā-ree-a-*gah*-tum. Variegated
(the flowers). C and E Europe.
volubile vol-*ew*-bi-lee. Twining. E Asia.

Acorus a-*ko*-rus or classically ă-ko-rus
Araceae. L. name for *Iris pseudacorus*. Aquatic
or marginal herbaceous perennials.
calamus kă-la-mus. Gk. *calamus* (a reed)
referring to the foliage. Sweet Flag. N
Hemisphere.
gramineus grah-*min*-ee-us. Grass-like (the
foliage). E Asia.

Acradenia ăk-ra-*deen*-ee-a *Rutaceae*. From
Gk. *akros* (at the tip) and *adenia* (a gland)
referring to the glands at the tips of the
carpels. Semi-hardy, evergreen shrub.
frankliniae frank-*lin*-ee-ie. After Lady

Acradenia (continued)
Franklin, wife of the Governor of
Tasmania in 1842. W Tasmania.

Acroclinium ak-ro-*klie*-nee-um. *Compositae*
Derivation uncertain; possibly from L. *acro*,
towards the top, and *clino*, prostrate, referring
to the nodding flower buds. Annual and
perennial herbs. Straw Flower, Everlastings.
roseum ro-see-um (= *Helipterum roseum*)
Rose-coloured (the flowers). SW
Australia.

Actaea ă-*tee*-a *Ranunculaceae*. From Gk. *aktea*
(elder) referring to the similar leaves.
Herbaceous perennials with poisonous fruits.
Baneberry.
rubra *rub*-ra. Red (the fruits). Red
Baneberry. N America.
spicata spee-*kaa*-ta. With flowers in spikes.
Herb Christopher. Europe, Asia.

Actinidia ăk-tin-*id*-ee-a *Actinidiaceae*. From
Gk. *aktinos* (a ray) referring to the radiating
styles. Deciduous, twining shrubs.
arguta ar-*gew*-ta. Sharply toothed (the
leaves). NE Asia.
deliciosa del-*liss*-i-o-să (= *A. chinensis*).
Delicious. Chinese Gooseberry, Kiwi
Fruit. China
kolomikta ko-lō-*mik*-ta. The native name.
China, Japan.
polygama po-*lig*-a-ma. Polygamous i.e.
with some flowers male, some female and
some hermaphrodite. Silver Vine. C Japan.
purpurea pur-*pewr*-ree-a. Purple (the
fruits). China.

Ada ah-da *Orchidaceae*. After Ada, sister of
Artemisia of Caria. Greenhouse orchid.
aurantiaca ow-răn-tee-*ah*-ka. Orange (the
flowers). Colombia.

Adam's Needle see *Yucca filamentosa*.

Adenocarpus a-deen-ō-*kar*-pus *Leguminosae*.
From Gk. *aden* (a gland) and *karpos* (a fruit)
referring to the sticky pods. Deciduous
shrub.
decorticans day-*kor*-ti-kănz. With peeling
bark. Spain.

Adenophora ă-den-*o*-fo-ra *Campanulaceae*.
From Gk. *aden* (a gland) and *phorea* (to bear)
referring to the glandular disc at the base of
the style. Perennial herbs.
liliifolia lee-lee-i-*fo*-lee-a. *Lilium*-leaved. C
Europe, N Asia.
tashiroi tăsh-ee-*rō*-ee. After Tashiro.

Adiantum ă-de-ăn-tum *Adiantaceae*. From Gk. *adiantos* (unwetted) referring to the way the fronds repel water. Tender ferns. Maidenhair Fern.

capillus-veneris ka-*pil*-lus-*ven*-e-ris. Venus's hair, referring to the delicate foliage. Common Maidenhair Fern. Worldwide.
caudatum kaw-*dah*-tum. With a tail (the pinnae). Trailing Maidenhair Fern. Old World tropics.
concinnum kon-*kin*-um. Elegant. S America.
formosum for-*mō*-sum. Beautiful. Australian Maidenhair Fern. Australia, New Zealand.
hispidulum his-*pid*-ew-lum. Finely bristly. Rough Maidenhair Fern. Old World tropics.
pedatum ped-*ah*-tum. Like a bird's foot (the fronds). N American Maidenhair Fern. N America, E Asia.
    'Japonicum' ja-*pon*-i-kum. Of Japan. Rose-fronded Maidenhair Fern.
raddianum rah-dee-*ah*-num. (= *A. cuneatum*). After Giuseppe Raddi (1770–1829). Delta Maidenhair Fern. Brazil.
trapeziforme tra-peez-ee-*for*-mee. Unequally four-sided (the segments of the fronds). Tropical America.
venustum ven-*us*-tum. Handsome. Kashmir Maidenhair Fern. Himalaya.

Adonis a-*dō*-nis *Ranunculaceae*. The Gk. name for Adonis, beautiful youth, and lover of Aphrodite, said to have been changed into this plant after death. Annual and perennial herbs.

aestivalis ie-sti-*vah*-lis. Of summer (flowering). Europe.
amurensis ăm-ew-*ren*-sis. Of the Amur region, NE Asia.
    'Plena' *play*-na. With double flowers.
annua ăn-ew-a (= *A. autumnalis*). Annual. Pheasant's Eye. S Europe, W Asia.
pyrenaica pi-ray-*nah*-i-ka. Of the Pyrenees. Pyrenees, Maritime Alps.
vernalis ver-*nah*-lis. Of spring (flowering). C and S Europe.

Adromischus ă-drō-*mis*-kus *Crassulaceae*. From Gk. *hadros* (stout) *mischos* (a stalk) referring to the stout, short pedicels. Tender succulents, S Africa.

cooperi koo-pa-ree. After Thomas Cooper (1815–1913). Plover Eggs.
cristatus kris-*tah*-tus. Crested (the leaves). Crinkle-leaf Plant.

Adromischus (continued)
maculatus măk-ew-*lah*-tus. Spotted (the leaves). Calico Hearts.

Aechmea iek-*mee*-a *Bromeliaceae*. From Gk. *aichme* (a point) referring to the pointed sepals. Tender, evergreen herbaceous perennials. Brazil, unless stated otherwise.

bracteata brăk-tee-*ah*-ta. With bracts (on the scape). Mexico to Colombia.
fasciata făs-kee-*ah*-ta. Banded (the leaves). Urn Plant.
fulgens ful-genz. Shining (the red bracts).
racineae ră-*seen*-ee-ie. After Racine, wife of M. B. Foster, see *Cryptanthus fosterianus*. Christmas Jewels.

Aeonium ie-*ō*-nee-um *Crassulaceae*. The L. name of one species. Tender or semi-hardy succulents.

arboreum ar-*bo*-ree-um. Tree-like. Gran Canaria.
    'Foliis Purpureis' *fo*-lee-is pur-*pewr*-ree-is. With purple leaves.
canariense ka-nah-ree-*en*-see. Of the Canary Islands.
domesticum see *Aichryson* × *domesticum*.
haworthii hay-*werth*-ee-ee. After Haworth, see *Haworthia*. Canary Islands.
simsii simz-ee-ee. After John Sims (1749–1831). Canary Islands.
tabuliforme tăb-ew-lee-*form*-ee. Table-shaped i.e flat-topped. Canary Islands.
undulatum un-dew-*lah*-tum. Wavy-edged (the leaves). Saucer Plant. Canary Islands.

Aerangis ah-e-*ran*-gis *Orchidaceae*. From Gk. *aer* (air), epiphytic orchids appearing to derive nourishment from the air. Tender Orchids. Africa, Madagascar and Mascarene Islands.

bilobum bi-*lō*-bum. Two lobed (the leaf apex). S Africa.
kotschyanum kot-shee-ah-num. After Kotschy. E Africa.

Aerides ah-*e*-ree-deez *Orchidaceae*. From Gk. *aer* (air), they are epiphytic, appearing to derive nourishment from the air. Tender orchids. Fox-tail Orchid.

crispa kris-pa. Finely wavy, referring to the fringed central lobe of the lip. Indonesia.
falcata făl-*kah*-ta. Sickle-shaped (the leaves). Burma.
japonica ja-*pon*-i-ka. Of Japan.
lawrenceae lo-*rens*-ee-ie. After Lady Lawrence wife of Sir Trevor Lawrence of Burford Lodge. Philippines.

Aerides (continued)
multiflora mul-tee-*flō*-ra. Many-flowered.
Himalaya, India, Burma.
odorata o-dō-*rah*-ta. Scented (the flowers).
Himalaya, India, SE Asia.
quinquevulnera kwing-kwee-*vul*-ne-ra.
With five marks (on the flower).
Philippines.

Aeschynanthus ie-skee-*năn*-thus *Gesneriaceae*.
From Gk. *aischyno* (shame) and *anthos* (a
flower) referring to the red flowers.
Greenhouse climbers. Basket Plant.
boschianus bosh-ee-*ah*-nus. After J. van der
Bosch (1807–54), Governor-General of
the Dutch East Indies. Java.
marmoratus mar-mo-*rah*-tus. Marbled (the
leaves). Zebra Basket Vine. SE Asia.
pulcher pul-ker. Pretty. Royal Red Bugler.
Java.
radicans rah-di-kănz. (= *A. lobbianus*).
With rooting stems. Lipstick Vine.
Malaysia, Java.

Aesculus ies-ku-lus *Hippocastanaceae*. L. name
for an oak with edible acorns. Deciduous
trees. Horse-chestnut, Buckeye.
californica kăl-i-*forn*-i-ka. Of California.
California Buckeye.
× *carnea* kar-nee-a. *A. hippocastanum* × *A.
pavia*. Flesh-coloured (the flowers). Red
Horse-chestnut.
flava flah-va. Yellow (the flowers). Yellow
Buckeye. United States.
hippocastanum hip-ō-kă-sta-num. The L.
name. Common Horse-chestnut. N
Greece, N Albania.
'Baumannii' bow-*mahn*-ee-ee. After A.
N. Baumann who found the original
double-flowered, sterile sport.
indica in-di-ka. Indian. Indian Horse-
chestnut. NW Himalaya, Afghanistan.
× *mutabilis* mew-*tah*-bi-lis. *A pavia* × *A.
sylvatica*. Changeable (the colour of the
flowers).
'Induta' in-*dew*-ta. Clothed, referring to
the densely hairy undersides of the
leaves.
parviflora par-vi-*flō*-ra. Small-flowered. SE
United States.
pavia pah-vee-a. After Peter Paaw (Petrus
Pavius), Dutch botanist, died 1616. Red
Buckeye. S United States.
turbinata tur-bin-*ah*-ta. Top-shaped (the
fruit). Japanese Horse-chestnut. Japan.

Aethionema ieth-ee-ō-*nee*-ma *Cruciferae*.
Possibly from Gk. *aethes* (unusual) and *nema*

Aethionema (continued)
(a thread) referring to the stamens. Sub-
shrubby perennials. Stone Cress.
armenum ar-*meen*-um. Of Armenia.
Turkey.
cordifolium ko-ri-di-*fo*-lee-um. With leaves
like *Coris*. W Asia.
grandiflorum grăn-di-*flō*-rum. Large-
flowered. W Asia.
iberideum i-be-*ri*-dee-um. Like *Iberis*.
Greece, Turkey.
oppositifolium op-o-si-ti-*fo*-lee-um. With
opposite leaves. Lebanon.

African Boxwood see *Myrsine africana*
African Daisy see *Arctotis*
African Hemp see *Sparmannia africana*
African Lily see *Agapanthus*
African Violet see *Saintpaulia*

Agapanthus ăg-a-*pănth*-us *Liliaceae*. From
Gk. *agape* (love) and *anthos* (a flower). Hardy
and semi-hardy perennial herbs. African Lily.
S Africa.
africanus ăf-ri-*kah*-nus. (= *A. umbellatus*).
African.
campanulatus kăm-păn-ew-*lah*-tus. Bell-
shaped (the flowers).
caulescens kaw-*les*-enz. With a stem.
inapertus in-a-*per*-tus. Closed, referring to
the narrow-mouthed flowers.
praecox prie-koks. Early (flowering).
orientalis o-ree-en-*tah*-lis. From the east
(of S Africa).

Agapetes ăg-a-*peet*-eez *Ericaceae*. From Gk.
agapetos (desirable). Evergreen, semi-hardy
shrub.
serpens ser-penz. Creeping. Himalaya.

Agastache a-*gah*-sta-kee *Labiatae*. From Gk.
agan (very much) and *stachys* (a spike)
referring to the numerous flower spikes.
Perennial herb. Giant Hyssop.
mexicana mek-si-*kah*-na. Of Mexico.

Agathis a-*gah*-this, classically ă-ga-this
Araucariaceae. From Gk. *agathis* (a ball of
thread) referring to the cones. Semi-hardy
conifer.
australis ow-*strah*-lis. Southern. Kauri Pine.
New Zealand.

Agave a-gah-vee *Agavaceae*. From Gk. *agave*
(noble) referring to the tall inflorescence of
A. americana. Tender and semi-hardy
succulents.
americana a-me-ri-*kah*-na. American.
Century Plant. Mexico.

Agave (continued)
'Marginata' mar-gin-*ah*-ta. Margined (the leaves).
angustifolia an-gust-i-*fo*-lee-a. Narrow-leaved. Cult.
attenuata a-ten-ew-*ah*-ta. Drawn out (the inflorescence). Mexico.
filifera fee-*li*-fe-ra. Bearing threads (on the leaf margins). Thread Agave. Mexico.
parviflora par-vi-*flō*-ra. Small-flowered. Arizona, N Mexico.
potatorum pō-tah-*to*-rum. Of drinkers, a drink is made from the plant. Mexico.
stricta strik-ta. Upright (the leaves). Mexico.
victoriae-reginae vik-*tor*-ree-ie-ray-*geen*-ie. After Queen Victoria. Mexico.

Ageratina a-ge-ra-*tee*-na *Compositae.* Diminutive of *Ageratum* q.v. Annuals, perennials and shrubs. E United States, S America.
altissima al-*tiss*-im-a (= *Eupatorium rugosum*). Very tall, tallest. E North America.
ligustrina lig-us-*tree*-na (= *Eupatorium ligustrinum*). Like *Ligustrum.* C America.

Ageratum a-*ge*-ra-tum *Compositae.* From Gk. *a* (without) and *geras* (age) referring to the long-lasting flowers. Annual herbs. Floss Flower.
conyzoides kon-ee-*zoi*-deez. Like *Conyza* a related genus. W Indies, Mexico, C America.
houstonianum hew-stōn-ee-*ah*-num. After William Houston (1695–1733), a Scottish physician who collected in C America. Mexico, C America.

Aglaonema a-glah-ō-*nee*-ma *Araceae.* From Gk. *aglaos* (bright) and *nema* (a thread) referring to the stamens. Tender, evergreen, foliage perennials.
commutatum kom-ew-*tah*-tum. Changeable. Philippines, Celebes.
'Treubii' *troy*-bee-ee. (= *A. treubii* hort.). After Melchior Treub, Director of the Botanic Garden, Buitenzorg (now Bogor) Java.
costatum kos-*tah*-tum. Ribbed, referring to the white midrib. Spotted Evergreen. Malay Peninsula.
crispum kris-pum. Finely wavy, perhaps referring to the irregular variegation. Painted Drop-tongue. Philippines.
modestum mod-*es*-tum. Modest, it is not variegated like many species. Chinese Evergreen. SE Asia.

Aglaonema (continued)
nitidum ni-tid-um. Shining (the leaves). SE Asia.
pictum pik-tum. Painted – the variegated leaves. Sumatra.

Agrostemma ăg-rō-*stem*-a *Caryophyllaceae.* From Gk. *agros* (a field) and *stemma* (a garland). Annual herb.
githago gi-*thah*-go. The L. name. Corn Cockle. E Mediterranean region.

Aichryson ie-*kris*-on *Crassulaceae.* The Gk. name for *Aenium arboreum.* Tender, succulent annuals and perennials.
× *domesticum* dom-*es*-ti-kum. *A. punctatum* × *A. tortuosum.* (= *Aeonium domesticum*). Cultivated. Youth and Old Age.
laxum lăks-um. Open (the inflorescence). Canary Islands.
villosum vil-*lō*-sum. Softly hairy. Azores, Madeira.

Ailanthus ie-*lan*-thus *Simaroubaceae.* From *ailanthos*, a native word meaning to reach for the sky. Deciduous tree.
altissima ăl-*tis*-si-ma. Tallest. Tree of Heaven. N China.

Air Plant see *Kalanchoe pinnata*

Ajuga a-*joo*-ga *Labiatae.* Derivation obscure. Perennial herbs.
genevensis gen-e-*ven*-sis. Of Geneva. S Europe.
pyramidalis pi-ra-mid-*ah*-lis. Pyramidal (the inflorescence). Europe.
reptans rep-tănz. Creeping. Bugle. Europe.

Akebia a-*kee*-bee-a *Lardizabalaceae.* From the Japanese name. Woody climbers. Chocolate Vine.
quinata kwi-*nah*-ta. In fives (the leaflets). Japan, China, Korea.
trifoliata tri-fo-lee-*ah*-ta. With three leaves (leaflets). China, Japan.

Albizia ăl-*biz*-ee-a *Leguminosae* (*Mimosoideae*). After F. del Albizzi who introduced the following. Semi-hardy tree.
julibrissin yoo-lee-*bris*-sin. From the native name. Pink Siris, Pink Mimosa, Silk Tree. W Asia.

Alcea ăl-*kee*-a *Malvaceae.* From *alkaia*, Gk. name for a kind of mallow. Herbaceous perennials or biennials.

Alcea (continued)
ficifolia fee-ki-*fo*-lee-a. *Ficus*-leaved. Fig-leaved Hollyhock. Cult.
rosea ro-see-a. (= *Althaea rosea*). Rose-coloured (the flowers). Hollyhock. Cult.

Alchemilla ăl-ke-*mil*-la *Rosaceae*. From the Arabic name, *alkemelych*. Herbaceous perennials. Lady's Mantle.
alpina ăl-*peen*-a. Alpine. Alpine Lady's Mantle. Europe.
conjuncta kon-*yunk*-ta. Joined, it resembles *A. alpina* but the leaflets are joined at the base. Alps.
erythropoda e-rith-rō-*pō*-da. Red-stalked (the leaves of some plants). E Europe, Caucasus.
mollis mol-is. Softly hairy (the leaves). E Carpathians.

Alder see *Alnus*
 Common see *A. glutinosa*
 Grey, see *A. incana*
 Italian, see *A. cordata*
Alder Buckthorn see *Rhamnus frangula*
Alecost see *Tanacetum balsamita*

Aletris a-*let*-ris *Liliaceae* (*Melianthaceae*). From Gk. *aleitris*, a female slave who ground corn, referring to the mealy perianth. Herbaceous perennials.
farinosa fă-ri-*nō*-sa. Mealy (the perianth). Unicorn Root. E United States.

Alexandrian Laurel see *Danae racemosa*
Algerian Statice see *Limonium bonduellii*

Alisma a-*lis*-ma *Alismataceae*. Gk. name for a water plant. Aquatic or marginal perennial herbs.
gramineum grah-*min*-ee-um. Grass-like (the leaves). N temperate regions.
lanceolatum lăn-kee-ō-*lah*-tum. Lance-shaped (the leaves). Europe.
natans see *Luronium natans*
plantago-aquatica plăn-*tah*-gō-a-*kwah*-ti-ka. Literally water plantain – the common name. N temperate regions.

Allamanda ăl-a-*măn*-da *Apocynaceae*. After Frederick Allamand, 18th century Swiss botanist who introduced *A. cathartica*. Tender, evergreen shrubs.
cathartica ka-*thar*-ti-ka. Purging. Golden Trumpet. S America.
schottii shot-ee-ee (= *A. neriifolia*) For H. W. Schott (1794–1865), botanist. S America

Allegheny Spurge see *Pachysandra procumbens*

Allium ă-lee-um *Liliaceae* (*Alliaceae*). The Gk. name for garlic, *A. sativum*. Perennial herbs
aflatunense ă-fla-tun-*en*-see. Of Aflatun, C Asia.
albo-pilosum see *A. christophii*.
ascalonicum hort. see *A. cepa* Aggregatum.
beesianum beez-ee-*ah*-num. After Bees Nursery, the introducers. China.
carinatum kăr-in-*ah*-tum. Keeled. C & S Europe.
 pulchellum pul-*kel*-lum. Pretty. SE Europe, W Asia.
cepa ke-pa. The L. Name. Onion. Cult.
 Aggregatum ag-re-*gah*-tum (= *A. ascalonicum* hort.). Clustered (the bulbs). Shallot. Cult.
 Proliferum prō-*li*-fe-rum. Proliferous referring to the production of bulbs in the flower heads. Tree Onion. Cult.
christophii kris-*tof*-ee-ee (= *A. albopilosum*). After? Stars of Persia. N Iran, C Asia.
cyaneum see-*ah*-ne-um. Blue (the flowers). NW China.
cyathophorum see-ă-tho-*fo*-rum. Cup-bearing, referring to the cup-shaped tube formed by the united filaments. C Asia.
 farreri fă-ra-ree. After Reginald Farrer, see *Viburnum farreri*. NW China.
fistulosum fis-tew-*lō*-sum. Hollow-stemmed. Welsh Onion. Cult.
flavum flah-vum. Yellow (the flowers). S Europe.
giganteum gi-*găn*-tee-um. Very large. Iran, Afghanistan, C Asia.
karataviense kă-ra-tah-vee-*en*-see. Of the Kara Tua, Kazakhstan. C Asia.
moly mo-lee. Gk. name of a herb. E Spain, SW France.
narcissiflorum nar-kis-i-*flō*-rum. Narcissus-flowered. SW Alps.
neapolitanum nee-ah-pol-i-*tah*-num. Of Naples. Mediterranean region.
oreophilum o-ray-o-fi-lum. (= *A. ostrowskianum*). Mountain-loving. C Asia, E Turkey, Caucasus.
ostrowskianum see *A. oreophilum*
paradoxum pă-ra-*doks*-um. Unusual, perhaps referring to the single leaf. Caucasus, N Iran.
porrum po-rum. The L. name. Leek. Cult.
roseum ro-see-um. Rose-coloured (the flowers, usually). Mediterranean region.
sativum sa-*tee*-vum. Cultivated. Garlic. Cult.
schoenoprasum skoyn-ō-*prah*-sum. From Gk. *schoinos* (a rush) and *prasum* (leek) referring to the rush-like leaves. Chives. Europe.

Allium (continued)
scorzonerifolium skorz-on-e-ree-i-*fo*-lee-um. *Scorzonera*-leaved.
siculum see *Nectaroscordium siculum.*
triquetrum tri-kwee-trum. Three-angled (the scape). W Mediterranean region.
tuberosum tew-be-*ro*-sum. Tuberous, referring to the rhizome. Chinese Chives. Himalaya, Assam, China.
zebdanense zeb-dan-*en*-see. Of Zebdani, Syria. E Mediterranean region.

Alloplectus ă-lō-*plek*-tus *Gesneriaceae.* From Gk. *allos* (diverse) and *pleco* (to plait) referring to the overlapping calyx lobes. Tender perennials.
capitatus kăp-i-*tah*-tus. In a dense head (the flowers). Velvet Alloplectus. N Andes.

Allspice, Californian see *Calycanthus occidentalis*
Carolina see *C. floridus*
Almond see *Prunus dulcis*
Dwarf Russian see *P. tenella*

Alnus ăl-nus *Betulaceae.* The L. name. Deciduous trees. Alder.
cordata kor-*dah*-ta. Heart-shaped (the leaves). Italian Alder. S Italy, Corsica.
glutinosa gloo-ti-*nō*-sa. Sticky (the young shoots and leaves). Common Alder. Europe, N Africa, W Asia.
incana in-*kah*-na. Grey. (the undersides of the leaves). Grey Alder. Europe, Caucasus.
japonica ja-*pon*-i-ka. Of Japan. NE Asia, Japan.

Alocasia ă-lō-*ka*-see-a *Araceae.* From Gk. *a* (without, not) and *Calocasia* a related genus. Tender foliage herbs.
cuprea kew-pree-a. Coppery (the leaves). Giant Caladium. Borneo.
lindenii see *Homalomena lindenii*
macrorrhiza măk-rō-*ree*-za. (= *A. indica*). With a large root. Giant Elephant's Ear. SE Asia.
micholitziana mich-o-litz-ee-*ah*-na. After W. Micholitz who collected for Sander in New Guinea. Philippines.
sanderiana sahn-da-ree-*ah*-na. After the Sander nursery. Kris Plant. Philippines.

Aloe ă-lō-ee *Liliaceae* (*Aloeaceae*). From the native name. Tender succulents.
arborescens ar-bo-*res*-enz. Tree-like. Candelabra Plant. S Africa.
aristata ă-ris-*tah*-ta. With a long, bristle-like tip (the leaves). Lace Aloe. S Africa.

Aloe (continued)
brevifolia brev-i-*fo*-lee-a. With short leaves. S Africa.
ciliaris ki-lee-*ah*-ris. Fringed with hairs. Climbing Aloe. S Africa.
ferox fe-rŏks. Fierce (the very spiny leaves). Cape Aloe. S Africa.
humilis hum-i-lis. Low-growing. S Africa.
striata stree-*ah*-ta. Striped (the leaves). S Africa.
variegata vă-ree-a-*gah*-ta. Variegated (the leaves). Partridge-breasted Aloe. S Africa.
vera ve-ra (= *A. barbadensis*). True. Mediterranean region (naturalised).

Alonsoa a-*lon*-zō-a *Scrophulariaceae.* After Alonzo Zanoni, Secretary of State of Colombia in the 18th century. Tender annuals. Mask Flower. Peru.
acutifolia a-kewt-i-*fo*-lee-a. With pointed leaves.
warscewiczii var-sha-*vich*-ee-ee. After Joseph Warscewicz (1812–66) who collected in S America.

Alopecurus ă-lo-pe-*kew*-rus *Gramineae.* From *alopekouros*, Gk. name for a grass with an inflorescence like a fox's tail. Perennial grass.
pratensis prah-*tayn*-sis. Growing in meadows. Meadow Foxtail. Europe, Caucasus, N Asia.

Aloysia ă-lō-*is*-ee-a *Verbenaceae.* After Maria Louisa, Princess of Parma, died 1819. Semi-hardy deciduous shrub.
triphylla tri-*fil*-la. (= *Lippia citriodora*). Three-leaved, the leaves are in whorls of three. Lemon Verbena. Chile.

Alsobia al-sō-*bee*-a *Gesneriaceae.* From Gk. *alsos*, forest, referring to the habitat. Perennial herbs. Tropical America.
dianthiflora dee-ănth-i-*flō*-ra (= *Episcia dianthiflora*). *Dianthus*-flowered. Lace Flower Vine. Mexico.

Alstroemeria ahl-strurm-*e*-ree-a *Liliaceae* (*Alstroemeriaceae*). After Baron Claus Alstroemer (1736–94). Hardy and semi-hardy perennial herbs. Chile.
aurea ow-ree-ă (= *A. auriantiaca*). Orange (the flowers).
ligtu lig-too. The Chilean name.

Alternanthera ăl-ter-nan-*the*-ra *Amaranthaceae.* From L. *alternans* (alternating) and *anthera* (an anther), alternate anthers are sterile. Tender foliage herbs.

Alternanthera (continued)
dentata den-*tah*-ta. Toothed. W Indies, S America.
ficoidea fee-*koi*-dee-a. Like *Ficus*. Mexico, C and S America.
amoena a-*moy*-na. Pleasant.
'Bettzickiana' bet-zik-ee-*ah*-na. After Bettzick.
'Versicolor' ver-*si*-ko-lor. Variously coloured (the leaves).

Althaea ăl-*thie*-a *Malvaceae*. The classical name for Gk. *althaine* (to heal) referring to medicinal properties. Perennial herb.
ficifolia see *Alcea ficifolia*.
officinalis o-fik-i-*nah*-lis. Sold as a herb. Marsh Mallow. Europe.
rosea see *Alcea rosea*.

Aluminium Plant see *Pilea cadierei*
Alum Root see *Heuchera*

Alyssum a-*lis*-sum *Cruciferae*. From Gk. *a* (not) and *lyssa* (madness), it was said to cure rabies. Madwort. Annual and perennial herbs.
alpestre ăl-*pes*-tree. Of the lower mountains. C Europe.
idaeum ee-*die*-um. Of Mt. Ida of ancient Greece. Crete.
maritimum see *Lobularia maritima*.
montanum mon-*tah*-num. Of mountains. Europe.
murale mew-*rah*-lee. Growing on walls. E Europe.
saxatile see *Aurinia saxatilis*
serpyllifolium ser-pil-li-*fo*-lee-um. Thyme-leaved. SW Europe.
spinosum speen-*ō*-sum. Spiny (the stems). S France, Spain.
wulfenianum wul-fen-ee-*ah*-num. After Wulfen. see *Wulfenia*. S Europe.

Amaranthus ăm-a-*răn*-thus *Amaranthaceae*. From Gk. *amarantos* (unfading) referring to the long-lasting flowers. Annual herbs.
caudatus kaw-*dah*-tus. With a tail, referring to the slender, drooping inflorescence. Love-lies-Bleeding. Tropics.
cruentus crew-ent-us. Blood-stained, referring to the flowers. Purple Amaranth. Americas.
hypochondriacus hi-pō-kon-*dri*-ă-cus. Of melancholy or sombre appearance; referring to colour of flowers. Widespread.
tricolor tri-ko-lor. Three-coloured (the leaves). Joseph's Coat. Tropics.
salicifolius să-lik-i-*fo*-lee-us. *Salix*-leaved. Fountain Plant.

Amaryllis ăm-a-*ril*-lis *Amaryllidaceae*. After a shepherdess in Gk. mythology. Semi-hardy, bulbous herb. The large-flowered bulbs commonly called *Amaryllis* are *Hippeastrum* hybrids.
belladonna bel-a-*don*-a. Beautiful lady. Belladonna Lily. S Africa.

Amberboa am-ber-*bō*-a (Compositae). From the Turkish name for *A. moschata*. Annual or biennial herbs. Med to C Asia.
moschata mos-*kah*-ta. (= *Centaurea moschata*) Musk scented. Sweet Sultan. W Asia.

Amelanchier ă-me-lan-kee-er *Rosaceae*. From the Provençale name for *A. ovalis*. Deciduous trees and shrubs. Serviceberry.
alnifolia ăl-ni-*fō*-lee-a. Alder-leaved. Alder-leaved Service Berry. NW America.
semi-integrifolia sem-i-in-teg-ri-fō-lee-a (= *A. florida*) Half-entire leaved (lit.) probably referring to toothed margins on upper part of the leaf.
canadensis kăn-a-*den*-sis. Of Canada or America. E N America.
lamarckii la-*mar*-kee-ee. (= *A. laevis*). After Lamarck (1744–1829), French naturalist. E N America.
ovalis ō-*vah*-lis. Oval (the leaves). Snowy Mespilus. C & S Europe.
spicata spee-*kah*-ta. In spikes (the flowers). E N America.
stolonifera sto-lō-*ni*-fe-ra. Producing stolons. E N America.

Ammannia a-*mah*-nee-a *Lythraceae*. After Paul Ammann (1634–91), German botanist. Aquarium plants.
baccifera bah-*ki*-fe-ra. Berry-bearing. Old World tropics.
senegalensis sen-e-gahl-*en*-sis. Of Senegal.

Amorpha a-*mor*-fa *Leguminosae*. (Papilionoideae). From Gk. *amorphos* (deformed) the flowers have only one petal. Deciduous shrub.
fruticosa froo-ti-*kō*-sa. Shrubby. False or Bastard Indigo. S United States.

Amorphophallus a-mor-fo-*făl*-lus *Araceae*. From Gk. *amorphos* (deformed) and *phallus* referring to the shape of the tubers. Tender perennial herbs.
bulbifer bul-bi-fer. Bulb-bearing (the leaves). Himalaya, India, Burma.
rivieri ri-vee-*e*-ree. After M. Riviere, head

Amorphophallus (continued)
gardener at Luxembourg Palace Gardens, Paris. SE Asia.

Ampelopsis ăm-pel-*op*-sis *Vitaceae*. From Gk. *ampelos* (vine) and *-opsis* indicating resemblance. Deciduous, woody climbers.
aconitifolia ă-kon-ee-ti-*fo*-lee-a. *Aconitum*-leaved. China.
brevipedunculata brev-ee-pe-dunk-ew-*lah*-ta. With short peduncles. NE Asia.
'Citrulloides' kit-rul-*loi*-deez. Like *Citrullus*, the water melon, with deeply divided leaves.
'Elegans' *ay*-le-gahnz. Elegant, the variegated leaves.
megalophylla meg-a-lō-*fil*-la. Large-leaved. W China.

Amsonia ăm-*son*-ee-a *Apocynaceae*. After Dr Charles Amson, 18th century Virginia physician. Blue Star. Herbaceous perennial.
orientalis ō-ree-en-*tah*-lis (= *Rhazya orientalis*) Eastern. SE Europe, W Asia.
tabernaemontana tă-ber-nie-mon-*tah*-nă. After Jakob Theodor von Bergzabern (Tabernaemontanus), 16th century physician and herbalist. E United States.

Anacampseros ăn-a-*kămp*-se-ros *Portulacaceae*. Gk. name for a herb reputed to restore love. Tender succulent.
rufescens roof-*e*-senz. Reddish (the leaves). S Africa.

Anacyclus ăn-a-*sik*-lus *Compositae*. From Gk. *an* (without) *anthos* (a flower) and *kuklos* (a ring) referring to the circle of ovaries surrounding the disc. Perennial herb.
pyrethrum pi-*ree*-thrum. From Gk. *pyr* meaning fire, an allusion to its herbal use as a rubifacient.
depressus day-*pres*-sus. Depressed (the habit). Mount Atlas Daisy. Morocco.

Anagallis ăn-a-*gah*-lis *Primulaceae*. Gk. name of a plant. Annual and perennial herbs.
arvensis ar-*ven*-sis. Of cultivated fields. Scarlet Pimpernel. Europe.
caerulea kie-*roo*-lee-a. Blue (the flowers).
monelli mon-el-ee. After ? SW Europe.
linifolia leen-i-*fo*-lee-a. (= *A. linifolia*). *Linum*-leaved.
tenella ten-*el*-la. Dainty. Bog Pimpernel. W Europe, N Africa.

Ananas ă-na-nas *Bromeliaceae*. From the native name. Tender evergreen perennial herbs.

Ananas (continued)
bracteatus brăk-tee-*ah*-tus. With bracts (on the fruit). Wild Pineapple. Brazil.
comosus kom-ō-sus. With a tuft of leafy bracts (on the fruit). Pineapple. Cult.

Anaphalis a-*nah*-fa-lis *Compositae*. From the Gk. name of a similar plant. Herbaceous perennials. Pearly Everlasting.
margaritacea mar-ga-ri-*tah*-kee-a. Pearl-like (the flower heads). Himalaya.
yedoensis yed-ō-*en*-sis. Of Yeddo (now Tokyo), Japan.
nepalensis nep-ă-len-sis (= *A. triplinervis*) Of Nepal. Himalaya, China.
monocephala mon-ō-*kef*-a-la. (= *A. nubigenus* hort.). One-headed.

Anchusa ăn-*kew*-sa *Boraginaceae*. From the L. name. Annual, biennial or perennial herbs.
azurea a-*zewr*-ree-a. (= *A. italica*). Deep blue (the flowers). S Europe.
caespitosa kie-spi-*tō*-sa. Tufted. Crete.
capensis ka-*pen*-sis. Of the Cape of Good Hope.

Andromeda ań-*drom*-e-da *Ericaceae*. After Andromeda of Gk. mythology. Evergreen shrubs. Bog Rosemary.
glaucophylla glow-kō-*fil*-la. Glaucous-leaved. N America.
polifolia pol-i-*fo*-lee-a. Grey-leaved. N temperate regions.

Androsace ăn-*dros*-a-kee *Primulaceae*. Gk. name for another plant from *aner* (a man) and *sakos* a shield. Alpine perennial herbs. Rock Jasmine.
alpina ăl-*peen*-a. Alpine. Alps.
carnea kar-nee-a. Flesh-coloured (the flowers). Alps. Pyrenees.
chamaejasme kăm-ie-*yăs*-mee. Dwarf jasmine. S Europe to N Russia.
helvetica hel-*vay*-ti-ka. Swiss. Alps.
lanuginosa lah-noo-gi-*nō*-sa. Woolly. Himalaya.
'Leichtlinii' liekt-*lin*-ee-ee. After Max Leichtlin (1831–1910).
primuloides preem-ew-*loi*-deez. Like *Primula*. Himalaya.
chumbyi chum-bee-ee. Of the Chumby Valley, N India.
pyrenaica pi-ray-*nah*-i-ka. Of the Pyrenees.
sarmentosa sar-men-*tō*-sa. Producing runners. Himalaya.
sempervivoides sem-per-vee-*voi*-deez. Like *Sempervivum*. Himalaya.
villosa vil-*lō*-sa. Softly hairy. Europe, W Asia.

Androsace (continued)
arachnoidea ă-răk-*noi*-dee-a. With hairs
like a spider's web.

Anemone a-*nem*-o-nee *Ranunculaceae*.
Derivation as for Adonis who was also
known as Naamen. His blood is said to have
given rise to the blood-red flowers of *A.
coronaria*. Perennial herbs.
alpina see *Pulsatilla alpina*
apennina ă-pen-*nee*-na. Of the Apennines.
S Europe.
blanda *blăn*-da. Pleasant. SE Europe, W
Asia.
coronaria ko-rō-*nah*-ree-a. Used in
garlands. Mediterranean region, C Asia.
× fulgens *ful*-genz. *A. hortensis* × *A.
pavonina*. Shining (the flowers). C
Mediterranean region.
halleri see *Pulsatilla halleri*
hortensis hor-*ten*-sis. Of gardens.
Mediterranean region.
hupehensis hew-pee-*hen*-sis. Of Hupeh
(now Hubei) China.
× hybrida hib-ri-da. *A. hupehensis* × *A.
vitifolia*. Hybrid.
× lesseri les-sa-ree. *A. multifida* × *A.
sylvestris*. After L. Lesser.
narcissiflora nar-kis-i-*flō*-ra. *Narcissus*-
flowered. S and C Europe, W Asia.
nemorosa nem-o-*rō*-sa. Of woods. Wood
Anemone. N Europe, NW Asia.
pulsatilla see *Pulsatilla vulgaris*
rivularis reev-ew-*lah*-ris. Growing by
streams. Himalaya, India, China.
vernalis see *Pulsatilla vernalis*
virginiana vir-jin-ee-*ah*-na. Of Virginia. N
America.

Anemonopsis a-nem-o-*nop*-sis *Ranunculaceae*.
From *Anemone* q.v. and Gk. -*opsis* indicating
resemblance. Perennial herb.
macrophylla măk-rō-*fil*-la. Large-leaved.
Japan.

Anethum a-*nay*-thum *Umbelliferae*. The Gk.
name. Annual herb.
graveolens gra-*vay*-o-lenz. Strong smelling.
Dill. SW Asia.

Angelica ăn-*gel*-i-ka *Umbelliferae*. From the
angelic medicinal properties. Biennial or
perennial herb.
archangelica ark-ăn-*gel*-i-ka. After the
Archangel Raphael. Angelica. Europe,
Asia.

Angel's Fishing Rod see *Dierama
pulcherrimum*

Angel's Tears see *Billbergia nutans, Narcissus
triandrus*
Golden see *Narcissus triandrus pallidulus*
Angel's Trumpet see *Brugmansia arborea*
Angel's Wings see *Caladium*

Angraecum ăn-*griek*-um *Orchidaceae*. From
angurek the Malayan name for epiphytic
plants. Greenhouse orchids.
distichum *dis*-ti-kum. Two-ranked (the
leaves). Tropical Africa.
eburneum ee-*burn*-ē-um. Ivory-white.
superbum soo-*perb*-um. Superb.
Madagascar.
eichlerianum iek-la-ree-*ah*-num. After
August Wilhelm Eichler (1839–1887),
German botanist. Tropical W Africa.
falcatum făl-*kah*-tum. Sickle-shaped (the
leaves). China, Japan.
sesquipedale ses-kwee-ped-*ah*-lee. 1½ feet
long (the spur of the flower). Star of
Bethlehem Orchid. Madagascar.

Anguloa ăn-gew-*lō*-a *Orchidaceae*. After Don
Francisco de Angulo, 18th century Spanish
botanist. Greenhouse orchids. Cradle
Orchid.
clowesii clowz-ee-ee. After the Rev. John
Clowes (1777–1846), who built up a
large collection of orchids. Colombia.
ruckeri ruk-a-ree. After Mr Rucker, an
orchid grower. Colombia.
uniflora ew-ni-*flō*-ra. One-flowered (as are
all species). Peru.

Anigozanthus a-nee-gō-*zăn*-thus
Haemadoreaceae. From Gk. *anoigo* (to open)
and *anthos* (a flower), referring to the widely
open flowers. Kangaroo's Paw. SW Australia.
flavidus *flah*-vi-dus. Yellow (the flowers).
manglesii mang-*galz*-ee-ee. After Robert
Mangles (died 1860) who introduced and
grew many W Australian plants,
presumably receiving seed from his two
brothers who both visited W Australia.
viridis *vi*-ri-dis. Green (the flowers).

Animated Oat see *Avena sterilis*

Anisodontea a-nee-sō-*dont*-ee-a *Malvaceae*.
From Gk. *anisos* (unequal) and *odon* (a
tooth). Tender shrub.
scabrosa skă-*brō*-sa. (= *Malvastrum
scabrosum*). Rough. Hairy Mallow. S
Africa.

Annona a-*no*-na *Annonaceae*. From the native
name. Tender evergreen shrubs.

Annona (continued)
cherimola ke-ree-*mō*-la. From *cherimoya* the native name. Custard Apple. N Andes.
muricata mew-ri-*kah*-ta. With sharp points (the fruit). Sour Sop. Tropical America.
reticulata ray-tik-ew-*lah*-ta. Netted (the carpels). Custard Apple. Tropical America.
squamosa skwah-*mō*-sa. Scaly (the fruit). Sweet Sop. Tropical America.

Anoectochilus a-neek-tō-*keel*-us *Orchidaceae*. From Gk. *anoektos* (open) and *cheilos* (a lip), the blades of the lip spread, giving it an open appearance. Warm greenhouse orchids.
argyroneurus arg-i-rō-*newr*-rus. Silver-veined (the leaves). Java.
discolor dis-ko-lor. Of two colours (the leaves). China.
elwesii el-*wez*-ee-ee. After H. J. Elwes, see *Galanthus elwesii*. Himalaya.
regalis ray-*gah*-lis. Royal. King Plant. Sri Lanka.

Anomatheca an-ō-ma-*theek*-a *Iridaceae* From Lat. *anomalus*, abnormal, and *theca*, case; the fruit capsule is covered with warts. Cormous perennials, closely related to *Lapeirousia*.
laxa lăk-sa (= *Lapeirousia laxa*) Loose (the inflorescence). S Africa.

Antennaria ăn-ten-ah-ree-a *Compositae*. From L. *antenna*, the pappus hairs of the male flowers have swollen tips resembling a butterfly's antennae. Perennial herb.
dioica dee-ō-*ee*-ka. Dioecious. Cat's Foot. Arctic and alpine N Hemisphere.

Anthemis ăn-them-is *Compositae*. The Gk. name for *Chamaemelum nobile*. Perennial herbs.
arabica see *Cladanthus arabicus*
marschalliana mar-shăl-ee-*ah*-na. (= *A. biebersteiniana*, *A. rudolphiana*). After Bieberstein, see *Cerastium biebersteinii*. Caucasus.
nobilis see *Chamaemelum nobile*
punctata punk-*tah*-ta. Spotted.
    cupaniana kew-pahn-ee-*ah*-na. After Francesco Cupani (1657–1711), Italian monk and naturalist. Italy.
sancti-johannis sank-tee-yō-*hăn*-is. For its flowering in mid-summer, referring to St John's day, June 24th. Bulgaria.
tinctoria tink-*to*-ree-a. Used in dyeing, a yellow dye can be extracted from the flowers. Dyer's Chamomile, Yellow Chamomile. Europe, W Asia.

Anthericum ăn-*the*-ri-kum *Liliaceae*. From *antherikon* Gk. name for asphodel. Perennial herbs.
liliago lee-lee-*ah*-go. Like *Lilium*. St Bernard Lily. S Europe.
ramosum rah-*mō*-sum. Branched (the inflorescence). W and S Europe.

Antholyza paniculata see *Crocosmia paniculata*

Anthriscus ăn-*thris*-kus *Umbelliferae*. L. name for a plant. Annual herb.
cereifolium kay-ree-*fo*-lee-um. Waxen-leaved. Chervil. SE Europe, W Asia.

Anthurium ăn-*thewr*-ree-um *Araceae*. From Gk. *anthos* (a flower) and *oura* (a tail) referring to the slender, tail-like spadix which bears the flowers. Tender perennial herbs.
andraeanum on-dray-*ah*-num. After Edouard Francis André (1840–1911). Tail Flower, Painter's Palette. Colombia.
crystallinum kris-tal-*lee*-num. Crystalline (the appearance of the leaves). Strap Flower. Colombia.
scherzerianum skairts-a-ree-*ah*-num. After Herr Scherzer of Vienna who discovered it. Flamingo Flower, Flame Plant. Guatemala, Costa Rica.

Anthyllis ăn-*thil*-lis *Leguminosae*. The Gk. name. Shrubs.
barba-jovis bar-ba-*yo*-vis. Jupiter's beard, referring to the silky hairy leaves and shoots. Jupiter's Beard. SW Europe, Mediterranean region.
hermanniae her-*mah*-nee-ie. From the resemblance to *Hermannia*. Mediterranean region.
montana mon-*tah*-na. Of mountains. S Europe.

Antigonon ăn-*ti*-go-non *Polygonaceae*. From Gk. *anti* (against) and *gonia*, an angle referring to its flexuous stems. Tender perennial climber.
leptopus lep-to-pus. Slender-stalked. Coral Vine. Mexico.

Antirrhinum ăn-tee-*ree*-num *Scrophulariaceae*. From Gk. *anti* (against) and *rhis* (a snout) from the appearance of the flower. Perennial herb.
asarina see *Asarina procumbens*
majus mah-yus. Larger. Snapdragon. Mediterranean region.

Anubias a-*new*-bee-as *Araceae*. From Anubis the ancient Egyptian god. Aquarium plants.

14   Aph-Ara

Anubias (continued)
afzelii ăf-*zee*-lee-ee. After Adam Afzelius
(1750–1837) who collected in W Africa.
Sierra Leone.
congensis kon-*gen*-sis. Of the Congo. W
Africa.
nana nah-na. Dwarf. Cameroons.

Aphelandra ăf-el-*ăn*-dra *Acanthaceae.* From
Gk. *apheles* (simple) and *aner* (male) referring
to one-celled anthers. Tender evergreen
shrubs.
aurantiaca ow-răn-tee-*ah*-ka. (= *A. nitens*).
Orange (the flowers). Mexico to NS
America.
    'Roezlii' *rurz*-lee-ee. After Benedikt
    Roezl (1824–85), a plant collector.
chamissoniana sham-i-son-ee-*ah*-na. After
Ludolf Adalbert von Chamisso
(1781–1838). Brazil.
squarrosa skwah-*rō*-sa. With the parts
spreading horizontally, referring to the
flower spikes. Saffron Spike, Zebra Plant.
Brazil.
    'Louisae' loo-*eez*-ie. After Queen
    Louise of Belgium.

Apium ă-pee-um *Umbelliferae.* The L. name
for celery and parsnip. Biennial herb.
graveolens gra-*vay*-o-lenz. Strong smelling.
Wild Celery. Widely distributed.
dulce dul-kee. Sweet. Celery.
rapaceum ra-*pah*-kee-um. Turnip-like, the
swollen stem. Celeriac.

Aponogeton a-pon-ō-*gay*-ton
Aponogetonaceae. From *Aquae Aponi* a
Roman healing spring and *geiton* (a
neighbour). Aquatic perennials.
crispus kris-pus. Crisped (the leaves). Sri
Lanka.
distachyos di-*stă*-kee-ŏs. With two spikes,
referring to the forked inflorescence.
Cape Asparagus, Cape Pondweed. S
Africa.
madagascariensis măd-a-găs-kay-ree-*en*-sis.
(= *A. fenestralis*). Of Madagascar.
Madagascar Lace Plant.
ulvaceus ul-*vah*-kee-us. Like *Ulva* (Sea
lettuce). Madagascar.
undulatus un-dew-*lah*-tus. Wavy-edged
(the leaves). SE Asia.

Aporocactus a-po-rō-*kăk*-tus *Cactaceae. From*
Gk. *aporos* (impenetrable) and *Cactus* q.v.
flagelliformis fla-gel-lee-*form*-is. Whip-like
(the slender stems). Rat's-tail Cactus.
Mexico.
× *mallisonii* see × *Heliaporus smithii.*

Apple see *Malus*
Apple of Peru see *Nicandra physalodes*
Apricot see *Prunus armeniaca*
    Japanese see *P. mume*

Aptenia ăp-*teen*-ee-a *Aizoaceae.* From Gk.
apten (unwinged), the capsules lack wings
distinguishing the genus from
Mesembryanthemum in which it was once
included. Tender succulent.
cordifolia kor-di-*fo*-lee-a. (=
Mesembryanthemum cordifolium). Heart-
shaped (the leaves). S Africa.

Aquilegia ă-kwi-*lee*-gee-a *Ranunculaceae.*
From L. *aquila* (an eagle) referring to the
shape of the petals. Perennial herbs.
Columbine.
alpina ăl-*peena.* Alpine. Alps.
bertolonii ber-to-*lō*-nee-ee. After Antonio
Bertoloni (1775–1869), Italian botanist.
S. Europe.
caerulea kie-*ru*-lee-a. Deep blue (the
flowers). Rocky mountains.
canadensis kăn-a-*den*-sis. Of Canada or NE
America. E N America.
chrysantha kris-*ănth*-a. With golden
flowers. SW United States, N Mexico.
discolor dis-ko-lor. Two-coloured (the
flowers). Spain.
flabellata flă-bel-*lah*-ta. Fan-shaped (the
leaflets). Japan.
glandulosa glăn-dew-*lō*-sa. Glandular.
Siberia.
longissima long-*gis*-si-ma. Longest,
referring to the very long spur. Texas.
scopulorum skop-ew-*lo*-rum. Growing on
cliffs. W United States.
vulgaris vul-*gah*-ris. Common. Europe.

Arabis *ă*-ra-bis *Cruciferae.* Derivation
obscure. Perennial herbs.
blepharophylla ble-fa-rō-*fil*-la. With fringed
leaves. California.
caucasica kaw-*kăs*-i-ka. Of the Caucasus.
SE Europe, W Asia.
ferdinandi-coburgii fer-di-*nahn*-dee-ee-kō-
burg-ee-ee. After King Ferdinand of
Bulgaria. Bulgaria.

Aralia a-*rah*-lee-a *Araliaceae.* From the
French-Canadian name *aralie.* Perennial
herbs and deciduous shrubs.
cachemerica kăsh-*meri*-ka. Of Kashmir.
Himalaya, Tibet.
elata ay-*lah*-ta. Tall. Japanese Angelica
Tree. NE Asia.
elegantissima see *Schefflera elegantissima*
sieboldii see *Fatsia japonica*

Aralia (continued)
spinosa spee-nō-sa. Spiny (the stem and petioles). Hercules Club. E United States.

Araucaria ă-row-*kay*-ree-a *Araucariaceae*. From the Araucani Indians who live where *A. araucana* grows. Hardy and tender conifers.
araucana ă-row-*kah*-na. As above. Monkey Puzzle, Chile Pine. Chile, Argentina.
heterophylla he-te-rō-*fil*-la. Variably-leaved. Norfolk Island Pine, Norfolk Islands.

Araujia a-*row*-hee-a. *Asclepiadaceae*. From the Brazilian name. Evergreen climber.
sericofera se-ri-*ko*-fe-ra. Silk-bearing, referring to the white hairs on the young shoots and beneath the leaves. Cruel Plant. S America.

Arbutus ar-bu-tus *Ericaceae*. The L. name. Evergreen trees and shrubs.
andrachne ăn-*drăk*-nee. The Gk. name. SE Europe.
× *andrachnoides* ăn-drăk-*noi*-deez. *A. andrachne* × *A. unedo*. Like *A. andrachne*. Greece.
menziesii men-*zeez*-ee-ee. After Menzies, see *Menziesia*. Madrona. WN America.
unedo ew-nee-do. The L. name. Strawberry Tree. Mediterranean Region, SW Ireland.

Arctostaphylos ark-tō-*stă*-fil-os *Ericaceae*. From Gk. *arctos* (a bear) and *staphyle* (a bunch of grapes). Evergreen shrubs.
manzanita măn-za-*neet*-a. The native name for this and similar species. Manzanita. California.
nevadensis nev-a-den-sis. Of the Sierra Nevada, California.
uva-ursi oo-va-ur-see. Bear's grape. Bearberry. N temperate regions.

Arctotis ark-tō-tis *Compositae*. From Gk. *arctos* (a bear) and *otus* (an ear), the pappus scales are said to resemble bear's ears. Annual and perennial herbs. African Daisy. S. Africa.
acaulis a-*kaw*-lis. Stemless.
breviscapa brev-ee-*skah*-pa. With a short scape.
fastuosa fas-tew-ō-sa (= *Venidium fastuosum*). Proud. Monarch of the Veldt, Namaqualand Daisy.
venusta vee-*nus*-tă. (= *A. stoechadifolia*). handsome; charming. Blue-eyed African Daisy.

Ardisia ar-*dis*-ee-a *Myrsinaceae*. From Gk.

Ardisia (continued)
Schefflera ardis (a point) referring to the pointed anthers. Tender, evergreen shrub.
crispa kris-pa. Finely wavy (the leaf margins). Coral Berry. Japan, S China.

Areca a-*ree*-ka *Palmae*. From the native name. Tender palm.
catechu kah-tee-koo. From the native name. Betel Nut Palm. Malaysia.

Arenaria ă-ray-*nah*-ree-a *Caryophyllaceae*. From L. *arena* (sand) many species grow in sandy places. Low growing perennial herbs.
balearica bă-lee-*ah*-ri-ka. Of the Balearic Islands. W Mediterranean region.
caespitosa hort. see *Sagina subulata*.
laricifolia see *Minuartia laricifolia*
montana mon-*tah*-na. Of mountains. W Europe.
purpurascens pur-pewr-*răs*-enz. Purplish (the flowers). SW Europe.
tetraquetra tet-ra-*kwee*-tra. Four-angled (the shoots). Pyrenees, Spain.

Argemone ar-ge-*mō*-nee *Papaveraceae*. Gk. name for a similar plant. Annual herbs.
grandiflora grăn-di-*flō*-ra. Large flowered. Mexico.
mexicana meks-ki-*kah*-na. Mexican. Devil's Fig. Prickly Poppy. Florida, C America.
platyceras plă-tee-*ke*-ras. With broad horns (on the fruit). Crested Poppy. Mexico.

Argyranthemum ar-jy-*ran*-thee-mum *Compositae*. From Gk. *argyros*, silver, and *anthemos*, flower. Perennial subshrubs. Macronesia.
frutescens froo-*tes*-enz (= *Chrysanthemum frutescens*). Shrubby. Paris Daisy. White Marguerite. Canary Islands.

Argyrocytisus ar-jy-rō-*si*-tis-us *Leguminosae* (*Papilionoideae*) From Gk. *argyros*, silver, and *Cytisus* q.v., in which this genus was previously included. Deciduous shrub. Morocco.
battandieri ba-ton-dee-e-ree (= *Cytisus battandieri*). After Jules Aimé Battandier, French botanist who studied Algerian plants. Pineapple-scented Broom.

Argyroderma ar-gi-rō-*der*-ma. *Aizoaceae*. From Gk. *argyros* (silver) and *derma* (skin) referring to the whitish leaves. Tender succulents.
octophyllum ok-tō-*fil*-lum. Eight leaved, which it rarely is. S Africa.

Arisaema ă-ris-*ie*-ma *Araceae. From Gk. aron* (arum) and *haema* (blood) referring to red-blotched leaves in some species. Perennial herbs.
candidissimum kăn-di-*dis*-si-mum. Most white (the spathe). W China.
dracontium dra-*kon*-tee-um. Dragon-like. S United States.
triphyllum tri-*fil*-lum. With three leaves (leaflets). Jack-in-the-Pulpit, Indian Turnip. E and S N America.

Arisarum a-*ris*-a-rum *Araceae.* From *arisaron* Gk. name for *A. vulgare.* Perennial herb.
proboscideum pro-bo-*ski*-dee-um. Like an elephant's trunk (the spadix). Mouse-tail Plant. S Italy, SW Spain.

Aristolochia a-ris-to-*lok*-ee-a *Aristolochiaceae.* From Gk. *aristos* (best) and *lochia* (childbirth) referring to supposed medicinal properties. Hardy and tender climbers. Birthwort.
littoralis lit-or-*ay*-lis (= *A. elegans*), Of the sea shore. Calico flower, Brazil.
macrophylla măk-rō-*fil*-la. (= *A. durior. A. sipho*). Large-leaved. Dutchman's Pipe. E United States.
sempervirens sem-per-vy-rens (= *A. altissima*). Evergreen.

Armeria ar-*me*-ree-a *Plumbaginaceae.* From the L. name for a *Dianthus.* Evergreen perennials. Thrift.
alliacea ăl-lee-ah-kee-a. (= *A. plantaginea*). *Allium*-like (the leaves). Jersey Thrift. W Europe.
caespitosa see-spit-*o*-sa (= *A. juniperifolia*) Tufted. Spain.
maritima ma-*ri*-ti-ma. Growing near the sea. Common Thrift. Widely distributed.
mauritanica mo-ri-*tah*-ni-ka. Of N Africa.
pseudarmeria sood-ar-*me*-ree-a. (= *A. latifolia*). False *Armeria.* Portugal.

Armoracia ar-mo-*rah*-kee-a *Cruciferae.* L. name for a related plant. Perennial herb.
rusticana rus-ti-*kah*-na. Of the country. Horse-radish. SE Europe.

Arnebia ar-*nee*-bee-a *Boraginaceae.* From the Arabian name. Annual and perennial herbs.
decumbens day-*kum*-benz. (= *A. cornuta*). Prostrate. N Africa, Russia, W and C Asia.
pulchra pul-kra. (= *A. echioides* hort.). Pretty. Prophet Flower. W Asia.

Arnica ar-ni-ka *Compositae.* From Gk. *arnakis*

Arnica (continued)
(lambskin) from the texture of the leaves. Perennial herb.
montana mon-*tah*-na. Of the mountains. Mountain Tobacco, Leopard's Bane. Europe.

Aronia a-*rō*-nee-a *Rosaceae. From aria* the Gk. name for *Sorbus aria.* Deciduous shrubs Chokeberry.
arbutifolia ar-bewt-i-*fo*-lee-a. *Arbutus*-leaved. Red chokeberry. E N America.
melanocarpa me-la-nō-*kar*-pa. With black fruits. Black Chokeberry. E N America.

Arrhenatherum ăren-*ă*-the-rum *Gramineae.* From Gk. *arren* (male) and *ather* (a bristle) referring to the bristles on the male flowers. Perennial grass.
elatius ay-*lah*-tee-us. Tall. Oat Grass. Europe, N Africa, W Asia.

Arrowhead see *Sagittaria*
Arrowhead Vine see *Syngonium angustatum, S. podophyllum*
Arrowroot see *Maranta arundinacea*

Artemisia ar-tay-*mis*-ee-a *Compositae.* After the Gk. goddess *Artemis.* Perennial herbs, shrubs and sub-shrubs.
abrotanum a-*brot*-a-num. The L. name. Southernwood. S Europe.
absinthium ăb-*sin*-thee-um. The L. name. Wormwood. Europe.
aborescens ar-bo-*res*-enz. Becoming tree-like. Mediterranean region.
dracunculus dra-*kun*-kew-lus. A small dragon; the words dragon and tarragon have the same derivation. French Tarragon. S Europe, Asia.
   inodorus in-o-*kō*-rus. Unscented Russian Tarragon.
frigida fri-gi-da. Of cold regions. W N America, Siberia.
lactiflora lăk-ti-*flō*-ra. Milk-flowered, referring to the creamy-white flowers. China.
ludoviciana loo-do-vik-ee-*ah*-na. Of Lousiana. W United States, Mexico.
schmidtiana shmit-ee-*ah*-na. After Schmidt. Japan.
stelleriana stel-la-ree-*ah*-na. After its discoverer Georg Wilhelm Steller (1709–46). Dusty Miller, Old Woman. NE Asia.
tridentata see *Seriphidium tridentatum.*

Arthropodium arth-rō-*pō*-dee-um *Liliaceae.* (*Asphodelaceae*) From Gk. *arthron* (a joint) and

Arthropodium (continued)
podion (a stalk) referring to the jointed pedicels. Perennial herb.
   cirrhatum ki-*rah*-tum. With tendrils, referring to the tendril-like appendages on the filaments. Rock Lily. New Zealand.

Artichoke, Chinese see *Stachys affinis*
   Globe see *Cynara scolymus*
   Jerusalem see *Helianthus tuberosus*
   Artillery Plant see *Pilea microphylla*

Arum ǎ-rum *Araceae*. From *aron* the Gk. name. Tuberous perennial herbs. The name *Arum* is also commonly applied to *Zantedischia*.
   cornutum see *Sauromatum guttatum*
   creticum kray-ti-kum. Of Crete.
   italicum ee-tǎ-li-kum. Of Italy. Europe, N Africa, W Asia.
   maculatum mǎk-ew-*lah*-tum. Spotted (the spathe). Cuckoo Pint, Lords and Ladies. Europe.
   nigrum ny-grum (= *A. orientale*). Black (the flowers). Eastern. SE Europe, W Asia.
   pictum pik-tum. Painted, referring to the white veined leaves. W Mediterranean region.

Aruncus a-*run*-kus *Rosaceae*. The Gk. name. Herbaceous perennial often called spiraea.
   dioicus dee-ō-*ee*-kus. (= *A. sylvester*). Dioecious. Goat's Beard. Europe, Asia, N America.

Arundinaria a-run-di-*nah*-ree-a *Gramineae*. From L. *arundo* (a reed). Bamboos.
   anceps see *Yushania anceps*
   gigantea jy-*gan*-tee-ǎ Unusually tall, or large. Switch cane. SE U.S.
   japonica see *Pseudosasa japonica*
   murielae see *Thamnocalamus spathacea*
   nitida see *Sinarundinaria nitida*
   pumila see *Pleioblastus humilis* var. *pumilus*
   vagans see *Sasaella ramosa*
   variegata see *Pleioblastus variegatus*
   viridistrata see *Pleioblastus auricoma*.

Arundo a-*run*-dō *Gramineae*. From L. *arundo* (a reed). Semi-hardy grass.
   donax do-nǎks. Gk. name for a kind of reed. Giant Reed. Mediterranean region.

Asarina a-*sah*-ri-na *Scrophulariaceae*. Spanish name for an *Antirrhinum*. Perennial herb.
   barclaiana bark-lay-*ah*-na (= *Maurandya barclaiana*). After Robert Barclay (1751–1830). Mexico.

Asarina (continued)
   erubescens e-roo-*bes*-enz (= *Maurandya erubescens*). Blushing (the pink flowers). Creeping gloxinia. Mexico.
   procumbens prō-*kumbenz*. (= *Antirrihinum asarina*). Prostrate. Spain, Portugal.
   scandens skǎn – denz (= *Maurandya scandens*). Climbing. Mexico.

Asarum a-*sah*-rum. *Aristolochiaceae*. The L. and Gk. name. Perennial herbs. Wild Ginger.
   canadense kǎn-a-den-see. Of Canada or NE America. E N America.
   caudatum kaw-*dah*-tum. With a tail (the calyx lobes are prolonged into slender tails). W N America.
   europaeum oy-rō-*pie*-um. Of Europe.

Asclepias a-*sklay*-pee-as *Asclepiadaceae*. From Gk. *Asklepios*, god of medicine, referring to medicinal properties. Herbaceous perennials and tender shrubs. Milkweed.
   curassavica ku-ra-*sah*-vi-ka. Of Curacao. S America.
   incarnata in-kar-*nah*-ta. Flesh pink (the flowers). N America.
   speciosa spe-kee-ō-sa. Showy. Canada, W United States.
   tuberosa tew-be-rō-sa. Tuberous (the root). N America, N Mexico.

Ascocentrum ǎs-kō-ken-trum *Orchidaceae*. From Gk. *ascos* (a bag) and *kentron* (a spur) referring to the large spur hanging from the lip. Greenhouse orchids.
   ampullaceum ǎm-pul-*ah*-kee-um. Flask-like (the lip). Himalaya, Burma.
   miniatum min-ee-*ah*-tum. Cinnabar red (the flowers). Java, Philippines.

Ash see *Fraxinus*
   Arizona see *F. velutina*.
   Common see *F. excelsior*.
   Manna see *F. ornus*.
   Narrow-leaved see *F. angustifolia*.
   White see *F. americana*.

Asimina as-*si*-mi-na *Annonaceae*. From the N American Indian name *assimin*. Deciduous shrub or small tree.
   tiloba tri-*lō*-ba. Three lobed (the calyx). Pawpaw. E N America. Not to be confused with the tropical American pawpaw, *Carica papaya*.

Asparagus a-*spǎ*-ra-gus *Liliaceae*. The classical name. Hardy and tender herbs.
   asparagoides a-spǎ-ra-*goi*-deez. *Asparagus*-

Asparagus (continued)
like, it was originally placed in another
genus. Smilax (of florists). S Africa.
densiflorus dens-i-*flō*-rus. Densely-
flowered. S Africa.
'Sprengeri' *spreng*-a-ree. After Carl L.
Sprenger (1846–1917), German
nurseryman who introduced it.
falcatus fål-*kah*-tus. Sickle shaped. Africa,
Sri Lanka.
officinalis o-fi-ki-*nah*-lis. Sold as a herb.
Garden Asparagus.
scandens skån-denz. Climbing. S Africa.
setaceus say-*tah*-kee-us. (= *A. plumosus*)
Bristled. Asparagus Fern. S Africa.

Aspasia a-*spah*-see-a *Orchidaceae*. After
Aspasia wife of Pericles, from Gk. *aspasies*
(delightful). Greenhouse orchids.
lunata loon-*ah*-ta. Crescent shaped, (? the
lip). Brazil.
variegata av-ree-a-*gah*-ta. Variegated (the
flowers). S America.

Aspen see *Populus tremula*.

Asperula a-*spe*-ru-la *Rubiaceae*. From L. *asper*
(rough) referring to the roughly hairy stems.
Annual and perennial herbs.
hexaphylla heks-a-*fil*-la. Six-leaved, the
leaves are in whorls of six. SW Alps.
hirta hir-ta. Hairy, Pyrenees.
lilaciflora li-lah-ki-*flō*-ra.(= *caespitosa*).
With lilac flowers. W Asia.
odorata see *Galium odoratum*.
orientalis o-ree-en-*tah*-lis. (= *A. azurea*).
Eastern. Europe, Asia.
suberosa soo-be-*rō*-sa. Corky-stemmed.
Greece, Bulgaria.

Asphodel see *Asphodelus*
White see *A. albus*
Yellow see *Asphodeline lutea*

Asphodeline a-sfod-e-*lee*-nee *Liliaceae*
(*Asphodelaceae*). Like *Asphodelus* q.v.
Perennial herbs.
liburnica li-*burn*-i-ka. (= *Asphodelus
liburnicus*). From Croatia (Liburnia) on
the Yugoslav coast. SE Europe.
lutea loo-tee-a. Yellow (the flowers).
Yellow Asphodel, King's Spear.
Mediterranean region.

Asphodelus a-*sfod*-e-lus *Liliaceae*
(*Asphodelaceae*) Gk. name for *A. ramosus*.
Perennial herbs. Asphodel.
albus ål-bus. White (the flowers). White
Asphodel. S Europe.

Asphodelus (continued)
liburnicus see *Asphodeline liburnica*
ramosus rah-*mō*-sus. (= *A. cerasiferus*).
Branched (the inflorescence). S Europe.

Aspidistra ă-spi-*di*-stra *Liliaceae*. From Gk.
aspideon (a small, round shield) referring to
the shape of the stigma. Tender, evergreen
herb.
elatior ay-*lah*-tee-or. Taller. Cast Iron
Plant. Japan.

Asplenium a-*splay*-nee-um *Aspleniaceae*.
From Gk. *a* (not) and *splen* (spleen) referring
to supposed medicinal properties. Hardy and
tender ferns. Spleenwort.
adiantum-nigrum ă-dee-ăn-tum-*nig*-rum.
Black Adiantum. Black Spleenwort. N
hemisphere.
bulbiferum bul-*bi*-fe-rum. Producing bulbs
(plantlets are borne on the leaves).
Mother Spleenwort, Hen and Chicken
Fern. New Zealand, Australia, India.
caudatum kaw-*dah*-tum. With a tail (the
pinnae). Tropics.
ceterach *ke*-te-rahk (= *Ceterach officinarum*).
From the Arabic name. Rusty back fern.
Europe and N Africa to the Himalaya.
daucifolium dow-ki-*fo*-lee-um. *Daucus*-
leaved. Mauritius, Reunion Islands.
falcatum fål-*kah*-tum. Sickle-shaped (the
pinnae).
flabellifolium fla-bel-lee-*fo*-lee-um. With
fan-shaped leaves. Necklace Fern. New
Zealand, Australia.
flaccidum *flăk*-ki-dum. Flaccid (the fronds).
Hanging Spleenwort. Australia,
Tasmania, Pacific Islands.
marinum ma-*reen*-num. Growing near the
sea. Sea Spleenwort. W Europe.
nidus *nee*-dus. A nest. Bird's Nest Fern.
Tropical Asia, Polynesia.
platyneuron plă-tee-*newr*-ron. With broad
veins. Ebony Spleenwort. E N America.
rhizophyllum ree-zō-*fil*-lum. Root-leaved,
the leaves root at the tip and hence the
plant 'walks'. Walking Fern. N. America.
ruta-muraria roo-ta-mew-*rah*-ree-a. Rue of
walls. Wall Rue. E United States, Europe,
Asia.
scolopendrium sko-lo-*pen*-dree-um (=
Phyllitis scolopendrium). From Gk.
skolopendra (a Millipede) referring to the
arrangement of the sori. Hart's-tongue
Fern. Europe.
trichomanes tri-kō-*mah*-neez. Gk. name for
a fern. Maidenhair Spleenwort. N
America, Europe, Asia.

Asplenium (continued)
viride vi-ri-dee. Green. Green Spleenwort.
N America, Europe, Asia.

Aster ă-ster *Compositae*. From L. *aster* (a star) referring to the flower heads. Herbaceous perennials.
acris see *A. sedifolius*
albescens ăl-*bes*-enz. (= *Microglossa albescens*). Whitish (the undersides of the leaves). Himalaya, W China.
alpinus ăl-*pee*-nus. Alpine, Europe, Asia.
amellus a-*mel*-lus. The L. name. Europe, W Asia.
ericoides e-ri-*koi*-deez. Like *Erica* (the foliage). N America, Mexico.
farreri fă-ra-ree. After Farrer, see *Viburnum farreri*. W China, Tibet.
× *frikartii* fri-*kart*-ee-ee. *A. amellus* × *A. thomsonii*. After Carl Ludwig Frikart (1879–1964).
natalensis see *Felicia rosulata*
nemoralis ne-mo-*rah*-lis. Of woods. N America.
novae-angliae no-vie-*ang*-glee-ie. Of New England. N America.
novi-belgii nō-vee-*bel*-gee-ee. Of New York (previously called New Belgium). Michaelmas Daisy. E N America.
pappei see *Felicia pappei*
puniceus pew-*ni*-kee-us. Reddish purple (the stems). E N America.
sedifolius say-di-*fo*-lee-us. (= *A. acris*). *Sedum*-leaved. S Europe.
spectabilis spek-*tah*-bi-lis. Spectacular. N America.
thomsonii tom-*son*-ee-ee. After Thomas Thomson (1817–78), Scottish physician and superintendant of Calcutta Botanic Garden. W Himalaya.

Asteranthera a-ste-ran-*the*-ra *Gesneriaceae*. From L. *aster* (a star) and *anthera* (an anther) referring to the star-like arrangement of the anthers. Semi-hardy, evergreen climber.
ovata ō-*vah*-ta. Ovate (the leaves). Chile, Argentina.

Astilbe a-*stil*-bee *Saxifragaceae*. From Gk. *a* (without) and *stilbe* (brilliance), the individual flowers are very small. Perennial herbs.
× *arendsii* ah-*rendz*-ee-ee. After Georg Arends of Ronsdorf (1862–1952), who hybridised astilbes.
chinensis chin-*en*-sis. Of China. China, Japan.
davidii dă-*vid*-ee-ee. After Armand David, see *Davidia*.

Astilbe (continued)
× *crispa* kris-pa. Finely wavy (the leaves).
grandis grăn-dis. Large. China.
japonica ja-*pon*-i-ka. Of Japan.
rivularis reev-ew-*lah*-ris. Growing near streams. Nepal.
× *rosea* ros-ee-a. *A. chinensis* × *A. japonica*. Rose-coloured (the flowers).
simplicifolia sim-pli-ki-*fo*-lee-a. With simple leaves. Japan.
thunbergii thun-*berg*-ee-ee. After Thunberg, see *Thunbergia*.

Astilboides a-stil-*boy*-deez *Saxifragaceae* From *Astilbe* q.v., and Gk. *-oides*, resembling (the flowers). A single species of herb formerly included in *Rodgersia*. N China.
tabularis tăb-ew-*lah*-ris (= *Rodgersia tabularis*). Table-like (the leaves).

Astrantia a-străn-tee-a *Umbelliferae*. Derivation obscure, possibly from L. *aster* (a star) referring to the star-like flowers. Herbaceous perennials, Masterwort.
carniolica kar-nee-o-li-ka. Of Carniola, N Yugoslavia. Europe.
major mah-yor. Larger. Europe.
maxima mahk-si-ma. Largest. S Europe, Caucasus.
minor min-or. Smaller. Europe.

Astrophytum a-*stro*-fi-tum *Cactaceae*. From Gk. *astron* (a star) and *phyton* (a plant) referring to the often star-shaped body.
asterias a-*ste-ree*-as. Star-like. Sea Urchin Cactus, Silver Dollar Cactus. S Texas, N Mexico.
capricorne kă-pri-*kor*-nee. Like a goat's horn. Goat's Horn Cactus. Texas, Mexico.
myriostigma mi-ree-ō-*stig*-ma. With many stigmas. Bishop's Cap Cactus. Mexico.
ornatum or-*nah*-tum. Ornamental. Bishop's Cap Cactus. Mexico.

Athrotaxis ăth-rō-*tăks*-is *Cupressaceae*. From G. *athroos* (crowded) and *taxis* (arrangement) referring to the densely arranged leaves. Evergreen conifers. Tasmania.
cupressoides kew-pres-*oi*-deez. *Cupressus*-like.
× *laxifolia* lăks-i-*fo*-lee-a. (*A. cupressoides* × *A. selaginoides*). Loose leaved, the leaves are more spreading than in *A. cupressoides*.
selaginoides se-lah-gi-*noi*-deez. Like *Lycopodium selago* (the foliage). King William Pine.

Athyrium a-*thi*-ree-um. *Athyriaceae*.

Athyrium (continued)
Derivation obscure possibly from Gk.
athyros (doorless) referring to the late
opening indusium of *A. filix-femina*. Lady
Fern.
   distentifolium dis-ten-ti-fo-lee-um. (= *A.
   alpestre*). With distended leaves. Alpine
   Lady Fern. Europe, Asia.
   filix-femina fi-liks-*fay*-mi-na. Lady fern,
   referring to the delicate fronds as
   compared with the male fern. (*Dryopteris
   filix-mas*). Common Lady Fern. N
   hemisphere.
   nipponicum nip-*on*-ik-um (= *A.
   goeringianum*). From Japan.
   'Pictum' *pik*-tum. Painted (the leaves).
   Japanese Painted Fern.

Atraphaxis ăt-ra-*făks*-is *Polygonaceae*. Gk.
name for a species of *Atriplex*. Deciduous
shrub.
   frutescens froo-*tes*-enz. Shrubby. SE
   Europe, Caucasus, C Asia.

Atriplex ah-tri-plex *Chenopodiaceae*. Gk.
name for *A. hortensis*. Herbs and shrubs.
   canescens kah-*nes*-enz. Grey-hairy (the
   leaves and shoots). Grey Sage Brush. W
   N America.
   halimus hă-li-mus. The Gk. name from
   halimos (maritime) it grows near the sea.
   Tree Purslane. S Europe.
   hortensis hor-*ten*-sis. Of gardens. Orache.
   Asia.

Aubergine see *Solanum melongena*

Aubrieta ō-bree-*aý*-ta *Cruciferae*. After
Claude Aubriet (1668–1743), French
botanical artist. Perennial herb.
   deltoidea del-*toi*-dee-a. Deltoid i.e. shaped
   like the Gk. letter delta (Δ) referring to
   the petals. SE Europe, W Asia.

Aucuba ow-*kew*-ba *Cornaceae*. From the
Japanese name. Evergreen shrub.
   japonica ja-*pon*-i-ka. Of Japan.
   'Crotonifolia' krō-ton-i-*fo*-lee-a. With
   leaves like croton i.e. *Codiaeum*.
   'Longifolia' long-gi-*fo*-lee-a. Long-
   leaved.
   'Nana Rotundifolia' *nah*-na ro-tun-di-
   fo-lee-a. Dwarf, round-leaved.
   'Picturata' pik-tew-*rah*-ta. Variegated.
   'Salicifolia' să-li-ki-*fo*-lee-a. *Salix*-
   leaved.

Auricula see *Primula auricula*

Aurinia ow-*rin*-ee-a *Cruciferae*. From L.
aureus (golden) referring to the flowers.
Sub-shrubby perennial.
   saxatilis săks-ah-ti-lis. (= *Alyssum saxatile*).
   Growing among rocks. S and C Europe,
   Turkey.

Australian Bluebell Creeper see *Sollya
heterophylla*
Australian Honeysuckle see *Banksia*
Austrian Copper Briar see *Rosa foetida*
'Bicolor'

Austrocedrus ow-stro-*ked*-rus *Cupressaceae*.
From L. *australis* (southern) and *Cedrus* q.v.
The only species is found in the Southern
Hemisphere. Evergreen conifer.
   chilensis chil-*en*-sis. (= *Libocedrus chilensis*)
   of Chile. Chile, Argentina.

Austrocylindropuntia cylindrica see *Opuntia
cylindrica*

Avena a-*vay*-na *Gramineae*. The classical
name for oat. Annual grass.
   candida hort. see *Helictotrichon sempervirens*
   sempervierens see *Helictotrichon sempervirens*
   sterilis ste-ri-lis. Barren i.e. not producing
   the oats of cultivation. Animated Oat. S
   Europe, W Asia.

Avocado Pear see *Persea americana*
Aylostera deminuta see *Rebutia deminuta*
Azalea see *Rhododendron*
   pontica see *R. luteum*

Azara a-*zah*-ra *Flacourtiaceae*. After J. N.
Azara (1731–1804) Spanish scientist.
Evergreen, semi-hardy shrubs.
   integrifolia in-teg-ri-*fo*-lee-a. With entire
   leaves. Chile, Argentina.
   microphylla mik-rō-*fil*-la. Small-leaved.
   Chile, Argentina.
   petiolaris pee-tee-ō-*lah*-ris. With a petiole,
   relatively long in this species. Chile.

Azolla a-*zol*-la *Azollaceae*. From Gk. *azo* (to
dry) and *ollua* (to kill), they are killed by
drying. Floating, aquatic ferns.
   caroliniana kă-ro-lin-ee-*ah*-na. Of
   Carolina. Mosquito fern. N America to
   W Indies.
   filiculoides fi-lik-ew-*loi*-deez. Like *Filicula*.
   N and S America.

B

Babiana băb-ee-*ah*-na *Iridaceae*. From *babiaans*, Afrikaans for baboon which is said to eat the plants. Semi-hardy, cormous herbs. Baboon Root.
   plicata pli-*kah*-ta. Pleated (the leaves).
   sambucina săm-bew-*kee*-na. Like *Sambucus* (the scent of the flowers).
   stricta strik-ta. Erect (the stems).

Baboon Root see *Babiana*
Baby Blue Eyes see *Nemophila*
Baby Rubber Plant see *Peperomia obtusifolia*
Baby's Tears see *Hypoestes phyllostachya*, *Soleirolia soleirolii*

Baccharis bă-ka-ris. *Compositae*. After Bacchus, god of wine. Deciduous and evergreen shrubs.
   halimifolia hă-li-mi-*fo*-lee-a. With leaves like *Atriplex halimus*. Bush Groundsel, Groundsel Tree. E N America.
   patagonica pă-*a*-gon-i-ka. Of Patagonia. S Chile, S Argentina.

Bachelor's Buttons see *Tanacetum parthenium*
   White see *Ranunculus aconitifolius*
   Yellow see *Ranunculus acris* 'Flore Pleno'
Badger's Bane see *Aconitum lycoctonum vulparia*
Balloon Flower see *Platycodon grandiflorus*
Balloon Vine see *Cardiospermum halicacabum*

Ballota ba-*lō*-ta *Labiatae*. Gk. name for *B. nigra*. Evergreen sub-shrub.
   pseudodictamnus soo-dō-dik-*tăm*-nus. False Dictamnus. Greece, Crete.

Balm of Gilead see *Populus* × *jackii* 'Gileadensis'.
Balsam see *Impatiens balsamina*
   Himalayan see *I. glandulifera*

Balsamita
   major see *Tanacetum balsamita*

Bamboo see *Arundinaria, Phyllostachys, Pleioblastus, Pseudosasa, Sasa, Sasaella, Sinarundinaria, Thamnocalamus*.
   Black see *Phyllostachys nigra*
Baneberry see *Actaea*
   Red see *A. rubra*

Banksia bank-see-a *Proteaceae*. After Sir Joseph Banks (1743–1820), botanist and patron of science. Australian Honeysuckle.

Banksia (continued)
   collina ko-*lee*-na. Growing on hills. SE Australia.
   integrifolia in-teg-ri-*fo*-lee-a. With entire leaves. E Australia.
   serrata se-*rah*-ta. Saw-toothed (the leaves). E Australia.

Banyan Tree see *Ficus benghalensis*

Baptisia băp-*tis*-ee-a *Leguminosae*. From Gk. bapto (to dye), the following have been used as substitutes for indigo (*Indigofera tinctoria*). Perennial herbs.
   australis ow-*strah*-lis. Southern. False Indigo. E United States.
   tinctoria tink-*to*-ree-a. Used in dyeing. Wild Indigo. E United States.

Barbados Gooseberry see *Pereskia aculeata*
Barbados Pride see *Caesalpinia pulcherrima*
Barberry see *Berberis*
   Common see *B. Vulgaris*
Barberton Daisy see *Gerbera jamesonii*

Barleria bar-le-*ree*-a *Acanthaceae*. After Jacques Barrelier (1606–73), French monk and botanist. Tender, evergreen shrub.
   lupulina lup-ew-*leen*-a. Hop-like (the flower spikes). Mauritius.

Barrenwort see *Epimedium*
Bartonia aurea see *Mentzelia lindleyi*
Basil see *Ocimum basilicum*
   Bush see *O. basilicum* 'Minimum'
Basket Grass see *Oplismenus hirtellus*
Basket Plant see *Aeschynanthus*

Bassia băss-ee-a *Chenopodiaceae* For Ferdinando Bassi (1710–1774), Italian botanist. Annual or (rarely) perennial herbs. Cosmopolitan.
   scoparia skō-*pah*-ree-a (= *Kochia scoparia*). Broom-like. Summer Cypress, Burning bush. Asia.
      tricophylla tri-ko-*fil*-a. With hair-like leaves. The form most commonly cultivated; the leaves turn deep red in late summer, hence 'Burning Bush'.

Bat Plant see *Tacca integrifolia*

Bauera bow-a-ra *Saxifragaceae*. After the brothers Franz (1758–1840) and Ferdinand (1760–1826) Bauer, Austrian botanical artists. Tender, evergreen shrub.
   rubioides roo-bee-*oi*-deez. Like *Rubia*, the trifoliolate, opposite leaves appear to be a

Bauera (continued)
whorl of six leaves as in *Rubia*. New South Wales.

Bauhinia bow-*hin*-ee-a. *Leguminosae*. After John and Caspar Bauhin, 16th century Swiss botanists. The two brothers are represented by the usually two-lobed leaves. Tender shrubs.
acuminata a-kew-min-*ah*-ta. Acuminate. Orchid Bush. SE Asia.
× *blakeana* blayk-ee-*ah*-na. (? *B. purpurea* × *B. variegata*). After Sir Henry and Lady Blake. Sir Henry Blake was Governor of the Hong Kong Botanic Garden from where it was described. China.
galpinii gal-*pin*-ee-ee (= *B. punctata*). After Ernest Edward Galpin (1848–1941), botanist. Spotted. Tropical Africa.
purpurea pur-*pewr*-ree-a. Purple (the flowers). Butterfly Tree. SE Asia.
variegata vă-ree-a-*gah*-ta. Variegated (the flowers). Purple Orchid Tree. India, China.

Bay Laurel see *Laurus nobilis*
Bayberry see *Myrica pensulvanica*
   California see *M. californica*
Bead Plant see *Nertera granadensis*
Bead Tree see *Melia azederach*
Bearberry see *Arctostaphylos uva-ursi*

Beaumontia bō-*mont*-ee-a *Apocynaceae*. After Lady Diana Beaumont (died 1831). Tender, woody climber.
grandiflora grăn-di-*flō*-ra. Large flowered. Herald's Trumpet. Himalaya.

Bear's Breeches see *Acanthus*
Beauty Berry see *Callicarpa*
Beauty Bush see *Kolkwitzia amabilis*
Bee Balm see *Monarda didyma*
Beech see *Fagus*
   Common see *F. sylvatica*
   Copper see *F. sylvatica* Purpurea
Beefsteak Plant see *Iresine herbstii*
Beetroot see *Beta vulgaris*

Begonia bay-*gon*-ee-a *Begoniaceae*. After Michael Bégon (1638–1710), Governor of French Canada. Tender perennial herbs and shrubs.
albo-picta ăl-bō-*pik*-ta. Painted with white (the leaves). Guinea-wing Begonia. Brazil.
× *argenteo-guttata* ar-gen-tee-ō-gu-*tah*-ta. *B. albo-picta* × *B. olbia*. Dotted with silver (the leaves). Trout Begonia. Brazil.
boliviensis bo-liv-ee-*en*-sis. Of Bolivia.

Begonia (continued)
bowerae bow-a-rie. After Bower. Miniature Eyelash Begonia. Mexico.
× *cheimantha* kie-*măn*-tha. *B. dregei* × *B. socotrana*. Winter-flowering. Christmas Begonia.
clarkei clark-ee-ee. After Major Trevor Clarke who grew it. Bolivia.
coccinea kok-*kin*-ee-a. Scarlet (the flowers). Angelwing Begonia. Brazil.
corallina ko-ra-*leen*-a. Coral-red (the flowers). Brazil.
daedalea see *B. strigillosa*
davisii day-*vis*-ee-ee. After Walter Davis (1847–1930) who collected for Veitch in S America. Peru.
diadema dee-a-*day*-ma. A crown, presumably referring to the red flower sheath. Borneo.
dregei dree-gee-ee. After Johann Franz Drege (1794–1881) a German botanist who collected in S Africa. Maple-leaf Begonia. S Africa.
× *erythrophylla* e-rith-rō-*fil*-la. (= *B. × feastii*). *B. hydrocotylifolia* × *B. manicata*. Red-leaved Beefsteak Begonia.
evansiana see *B. grandis* × *feastii* see *B. × erythrophylla*
fuchsioides fuk-see-*oi*-deez. *Fuchsia*-like. Fuchsia Begonia. Venezuela.
grandis grăn-dis. (= *B. evansiana*). Large. Hardy Begonia. SE Asia.
haageana see *B. scharffii*
luxurians luk-*sewr*-ree-ănz. Luxuriant. Palm-leaf Begonia Brazil.
maculata măk-ew-*lah*-ta. Spotted (the leaves). Brazil.
manicata măn-i-*kah*-ta. Long-sleeved, referring to the long petioles with a ruff of hairs at the apex. Mexico, Guatemala.
masoniana may-son-ee-*ah*-na. After Mr L. Maurice Mason who introduced it from Singapore in 1957. Iron Cross Begonia. China.
metallica me-tă-li-ka. Metallic (the appearance of the leaves). Metallic-leaf Begonia. Brazil.
rex reks. King. Assam.
Rex-cultorum reks-kul-*to*-rum. The cultivated *B. rex* which is grown mainly as hybrids.
'Richmondensis' rich-mond-*en*-sis. Of Richmond.
scharffii sharf-ee-ee. (= *B. laageana*). After Carl Scharff who collected in Brazil about 1888. Elephant's-ear Begonia. Brazil.
semperflorens sem-per-*flō*-renz. Everflowering. Brazil, Argentina.
Semperflorens-cultorum sem-per-*flō*-

Begonia (continued)
renz-kul-*to*-rum. The cultivated *B.*
semperflorens which is largely grown as
hybrids.
serratipetala se-rah-tee-*pe*-ta-la. With
toothed petals. New Guinea.
socotrana so-ko-*trah*-na. Of Socotra.
strigillosa stri-gi-lō-sa. (= *B. daedalea*).
With short, appressed bristles. C
America.
Tuber-hybrida *tew*-ber-*hib*-ri-da.
Tuberous hybrid. The commonly grown
tuberous Begonias.

Belamcanda bel-am-*kăn*-da *Iridaceae*. From
the native name. Perennial herb.
chinensis chin-en-sis. Of China. Leopard
Flower. Himalaya, China, SE Asia, Japan.

Belladonna Lily see *Amaryllis belladonna*

Bellevalia bel-*vah*-lee-a *Liliaceae*
(*Hyacinthaceae*). After Pierre Riche de
Belleval (1564–1632). Bulbous perennial.
pycnantha pik-nan-*tha* (= *Muscari
paradoxum*). With flowers crowded
together. E Turkey, Caucasus.

Bellflower see *Campanula*
   Adriatic see *C. garganica*
   Chimney see *C. pyramidalis*
   Clustered see *C. glomerata*
   Giant see *C. latifolia, Ostrowskia magnifica*
   Italian see *C. isophylla*
   Milky see *C. lactiflora*
   Spurred see *C. alliariifolia*

Bellis bel-is *Compositae*. From L. *bellus*
(pretty). Perennial herb.
perennis pe-*re*-nis. Perennial. Daisy.
Europe, W Asia.

Bellium bel-ee-um *Compositae*. From *Bellis*
q.v. which they resemble. Annual and
perennial herbs.
bellidioides bel-i-dee-*oi*-deez. Like *Bellis.*
W Mediterranean region.
minutum mi-*new*-tum. Small. Greece, W
Asia.

Bells of Ireland see *Moluccella laevis*
Bellwort see *Uvularia*
Beloperone guttata see *Justicea brandegeana*

Berberidopsis ber-be-ri-*dop*-sis *Flacourtiacea.*
From *Berberis* q.v. and Gk. *-opsis* indicating
resemblance. The two genera are not related.
Evergreen, woody climber.

Berberidopsis (continued)
corallina ko-ra-*lee*-na. Coral-red (the
flowers). Coral Plant. Chile.

Berberis ber-be-ris *Berberidaceae*. From the
Arabic name. Deciduous and evergreen
shrubs.
aggregata ăg-re-*gah*-ta. Clustered (the
flowers). W China.
buxifolia buks-i-*fo*-lee-a. *Buxus*-leaved.
Chile, Argentina.
candidula kăn-*did*-ew-la. Diminutive of
candida (white), the leaves are white
beneath. China.
darwinii dar-*win*-ee-ee. After Charles
Darwin who discovered it in 1835. Chile.
gagnepainii găn-ya-*păn*-ee-ee. After
Francois Gagnepain (1866–1952). China.
hookeri *huk*-a-ree. After W. J. Hooker.
Himalaya.
ilicifolia ee-li-ki-*fo*-lee-a. *Ilex*-leaved.
Chile.
julianae yoo-lee-*ah*-nie. Name given by
Schneider, after his wife Juliana. C China.
koreana ko-ree-*ah*-na. Of Korea.
linearifolia lin-ee-ah-ri-*fo*-lee-a. With
linear leaves. Argentina, Chile.
× *lologensis* lo-lo-*gen*-sis. *B. darwinii* × *B.
linearifolia*. Of Lolog, Argentina.
× *ottawensis* o-ta-*wen*-sis. *B. thunbergii* × *B.
vulgaris*. Of Ottawa, where it was raised.
panlanensis păn-lăn-*en*-sis. Of Pan-lan,
China.
× *rubrostilla* rub-rō-*stil*-la. From L. *rubra*
(red) and *stilla* (a drop of liquid) referring
to the red fruits.
sargentiana sar-jen-tee-*ah*-na. After
Sargent, see *Prunus sargentii*. China.
× *stenophylla* sten-ō-*fil*-la. *B. darwinii* × *B.
empetrifolia*. Narrow-leaved.
thunbergii thun-*berg*-ee-ee. After
Thunberg, see *Thunbergia*. Japan, China.
valdiviana văl-di-vee-*ah*-na. Of Valdivia,
Chile.
verruculosa ve-roo-kew-*lō*-sa. With small
warts (on the shoots). China.
vulgaris vul-*gah*-ris. Common. Common
Barberry. Europe, N Africa, temperate
Asia.
wilsoniae wil-*so*-nee-ie. After Ernest
Wilson's wife, Ellen. W China.

Bergenia ber-*gen*-ee-a *Saxifragaceae*. After
Karl August von Bergen (1704–68).
Herbaceous perennials.
ciliata ki-lee-*ah*-ta. Fringed with hairs (the
leaves). Himalaya.
ligulata see *B. ciliata*

Bergenia (continued)
cordifolia kor-di-*fo*-lee-a. With heart-shaped leaves. NE Asia.
crassifolia kră-si-*fo*-lee-a. Thick-leaved. NE Asia.
purpurascens pur-pew-*răs*-enz. (= *B. delavayi*). Purplish (the young leaves). Himalaya, China.
x *schmidtii shmit*-ee-ee. *B. ciliata* x *B. crassifolia*. After Ernst Schmidt.
stracheyi stray-kee-ee. After Lieutenant-General Sir Richard Strachey (1817–1908) who collected in the Himalaya. Himalaya.

Bergeranthus ber-ga-*răn*-thus *Aizoaceae*. After Alwyn Berger (1871–1931), German botanist and Bk. *anthos* (a flower). Tender succulents. S Africa.
multiceps mul-tee-keps. With many heads.
scapiger skah-pi-ger. With a scape.
vespertinus ves-per-*teen*-us. With the flower opening in the evening.

Bertolonia ber-to-*lō*-nee-a *Melastomataceae*. After Antonio Bertoloni (1775–1869), Italian botanist. Tender herbs. Brazil.
maculata măk-ew-*lah*-ta. Spotted (the leaves).
marmorata mar-mo-*rah*-ta. Marbled (the leaves).
   aenea ie-nee-a. Bronze (the leaves).

Bessera *bes*-a-ra *Liliaceae (Alliaceae)*. After Wilibald von Besser (1784–1842), Austrian botanist. Tender cormous herb.
elegans ay-le-gahnz. Elegant. Coral Drops. Mexico.

Beta *bay*-ta *Chenopodiaceae*. The L. name. Annual or biennial herb.
vulgaris bul-*gah*-ris. Common. Beetroot, Sugarbeet. Cult.

Betula *bet*-ew-la *Betulaceae*. The L. name. Deiduous trees. Birch.
albo-sinensis ăl-bō-si-*nen*-sis. The Chinese *B. alba* (= *B. pendula*). W China.
septentrionalis sep-ten-treeō-*nah*-lis. Northern.
alleghaniensis ăl-eg-*gan*-ee-en-sis (= *B. lutea*). From the Allegheny mountains. Yellow Birch. Eastern North America.
ermanii er-*mahn*-ee-ee. After Adolph Erman. NE Asia.
grossa grō-sa. Very large. Japanese Cherry Birch. Japan.
lenta len-ta. Tough but flexible. Sweet Birch. EN America.
lutea see *B. alleghaniensis*.

Betula (continued)
nana nah-na. Dwarf. Dwarf Birch. N Europe, N America.
nigra ni-gra. Black (the bark). River Birch. E United States.
papyrifera pă-pi-*ri*-fe-ra. Paper-bearing, referring to the papery bark. N America.
pendula pen-dew-la. Pendulous (the branchlets). Silver Birch. Europe, N Asia.
   'Dalecarlica' dă-lee-*kar*-li-ka. Of Dalarno (Dalecarlia) Sweden.
   'Tristis' *tris*-tis. Sad, referring to the weeping habit.
utilis ew-ti-lis. Useful, the wood is used for timber. Himalayan Birch. Himalaya, W China.
   jacquemontii zhahk-a-*mont*-ee-ee. After Victor Jacquemont (1801–32), French naturalist. W Himalaya.

Bifrenaria bi-fray-*nah*-ree-a *Orchidaceae*. From L. *bis* (twice) *frenum* (a strap) referring to the two strap-like structures which connect the pollinia and the glands.
harrisoniae hă-ri-*so*-nee-ie. After Mrs Arnold Harrison of Aigburth who grew it and painted it for Hooker's Exotic Flora. Brazil.

Big Tree see *Sequoiadendron giganteum*

Bignonia big-*nō*-nee-a *Bignoniaceae*. After Abbé Jean Paul Bignon (1662–1743). Evergreen woody climber.
capreolata kă-pree-ō-*lah*-ta. Bearing tendrils (from the ends of the petioles). Cross Vine. SE United States.

Bilberry see *Vaccinium myrtillus*

Billardiera bi-lar-dee-*e*-ra *Pittosporaceae*. After J. J. H. de Labillardière, French botanist who worked on the Australian flora. Semi-hardy, evergreen climber.
longiflora long-gi-*flō*-ra. Long-flowered. Tasmania.

Billbergia bil-*berg*-ee-a *Bromeliaceae*. After J. G. Billberg (1772–1844), Swedish botanist. Tender, epiphytic herbs.
amoena a-*moy*-na. Pleasant. Brazil.
   rubra rub-ra. Red (the leaves).
horrida ho-ri-da. Spiny (the leaves).
iridifolia ee-ri-di-*fo*-lee-a. *Iris*-leaved. Brazil.
nutans new-*tănz*. Nodding (the flowers). Angel's Tears, Friendship Plant. Brazil.
pyramidalis pi-ra-mi-*dah*-lis. Pyramidal (the inflorescence). Summer Torch. Brazil.

Billbergia (continued)
venezuelana ven-ez-way-*lah*-na. Of
Venezuela.
vittata vi-*tah*-ta. Banded (the leaves).
Brazil.
× *windii* vin-dee-ee. *B. decora* × *B. nutans*.
After Wind, a gardener.
zebrina ze-*bree*-na. Banded (the leaves).
Brazil.

Birch see *Betula*
   Dwarf see *B. nana*
   Himalayan see *B. utilis*
   Japanese Cherry see *B. grossa*
   Paper see *B. papyrifera*
   River see *B. nigra*
   Silver see *B. pendula*
   Sweet see *B. lenta*
   Yellow see *B. alleghaniensis*
Bird of Paradise Flower see *Strelitzia
reginae*
Bird's Eyes see *Gilia tricolor*
Bird's Foot Trefoil see *Lotus corniculatus*
Birthwort see *Aristolochia*
Bishop's Cap see *Mitella*
Bishop's Wort see *Stachys macrantha*
Bistort see *Polygonum bistorta*
Black-eyed Susan see *Rudbeckia hirta,
Thunbergia alata*
Black Gum see *Nyssa sylvatica*
Black Snakeroot see *Cimicifuga racemosa*
Blackthorn see *Prunus spinosa*
Blackwood see *Acacia melanoxylon*
Bladder-nut see *Staphylea*
Bladder Senna see *Colutea arborescens*
Bladderwort see *Utricularia*
Blanket Flower see *Gaillardia*
Blazing Star see *Mentzelia lindleyi*

Blechnum blek-num *Blechnaceae*. From
blechnon Gk. name for a fern. Hardy and
tender ferns.
brasiliense bra-zil-ee-*en*-see. Of Brazil. Rib
Fern, Brazil Tree Fern. Brazil, Peru.
capense ka-*pen*-see. Of the Cape of Good
Hope. Palm-leaf Fern.
discolor dis-ko-lor. Two-coloured, the
leaves are dark green above and
cinnamon-hairy beneath. Crown Fern.
Australia, New Zealand.
gibbum gi-bum. (= *B. moorei* hort.).
Swollen on one side. New Caledonia, S
Pacific Islands.
occidentale ok-ki-den-*tah*-lee. Western.
Hammock Fern. Tropical America.
orientale o-ree-en-*tah*-lee. Eastern.
Himalaya to Australia.
penna-marina pen-a-ma-*reen*-a. Sea pen,
presumably from the resemblance of the

Blechnum (continued)
fronds to this marine animal. Dwarf Hard
Fern. New Zealand, Australia, S
America.
spicant spee-kant. Tufted. Hard Fern.
Europe, W Asia, Japan, W N America.

Bleeding Heart see *Dicentra spectabilis*
Bleeding-heart Vine see *Clerodendrum
thomsoniae*

Bletilla ble-*til*-la *Orchidaceae*. A diminutive of
Bletia, a related genus, named after Louis
Blet, 18th century Spanish apothecary. Semi-
hardy orchid.
striata stree-*ah*-ta. Striped, the ribbed
leaves. China, Japan.

Blood Flower see *Haemanthus katharinae*
Blood Lily see *Haemanthus*
Bloodleaf see *Iresine herbstii*
Bloodroot see *Sanguinaria*
Blue-eyed Mary see *Collinsia verna,
Omphalodes verna*
Blue Flowered Torch see *Tillandsia lindenii*
Blue Lace Flower see *Trachymene caerulea*
Blue Lips see *Collinsia grandiflora*
Blue Thimble Flower see *Gilia capitata*
Blue Trumpet Vine see *Thunbergia
grandiflora*
Bluebell see *Hyacinthoides non-scripta*
   (Scotland) see *Campanula rotundifolia*
   Spanish see *Hyacinthoides hispanica*
Blueberry, Box see *Vaccinium ovatum*
   Highbush see *V. corymbosum*
Bluets see *Hedyotis caerulea*
Blushing Bromeliad see *Neoregelia carolinae*
Boat Lily see *Rhoeo spathacea*

Boenninghausenia bur-ning-how-*zen*-ee-a
Rutaceae. After von Boenninghausen
(1785–1864), German botanist. Deciduous
sub-shrub.
albiflora ăl-bi-*flō*-ra. White-flowered. E
Asia.

Bog Arum see *Calla palustris*
Bog Bean see *Menyanthes trifoliata*
Bog Myrtle see *Myrica gale*
Bog Rosemary see *Andromeda*

Bomarea bō-*mah*-ree-a *Liliaceae*
(*Alstroemeriaceae*). After Jacques Christophe
Valmont de Bomare, French patron of
science. Tender, twining herb.
caldasii kăl-dă-see-ee. (× *B. kalbreyeri*).
After Francisco José de Caldas
(1771–1816), botanical explorer. S
America.

Bongardia bon-*gard*-ee-a *Berberidaceae*. After
August Bongard (1786–1839). German
botanist. Herbaceous perennial.
  chrysogonum kris-*o*-go-num. A golden star,
  referring to the flowers. W and C Asia.

Borago bo-*rah*-gō *Boraginaceae*. Possibly from
L. *burra* (a hairy garment) referring to the
leaves. Annual and perennial herbs.
  officinalis o-fi-ki-*nah*-lis. Sold as a herb.
  Borage. Europe, N Africa.
  pygmaea pig-*my*-a (= *B. laxifolia*). Dwarf.
  Europe.

Borecole see *Brassica oleracea* Acephala

Boronia bo-*rō*-nee-a *Rutaceae*. After
Francesca Borone (1769–94). Tender,
evergreen shrubs.
  elatior ay-*lah*-tee-or. Taller. W Australia.
  heterophylla he-te-rō-*fil*la. With variable
  leaves sometimes simple, sometimes
  pinnate. W Australia.
  megastigma meg-a-*stig*-ma. With a large
  stigma. W Australia.
  serrulata se-ru-*lah*-ta. With small teeth (the
  leaves). SE Australia.

Borzicactus see *Matucana aurantiacus*

Boston Ivy see *Parthenocissus tricuspidata*
Bottle-brush see *Callistemon*
  Crimson see *C. citrinus*
Bottle Gourd see *Lagenaria siceraria*

Bougainvillea boo-gan-*vil*-lee-a
Nyctaginaceae. After Louis Antoine de
Bougainville (1729–1811), explorer and
scientist. Tender climbers.
  × *buttiana* būt-ee-*ah*-na. *B. glabra* × *B.
  glabra* blǎ-bra. Glabrous. Brazil.
  peruviana. After Mrs R. V. Butt who found
  it in 1910.
  spectabilis spek-*tah*-bi-lis. Spectacular.
  Brazil.

Bouncing Bet see *Saponaria officinalis*

Bouvardia boo-*var*-dee-a *Rubiaceae*. After Dr
Charles Bouvard (1572–1658). Tender,
evergreen shrubs.
  longiflora long-gi-*flō*ra. Long-flowered.
  Sweet Bouvardia. Mexico.
  ternifolia tern-i-*fo*-lee-a. With leaves in
  threes. Scarlet Trompetilla. Texas,
  Miexico.

Bower Plant see *Pandorea jasminoides*

Bowles' Golden Grass see *Milium effusum*
'Aureum'
Box see *Buxus*
  Common see *B. sempervirens*
Box elder see *Acer negundo*
Box Thorn sec *Lycium barbarum*

Brachycome brǎ-kee-*kō*-mee *Compositae*.
From Gk. *brachys* (short) and *kome* (hair)
referring to the short pappus bristles. Annual
and perennial herbs.
  iberidifolia i-be-ri-di-*fo*-lee-a. With leaves
  like *Iberis*. Swan River Daisy. W and S
  Australia.
  rigidula ri-*gi*-dew-la. Rigid. SE Australia,
  Tasmania.

Brachyglottis brǎ-kee-*glo*-tis *Compositae*
From Gk. *brachys* (short) and *glotta* (a
tongue) referring to the short ray florets.
Semi-hardy, evergreen shrub.
  compacta com-*pǎk*-tǎ (= *Senecio compactus*).
  Compact. New Zealand (North Island).
  Dunedin Hybrids *dun*-ee-din (= *Senecio
  greyi* hort.) Hybrids between *B. compacta*,
  B. greyi and *B. laxifolia*; originating at the
  Dunedin Botanic Gardens, New Zealand,
  between 1910–13.
  greyi gray-ee (= *Senecio greyi*). After Sir
  George Grey (1812–98), Prime Minister
  of New Zealand. Plants offered under this
  name are usually Dunedin Hybrids. New
  Zealand (North Island).
  laxifolia lǎks-i-*fo*-lee-a (= *Senecio laxifolius*).
  Loose-leaved. New Zealand (South
  Island).
  repanda re-*pǎn*-da. Wavy-margined (the
  leaves). New Zealand.

Brake, Australian see *Pteris tremula*
  Cretan see *P. cretica*
  Sword see *P. ensiformis*
Brandy Bottle see *Nuphar lutea*
Brasiliopuntia brasiliensis see *Opuntia
brasiliensis*
Brass Buttons see *Cotula coronopifolia*
Brassaia actinophylla see *Schefflera actinophylla*

Brassavola bra-*sah*-vo-la *Orchidaceae*. After
Antonio Musa Brassavola (1500–55), Italian
botanist. Greenhouse orchids.
  cordata kor-*dah*-ta. heart-shaped (the lip).
  W Indies.
  digbyana see *Rhyncolaelia digbyana*
  fragrans frah-granz. Fragrant (the flowers).
  Brazil.
  glauca see *Rhyncolaelia glauca*.
  nodosa nō-*dō*-sa. With conspicuous nodes.
  Lady of the Night. C and S America.

Brassavola (continued)
tuberculata tew-ber-kew-*lah*-ta (= *B. perrinii*). Warty. Brazil.

Brassia brahs-ee-a Orchidaceae. After William Brass, plant collector and botanical artist, (died 1783). Greenhouse orchids.
caudata kaw-*dah*-ta. Prolonged into a tail (the sepals). Tropical America.
elegantula ay-le-*gănt*-ew-la. Elegant. Mexico.
lawrenceana lo-rens-ee-*ah*-na. After Mrs Lawrence. NS America.
longissima long-*gis*-si-ma. (= *B. lawrenceana longissima*). Longest (the sepals). Costa Rica.
maculata măk-ew-*lah*-ta. Spotted (the petals). W Indies, C America.
verrucosa ve-roo-*kō*-sa. Warty (the lip). Mexico to S America.

Brassica bră-si-ka Cruciferae. L. name for cabbage. Annual and biennial herbs.
napus nah-pus. L. name for turnip. Rape. Cult.
Napobrassica nah-pō-*bră*-si-ka. From *napus* and *brassica*. Swede, Rutabaga.
oleracea o-le-*rah*-kee-a. Vegetable-like. Wild Cabbage. W Europe.
Acephala a-*kef*-a-la. Without a head. Ornamental Kale, Borecole.
Botrytis *bot*-ri-tis. Like a bunch of grapes. Cauliflower, Perennial Broccoli.
Capitata kă-pi-*tah*-ta. In a dense head (the leaves). Cabbage.
Costata see *B. oleracea* Tronchuda.
Gemmifera gem-*i*-fe-ra. Bearing buds. Brussels Sprouts.
Gongylodes gon-gi-*lō*-deez. Swollen (the stem). Kohl Rabi.
Italica ee-*tă*-li-ka. Of Italy. Sprouting Broccoli.
Tronchuda tron-*koo*-da. The Portuguese name. Couve Tronchuda, Portuguese Cabbage.
rapa rah-pa. The L. name. Turnip. Cult.
Pekinensis pee-kin-*en*-sis. Of Peking. Chinese Cabbage.
Perviridis per-*vi*-ri-dis. Very green. Tendergreen.

× **Brassocattleya** brah-sō-*kăt*-lee-a Orchidaceae. Intergeneric hybrids, from the names of the parents. *Brassavola* × *Cattleya*. Greenhouse orchids.

× **Brassolaelia** brah-sō-*lie*-lee-a Orchidaceae. Intergeneric hybrids, from the names of the

× **Brassolaelia** (continued)
parents. *Brassavola* × *Laelia*. Greenhouse Orchids.

× **Brassolaeliocattleya** brah-sō-lie-lee-ō-*kăt*-lee-a Orchidaceae. Intergeneric hybrids, from the names of the parents. *Brassavola* × *Cattleya* × *Laelia*. Greenhouse orchids.

Brazilian Edelweiss see *Sinningia leucotricha*
Brazilian Plume see *Justicea carnea*
Bridal Wreath see *Francoa sonchifolia, Spiraea* 'Arguta'
Bridewort see *Spiraea salicifolia*

Brimeura bri-*mewr*-ra Liliaceae (Hyacinthaceae). After Maria de Brimeur, a lover and grower of flowers in the time of Clusius. Bulbous perennial.
amethystinus ă-me-*this*-ti-nus. (= *Hyacinthus amethystinus*). Violet (the flowers). SW Europe, NW Yugoslavia.

Briza bree-za Gramineae. L. name of a grass grown for food. Annual and perennial grasses.
maxima mahk-si-ma. Larger. Pearl Grass. Mediterranean region.
media me-dee-a. Intermediate. Quaking Grass. Europe, W Asia.
minor mi-nor. (= *B. minima*). Smaller. Europe, W. Asia.

Broad Bean see *Vicia faba*
Broad-leaved Kindling Bark see *Eucalyptus dalrympleana*
Broccoli see *Brassica oleracea* Botrytis
Sprouting see *B. oleracea* Italica

Brodiaea brō-dee-*ie*-a Liliaceae (Alliaceae). After James Brodie (1744–1824), Scottish botanist. Semi-hardy, cormous perennials.
californica kăl-i-*for*-ni-ka. Of California.
coronaria ko-rō-*nah*-ree-a. Of garlands. Triplet Lily. WN America.
elegans ay-le-gahnz. Elegant. WN America.
ida-maia see *Dichelostemma ida-maia*
laxa see *Triteleia laxa*
stellaris ste-*lah*-ris. Star-like (the flowers). California.
uniflora see *Ipheion uniflorum*

Broom see *Cytisus, Genista*
Common see *Cytisus scoparius*
Dalmatian see *Genista sylvestris*
Genoa see *Genista januensis*
Montpelier see *Genista monspessulanus*
Mt Etna see *Genista aetnensis*

Broom (continued)
Pineapple-scented see *Argyrocytisus battandieri*
White Spanish see *Cytisus multiflorus*

Broussonetia broo-son-*ay*-tee-a *Moraceae*. After T.N.V. Broussonet (1761–1807), French naturalist. Deciduous tree.
papyrifera pǎ-pi-*ri*-fe-ra. Paper bearing, the bark is used to make paper in Japan. Paper Mulberry. China, Japan.

Browallia brō-*ah*-lee-a *Solanaceae*. After John Browall (1707–55), Swedish botanist. Tender annuals.
americana ǎm-me-ri-*kah*-na (= *B. demissa*) Of America. C America.
speciosa spek-ee-*ō*-sa. Showy. Sapphire Flower. Colombia.
viscosa vis-*kō*-sa. Sticky (the young growths and calyx). Peru.

Bruckenthalia bruk-an-*thahl*-ee-a *Ericaceae*. After Samuel and Michael von Bruckenthal, 18th century Austrian noblemen. Evergreen, heath-like shrub.
spiculifolia speek-ew-lee-*fo*-lee-a. With spiky leaves. Spike Heath. E Europe, W Asia.

Brugmansia brǔg-*mǎn*-see-a *Solanaceae*. Shrubs and trees, formerly included in *Datura*, from which the species differ in having pendulous flowers.
arborea ar-*bo*-ree-a. Tree-like. Angel's Trumpet. S America.
sanguinea san-*gwin*-ee-a. Blood-red (the corolla). S America.
suaveolens swah-*vee*-o-lenz. Sweetly scented. Brazil.

Brunfelsia brun-*fel*-see-a *Solanaceae*. After Otto Brunfels (1489–1534), German monk and botanist. Tender, evergreen shrubs.
americana a-me-ri-*kah*-na. American. Lady of the Night. W Indies.
latifolia la-tee-*fo*-lee-a. Broad-leaved. Brazil.
pauciflora paw-si-*flō*-ra. Few-flowered. Yesterday, Today and Tomorrow. Brazil.
calycina kǎ-li-*kee*-na. (= *B. calycina*). With a well-developed calyx.
undulata un-dew-*lah*-ta. Undulate (the corolla). White Raintree. Jamaica.

Brunnera brun-a-ra *Boraginaceae*. After Samuel Brunner (1790–1844), Swiss botanist. Perennial herb.

Brunnera (continued)
macrophylla mǎk-rō-*fil*-la. Large-leaved. Caucasus, Siberia.

Brussels Sprouts see *Brassica oleracea* Gemmifera
Bryophyllum see *Kalanchoe*
Buck Bean see *Menyanthes trifoliata*
Buckeye see *Aesculus*
  California see *A. californica*
  Red see *A. pavia*
  Yellow see *A. flava*
Buckthorn, Common see *Rhamnus cathartica*

Buddleja būd-lee-a *Loganiaceae*. After the Rev. Adam Buddle (1660–1715). Deciduous shrubs. Butterfly Bush.
alternifolia ǎl-tern-i-*fo*-lee-a. With alternate leaves, they are opposite in most species. China.
colvilei kol-*vil*-ee-ee. After Sir James Colvile (died 1890). Himalaya.
crispa kris-pa. Finely wavy (the leaves). Himalaya.
davidii dǎ-*vid*-ee-ee. After Armand David who discovered it, see *Davidia*. China.
fallowiana fǎ-lō-ee-*ah*-na. After George Fallow (1890–1915), gardener at the Royal Botanic Garden, Edinburgh. China.
globosa glo-*bō*-sa. Spherical (the inflorescence). Chile, Peru.
× *weyeriana* vay-a-ree-*ah*-na. *B. davidii magnifica* × *B. globosa*. After Van de Weyer who raised it in 1914.

Bugbane see *Cimicifuga*
Bugle see *Ajuga reptans*

Buglossoides bew-glos-*oi*-deez *Boraginaceae*. Like *Buglossum*. Perennial herb.
purpurocaerulea pur-pew-rō-kie-*ru*-lee-a. (= *Lithospermum purpurocaeruleum*). Purple-blue (the flowers). Europe, W Asia.

Bulbine bul-*bee*-nay *Liliaceae* (*Asphodelaceae*). From *bolbine* Gk. name for a bulbous plant. Perennial herb.
alooides al-ō-*oy*-deez. Aloe-like (the rosettes). S. Africa.
semibarbata see *Bulbinopsis semibarbata*.

Bulbinella bul-bi-*nel*-a *Liliaceae* (*Asphodelaceae*). Diminutive of *Bulbine* q.v. Perennial herb.
hookeri hook-a-ree. After J. D. Hooker. New Zealand.

Bulbinopsis bul-been-*op*-sis *Liliaceae (Asphodelaceae)*. From *Bulbine*, and *opsis*, meaning appearance. Stemless herbs allied to *Bulbine*. E. Australia.
   semibarbata se-mee-bar-*bah*-ta. Half bearded, only the inner three filaments, out of six, are hairy. Australia.

Bulbocodium bul-bō-*kō*-dee-um *Liliaceae*. From Gk. *bolbos* (a bulb) and *kodion* (wool). Cormous, perennial herb.
   vernum ver-num. Of spring (flowering). Spring Meadow Saffron. Pyrenees, Alps to Russia.

Bulbophyllum bul-bō-*fil*-lum *Orchidaceae*. From Gk. *bolbos* (a bulb) and *phyllon* (a leaf) referring to the leaves growing from the top of the pseudobulb. Greenhouse orchids.
   barbigerum bar-*bi*-ge-rum. Bearded (the lip). Tropical Africa.
   careyanum kair-ree-*ah*-num. After Dr Carey of Serampore who sent it to Liverpool Botanic Garden before 1824. Himalaya.
   collettii ko-*let*-ee-ee. After Gen. Sir Henry Collett of the Indian Army. Burma.
   dayanum day-ah-num. After John Day (1824–88), orchid grower. Assam.
   dearei dear-ree-ee. After Col. Deare of Engelfield Green. Borneo, Philippines.
   ericssonii e-rik-*son*-ee-ee. After Ericsson, a Swedish plant collector who discovered it. Borneo.
   medusae may-*dew*-sie. Of Medusa, likening the many-flowered umbel to Medusa's head. SE Asia.
   odoratissimum o-dō-ra-*ti*-si-mum. Most fragrant. Himalaya, China.
   ornatissimum or-na-*ti*-si-mum. Most ornamental. Himalaya, Philippines.
   umbellatum um-bel-*ah*-tum. The flowers appear to be in umbels. Himalaya.

Bull Bay see *Magnolia grandiflora*
Bunny Rabbits see *Linaria maroccana*

Buphthalmum but-*thahl*-mum *Compositae*. From Gk. *bous* (an ox) and *ophthalmos* (an eye) referring to the flowers.
   salicifolium să-li-ki-*fo*-lee-um. *Salix*-leaved. C Europe.
   speciosissimum see *Telekia speciosissima*
   speciosum see *Telekia speciosa*

Bupleurum boo-*plur*-rum *Umbelliferae*. From Gk. *boupleuros* (ox rib) referring to another plant. Evergreen shrub. Thorow-wax.
   fruticosum froo-ti-*kō*-sum. Shrubby (most

Bupleurum (continued)
   species are herbaceous). S. Europe, Mediterranean region.

Burhead see *Echinodorus*
Burning Bush see *Dictamnus albus, Bassia scoparia*
Burnet see *Sanguisorba*
Bush Clover see *Lespedeza*
Bush Groundsel see *Baccharis halimifolia*
Busy Lizzie see *Impatiens walleriana*
Butcher's Broom see *Ruscus aculeatus*

Butia bew-tee-a *Palmae*. The native name. Tender palms.
   capitata kā-pi-*tah*-ta. In a dense head (the leaves). Jelly Palm. Brazil, Argentina.
   yatay yah-tay. The native name. Yatay Palm. Argentina.

Butomus boo-to-mus *Butomaceae*. From Gk. *bous* (an ox) and *temmo* (to cut), the sharp-edged leaves prevent it being used for fodder. Aquatic perennial herb.
   umbellatus um-bel-*ah*-tus. With flowers in umels. Flowering Rush. Europe, Asia.

Buttercup see *Ranunculus*
   Persian see *R. asiaticus*
Butterfly Bush see *Buddleja*
Butterfly Flower see *Schizanthus*
Butterfly Lily see *Hedychium coronarium*
Butterfly Tree see *Bauhinia purpurea*
Butter-nut see *Juglans cinerea*
Butterwort see *Pinguicula*
Button Snake Root see *Liatris pycnostachya*
Buttons-on-a-String see *Crassula rupestris*

Buxus buls-us *Buxaceae*. The L. name for *B. sempervirens*. Evergreen shrubs and trees. Box.
   balearica bā-lee-*ah*-ri-ka. Of the Balearic Islands. Balearic Islands, SW Spain.
   microphylla mik-rō-*fil*-la. Small-leaved. Cult.
      koreana ko-ree-*ah*-na. Of Korea. Korea, China.
      sinica sin-i-ka. Of China.
   sempervirens sem-per-*vi*-renz. Evergreen. Common Box. Europe, N Africa, W Asia.

C

Cabbage see *Brassica oleracea* Capitata
   Chinese see *B. rapa* Pekinensis
   Portuguese see *B. oleracea* Tronchuda

Cabbage (continued)
Wild see *B. oleracea*
Cabbage Gum see *Eucalyptus pauciflora*
Cabbage Tree see *Cordyline australis*

Cabomba ka-*bom-ba Nymphaeaceae*. From the
native name. Aquatic herbs. Fanwort.
   aquatica a-*kwah*-ti-ka. Growing in water.
   NE South America.
   caroliniana kǎ-ro-lin-ee-*ah*-na. Of
   Carolina. SE United States.

Cactus Gk. name for another spiny plant.
Now used only as a common name, see
right.

Cactus, Baseball see *Euphorbia obesa*
   Bird's-nest see *Mammillaria camptotricha*
   Bishop's Cap see *Astrophytum
   myriostigma, A. ornatum*
   Cane see *Opuntia cylindrica*
   Chain see *Rhipsalis paradoxa*
   Chain-link see *Opuntia imbricata*
   Christmas see *Schlumbergera × buckleyi*
   Cinnamon see *Opuntia microdasys rufida*
   Claw see *Schlumbergera truncata*
   Cob see *Echinopsis*
   Coral see *Mammillaria heyderi*
   Corncob see *Euphorbia mammillaris*
   Cotton-pole see *Opuntia vestita*
   Crab see *Schlumbergera truncata*
   Crown see *Rebutia*
   Dumpling see *Lophophora williamsii*
   Easter see *Hatiora gaertneri*
   Eve's Pin see *Opuntia subulata*
   Feather see *Mammillaria plumosa*
   Fire Crown see *Rebutia senilis*
   Fire Cracker see *Cleistocactus
   smaragdiflorus*
   Goat's Horn see *Astrophytum capricorne*
   Gold Lace see *Mammillaria elongata*
   Golden Ball see *Parodia leninghausii*
   Golden Barrel see *Echinocactus grusonii*
   Indian Fig see *Opuntia ficus-indica*
   Lamb's Tail see *Echinocereus schmollii*
   Mistletoe see *Rhipsalis baccifer*
   Old Man see *Cephalocereus senilis*
   Old Woman see *Mammillaria hahniana*
   Orchid see *Nopalxochia ackermanii*
   Peanut see *Echinopsis chamaecereus*
   Peruvian Apple see *Cereus uruguayanus*
   Plain see *Gymnocalycium mihanovichii*
   Rat's-tail see *Aporocactus flagelliformis*
   Red Crown see *Rebutia minuscula*
   Ribbon see *Pedilanthes tithymaloides*
   Scarlet Ball see *Parodia haselbergii*
   Sea Urchin see *Astrophytum asterias,
   Echinopsis*
   Silver Ball see *Parodia scopa*

Cactus (continued)
   Silver Cluster see *Mammillaria prolifera*
   Silver Dollar see *Astrophytum asterias*
   Snowball see *Mammillaria bocasana*
   Snowdrop see *Lepismium houlletianum*
   Spider see *Gymnocalycium denudatum*
   Star see *Astrophytum*
   Sun see *Helocereus speciosus*
   Teddy Bear see *Opuntia bigelovii*
   Toothpick see *Stetsonia coryne*
   White Torch see *Echinopsis spachiana*

Caesalpinia kie-sal-*pee*-nee-a *Leguminosae*.
After Andreas Caesalpini (1519–1603,
Italian botanist. Tender trees.
   gilliesii gi-*leez*-ee-ee. After John Gillies
   (1747–1836). Argentina, Uruguay.
   pulcherrima pul-*ke*-ri-ma. Very Pretty.
   Barbados Pride. W Indies.

Caladium ka-*lah*-dee-um *Araceae*. From
kaladi the native name. Tender herbs. Angel's
Wings, Elephant's Ears, Heart of Jesus.
   bicolor bi-*ko*-lor. Two-coloured (the
   leaves). Brazil.
   × *hortulanum* see *c. bicolor*
   'Candidum' *kǎn*-di-dum. White (the
   leaves).

Calamintha kǎl-a-*min*-tha *Labiatae*. The Gk.
name from *kallos* (beautiful) and *minthe*
(mint). Perennial herbs. Calamint.
   alpina see *Acinos alpinus*.
   grandiflora grǎn-di-*flō*-ra. Large-flowered. S
   Europe.
   nepeta ne-*pe*-ta. From *Nepeta* a related
   genus. Mediterranean region.

Calamondin see × *Citrofortunella microcarpa*

Calandrinia kǎ-lan-*dreen*-ee-a *Portulacaceae*.
After Jean Louis Calandrina (1703–58).
Annuals. Rock Purslane.
   ciliata ki-lee-*ah*-ta. Fringed with hairs.
   Peru, Ecuador.
   umbellata um-bel-*ah*-ta. With flowers in
   umbels. Peru, Chile.

Calanthe ka-*lǎn*-thee *Orchidaceae*. From Gk.
kalos (beautiful) and *anthos* (a flower).
Greenhouse orchids.
   furcata fur-*kah*-ta. (= *C. veratrifolia*). Cleft
   (the lip). SE Asia to Australia.
   masuca ma-*soo*-ka. A native name.
   Himalaya.
   vestita ves-*tee*-ta. Clothed, referring to the
   hairy stem. SE Asia.

Calathea ka-*lah*-thee-a *Marantaceae*. From

Calathea (continued)
Gk. *kalathos* (a basket), the flowers are
clustered as if in a basket. Tender, evergreen
herbs.
 bachemiana bah-kem-ee-*ah*-na. After Herr
 Bachem, Burgermeister of Cologne.
 Brazil.
 lancifolia lǎn-ki-*fo*-lee-a. With lance-
 shaped leaves. Rattlesnake Plant. Tropical
 America.
 lindeniana lin-den-ee-*ah*-na. After J. J.
 Linden, Belgian horticulturist. Brazil.
 louisae loo-*eez*-ie. After Queen Louisa of
 Belgium.
 makoyana mǎk-oy-*ah*-na. After Jacob
 Makoy (1790–1873), nurseryman of
 Liège. Cathedral Windows, Peacock Plant.
 majestica ma-*jest*-ik-a (= *C. ornata*).
 Majestic. N. South America.
 picturata pik-tew-*rah*-ta. Variegated. Brazil.
 zebrina zeb-*ree*-na. Striped. Zebra Plant.
 Brazil.

Calceolaria kǎl-kee-ō-*lah*-ree-a
Scrophulariaceae. After F. Calceolari, 16th
century Italian botanist. The name in L. also
means slipper-like, referring to the flowers.
Herbs and shrubs. Slipperwort.
 biflora bi-*flō*-ra. Two-flowered. Chile,
 Argentina.
 darwinii dar-*win*-ee-ee. After Charles
 Darwin who discovered it. Patagonia.
 × Herbeohybrida herb-ee-ō-*hib*-ri-da.
 Herbaceous hybrid.
 integrifolia in-teg-ri-*fo*-lee-a. With entire
 leaves. Chile.
 mexicana meks-i-*kah*-na. Of Mexico.
 Mexic, C America.
 tenella ten-*el*-la. Dainty. Chile.
 tripartita tri-*part*-ee-ta. Three-parted (the
 leaf segments). W South America.

Calendula ka-*len*-dew-la *Compositae*. From L.
calendae (the first day of the month) referring
to the long flowering period. Annual herb.
 officinalis o-fi-ki-*nah*-lis. Sold as a herb. Pot
 Marigold. S Europe.

Calico Bush see *Kalmia latifolia*
Calico Flower see *Aristolochia littoralis*
Calico Hearts see *Adromischus maculatus*
California Bluebell see *Phacelia
campanularia*
California Geranium see *Senecio petasites*
California Laurel see *Myrica californica*
California Nutmeg see *Torreya californica*

Calla kǎ-la *Araceae*. From Gk. *kallos*
(beautiful). Herbaceous perennial.

Calla (continued)
 palustris pa-*lus*-tris. Growing in bogs. Bog
 Arum. N America, Europe, Asia.

Callicarpa kǎ-lee-*kar*-pa *Verbenaceae*. From
Gk. *kallos* (beautiful) and *karpos* (a fruit).
They are grown for their attractive fruits.
Deciduous shrubs. Beauty Berry.
 bodinieri bo-din-ee-*e*-ree. After Emile
 Marie Bodinieri (1842–1901), French
 missionary and plant collector in China.
 China.
 giraldii see *C. bodinieri*
 dichotoma di-*ko*-to-ma. With forking
 shoots. China, Japan.
 japonica ja-*pon*-i.ka. Of Japan.
 'Leucocarpa' loo-kō-*kar*-pa. With
 white fruits.
 rubella roo-*bel*-a. Reddish (the fruits). SE
 Asia.

Callirhoe kǎ-lee-*rō*-ee *Malvaceae*. After the
daughter of Achelous in Gk. mythology.
Annual and perennial herbs. Poppy Mallow.
 digitata di-gi-*tah*-ta. Fingered, the leaves
 have finger-like lobes. United States.
 involucrata in-vol-oo-*krah*-ta. With an
 involucre (around the flowers). United
 States.
 leiocarpa ly-ō-*kah*-pa (= *C. pedata*). From
 Gk. *leio* smooth, and *carpos*, fruit. Texas
 and Oklahoma.

Callisia ka-*lis*-ee-a *Commelinaceae*. From Gk.
kallis (beauty). Tender, evergreen herbs.
Inch Plant. Mexico.
 elegans ay-le-gahnz. (= *Setcreasia striata
 hort.*). Elegant. Striped Inch Plant.
 fragrans frah-granz. Fragrant (the flowers).
 navicularis nah-vik-ew-*lah*-ris Boat-shaped
 (the leaves). Chain plant. Mexico, Peru.

Callistemon ka-*lee*-stay-mon *Myrtaceae*. From
Gk. *kallos* (beautiful) and *stemon* (a stamen).
The stamens are the showy part of the flower.
Evergreen, semi-hardy shrubs. Bottle-
brush.
 citrinus ki-*tree*-nus. Lemon-scented (the
 leaves). Crimson Bottle-brush. SE
 Australia.
 linearis lin-ee-*ah*-ris. Linear (the leaves).
 New South Wales.
 rigidus ri-gi-dus. Rigid (the leaves). E
 Australia.
 salignus sa-*lig*-nus. Willow-like (the leaves).
 SE Australia.
 speciosus spe-kee-ō-sus. Showy. W
 Australia.

Callistemon (continued)
subulatus sub-ew-*lah*-tus. Awl-shaped (the leaves). SE Australia.

Callistephus ka-*lee*-ste-fus *Compositae*. From Gk. *kallos* (beautiful) and *stephanus* (a crown) referring to the showy, terminal flower heads.
chinensis chin-*en*-sis. Of China. China Aster.

Callitriche ka-*lee*-tri-kee *Callitrichaceae*. From Gk. *kallos* (beautiful) and *trichos* (hair) referring to the delicate foliage. Aquatic herbs. Water Starwort.
hamulata hahm-ew-*lah*-ta. Hooked (the ends of the leaves) Europe.
hermaphroditica her-mă-frŏ-*dee*-ti-ka. (= *C. autumnalis*). Hermaphrodite. Europe, N Africa.
palustris pa-*lus*-tris. Growing in bogs. N America, Europe, Asia.
stagnalis stăg-*nah*-lis. Growing in still water. S Europe, N Africa.

Calluna ka-*loo*-na *Ericaceae*. From L. *kalluno* (to cleanse), it was used to make brooms. Evergreen shrub.
vulgaris vul-*gah*-ris. Common. Heather, Ling. Europe, W Asia.

Calocedrus kă-lo-*ked*-rus *Cupressaceae*. From Gk. *kallos* (beautiful) and *Cedrus* q.v. Evergreen conifer.
decurrens day-ku-renz. (= *Libocedrus decurrens*). With the leaf margin running gradually into the stem. Incense Cedar. WN America.

Calochortus kă-lo-*kor*-tus *Liliaceae* (*Calochortaceae*). From Gk. *kallos* (beautiful) and *chortos* (grass) referring to the slender leaves. Semi-hardy bulbous herbs. Mariposa Lily.
albus ăl-bus. White (the flowers). California.
amabilis a-*mah*-bi-lis. Beautiful. NW California.
barbatus bar-*bah*-tus. Bearded (the petals). Mexico.
coeruleus kie-*ru*-lee-us. Blue (the flowers). California.
luteus loo-tee-us. Yellow (the flowers). California.
uniflorus ew-ni-*flŏ*-rus. One-flowered. California, Oregon.

Calomeria kă-lo-*me*-ree-a *Compositae*. From

Calomeria (continued)
Gk. *kallos* (beautiful) and *meris* (a part). Tender biennial.
amaranthoides ăm-a-rănth-*oi*-deez. (= *Humea elegans*). Like *Amaranthus*. Plume Bush. SE Australia.

Caltha kăl-tha *Ranunculaceae*. L. name for a yellow-flowered plant. Aquatic and marginal perennials.
asarifolia see *C. palustris*.
chelidonii see *C. leptosepala*.
leptosepala lep-tŏ-*se*-pa-la. With slender sepals. W N America.
palustris pa-lus-tris. Growing in bogs. King Cup, Marsh Marigold. N America, Europe, Asia.
   'Plena' *play*-na. With double flowers.
polypetala see *C. palustris*.

Calycanthus kă-lee-*kanth*-us *Calycanthaceae*. From Gk. *kalyx* (calyx) and *anthos* (a flower), the sepals and petals are similar. Deciduous shrubs.
fertilis fer-ti-lis. Fertile. SE United States.
floridus flo-ri-dus. Flowering. Carolina Allspice. SE United States.
occidentalis ok-ki-den-*tah*-lis. Western. Californian Allspice. California.

Camassia ka-mă-see-a *Liliaceae* (*Hyacinthaceae*). From the N American Indian name *quamash* or *camass*. Bulbous herbs.
cusickii kew-*sik*-ee-ee. After W. C. Cusick. Oregon.
leichtlinii liet-*lin*-ee-ee. After Max Leichtlin (1831–1910). Oregon.
quamash kwah-măsh. The native name, see above. W N America.
scilloides skil-*oi*-deez. Like *Scilla*. E N America.

Camellia ka-*mel*-lee-a *Theaceae*. After George Joseph Kamel Camellus) (1661–1706), pharmacist who studied the Philippines flora. Evergreen shrubs.
cuspidata kus-pi-*dah*-ta. With a stiff point (the leaves). W China.
japonica ja-*pon*-i-ka. Common Camellia. Japan, Korea.
reticulata ray-tik-ew-*lah*-ta. Net-veined (the leaves). China.
sasanqua sa-*sang*-kwa. The Japanese name. Japan.
sinensis si-nen-sis. Of China. Tea Plant.
× *williamsii* wil-*yămz*-ee-ee. *C. japonica* × *C. saluenensis*. After J. C. Williams of

Camellia (continued)
Caerhays who raised many fine forms of this hybrid.

Campanula kăm-*pahn*-ew-la *Campanulaceae*. From L. *campana* (a bell) referring to the shape of the flowers. Perennial herbs. Bellflower.
alliariifolia ă-lee-ah-ree-i-*fo*-lee-a. With leaves like *Alliaria petiolata* (Garlic Mustard). Spurred Bellflower. Caucasus, Turkey.
allionii see *C. alpestris*.
alpestris ăl-*pes*-tris. (= *C. allionii*). Of the lower mountains. SW Alps.
arvatica ar-*vă*-ti-ca. Of Arvas, N Spain.
aucheri ow-ka-ree. After P. M. R. Aucher-Eloy (1792–1836). W Asia.
barbata bar-*bah*-ta. Bearded (the corolla). Alps, C Europe, Norway.
bononiensis bo-nō-nee-*en*-sis. Of Bologne, N Italy. C and E Europe, W Asia.
carpatica kar-*pă*-ti-ka. Of the Carpathians.
cochleariifolia kok-lee-ah-ree-i-*fo*-lee-a. With leaves like *Cochlearia*. S and E Europe.
collina ko-*leen*-a. Growing on hills. Caucasus.
excisa eks-*kee*-sa. Cut away (the deep corolla sinuses). Alps.
garganica gar-*gah*-ni-ka. Of Monte Gargano, Italy. Adriatic Bellflower. SE Italy, W Greece.
glomerata glo-me-*rah*-ta. Clustered (the flowers). Clustered Bellflower. Europe, W Asia.
isophylla ee-sō-*fil*-la. With equal-sized leaves. Italian Bellflower. NW Italy.
lactiflora lăk-ti-*flō*-ra. Milky-flowered. Milky Bellflower. Caucasus, W Asia.
latifolia lah-tee-*fo*-lee-a. Broad-leaved. Giant Bellflower. Europe, W Asia.
latiloba lă-i-*lō*-ba. Broad-lobed (the flowers) Siberia.
medium *may*-de-um. Classical name for a similar plant from Media, W Asia. Canterbury Bells. SE France, Ialy.
'Calycanthema' kă-lee-kăn-*thee*-ma. Flowering in the calyx, the calyx resembles the corolla. Cup and Saucer Canterbury Bells.
morettiana mo-ret-ee-*ah*-na. After G. Moretti (1782–1853). Dolomites (Italy).
persicifolia per-si-ki-*fo*-lee-a. With leaves like *Prunus persica* (Peach). Europe to N Asia.
portenschlagiana por-tan-shlahg-ee-*ah*-na. After Franz von Portenschlag-

Campanula (continued)
Ledermeyer (1777–1827), Austrian botanist. W Yugoslavia.
poscharskyana po-shar-skee-*ah*-na. After Gustav Adolf Poscharsky (1832–1914). W Yugoslavia.
pulla *pul*-la. Dark (the flowers). NE Alps (Austria).
× *pulloides* pul-*loi*-deez. *C. carpatica* × *C. pulla*. Like *C. pulla*.
punctata punk-*tah*-ta. Spotted (the corolla). E Asia.
pyramidalis pi-ra-mi-*dah*-lis. Pyramidal (the inflorescence). Chimney Bellflower. N Italy, Albania, Yugoslavia.
raineri *ray*-na-ree. After Rainer. SE Alps.
rapunculoides ra-punk-ew-*loi*-deez. Like *C. rapunculus*. Europe, Asia.
rotundifolia ro-tund-i-*fo*-lee-a. With rounded (basal) leaves). Harebell, Bluebell (Scotland). Europe, Asia, N America.
sarmatica sar-*mă*-ti-ka. Of Sarmatia (E Europe). Caucasus.
thyrsoides thur-*soi*-deez. Staff-like (the inflorescence). Alps, Balkans.
trachelium tra-*kay*-lee-um. From Gk. *trachelos* (neck) referring to supposed medicinal properties. Throatwort. Europe, Asia.
zoysii zoys-ee-ee. After Karl von Zoys, Austrian botanist. SE Alps.

Campernelle see *Narcissus* × *odorus*
Campion see *Silene*
   Alpine see *Lychnis alpina*
   Moss see *Silene acaulis*
   Rose see *Lychnis coronaria*
   Sea see *Silene vulgaris maritima*

Campsis *kămp*-sis *Bignoniaceae*. From Gk. *kampe* (something bent) referring to the curved stamens. Deciduous woody climbers.
grandiflora grăn-di-*flō*-ra. Large-flowered. China.
radicans rah-di-kănz. With rooting stems. Trumpet Vine. SE United States.
× *tagliabuana* tăg-lee-ă-bew-*ah*-na. *C. grandiflora* × *C. radicans*. After the Tagliabue brothers, Italian nurserymen.

Camptosorus see *Asplenium*

Canada Lily see *Lilium canadense*
Canary Creeper see *Tropaeolum speciosum*
Canary Grass see *Phalaris canariensis*
Candelabra Plant see *Aloe arborescens*
Candle Plant see *Plectranthus oertendahlii*, *Senecio articulatus*

Canna ka-na *Cannaceae*. From Gk. *kanna* (a reed). Tender herbaceous perennials.
edulis see *C. indica*
× *generalis* gen-e-*rah*-lis. Normal i.e. the commonly grown form.
indica in-di-ka. (= *C. edulis, C. lutea*). Indian. Indian Shot. Tropical America.

Cannabis kă-na-bis *Cannabaceae*. The L. and Gk. name. Annual herb.
sativa sa-*tee-va*. Cultivated. Hemp. C. Asia.

Canterbury Bells see *Campanula medium*
Cup and Saucer see *C. medium* 'Calycanthema'

Cantua kăn-tew-a *Polemoniaceae*. From the Peruvian name. Tender evergreen shrub.
buxifolia buks-i-*fo*-lee-a. *Buxus*-leaved. Peru.

Cape Asparagus see *Aponogeton distachyos*
Cape Cowslip see *Lachenalia*
Cape Gooseberry see *Physalis peruviana*
Cape Honeysuckle see *Tecomaria capensis*
Cape Ivy see *Senecio macroglossus*
Cape Jasmine see *Gardenia augusta*
Cape Leadwort see *Plumbago auriculata*
Cape Pondweed see *Aponogeton distachyos*
Caper Spurge see *Euphorbia lathyris*

Capsicum kăp-si-kum *Solanaceae*. From Gk. *kapto* (to bite) referring to the hot taste. Tender annuals.
annuum ăn-ew-um. Annual. Sweet Pepper, Christmas Pepper. Cult.
frutescens froo-*tes*-enz. Shrubby. Chilli Pepper. Tropical America.

Caragana kă-ra-*gah*-na *Leguminosae*. From *Caragan*, the Mongolian name for *C. arborescens*. Deciduous tree.
arborescens ar-bo-*res*-enz. Tree-like. Pea Tree. Siberia, Mongolia.
'Lorbergii' lor-*berg-ee-ee*. After Lorberg's nursery, Germany, where it was raised.

Caraway see *Carum carvi*

Cardamine kar-*dah*-mi-nee *Cruciferae*. From the Gk. name of a related plant. Perennial herbs.
californica kăl-i-*for*-ni-ka. (= *Dentaria californica*). Of California. W United States.
enneaphyllos en-ee-a-*fil*-los. (=*Dentaria enneaphyllos*). With nine leaves (leaflets). E Europe.
heptaphylla hep-ta-*fil*-la. (= *Dentaria*

Cardamine (continued)
heptaphylla). With seven leaves (leaflets). W and C Europe.
laciniata la-kin-ee-*ah*-ta. Deeply cut (the leaves). E United States.
lyrata li-*rah*-ta. Lyre-shaped (the leaves). E Asia.
pentaphyllos pen-ta-*fil*-los. (= *Dentaria pentaphyllos*). With five leaves (leaflets). W and C Europe.
pratensis prah-*tayn*-sis. Of meadows. Ladies' Smock, Cuckoo Flower. Europe, Asia, N America.

Cardamon see *Elettaria cardamomum*
Cardinal Flower see *Lobelia cardinalis, Sinningia cardinalis*

Cardiocrinum kar-dee-ō-*kree*-num *Liliaceae*. From Gk. *kardio* (heart) and *krinon* (a lily) referring to the heart-shaped leaves. Bulbous herbs.
cordatum kor-*dah*-tum. Heart-shaped (the leaves). Japan.
giganteum gi-*găn*-tee-um. Very large. Himalaya, Burma.

Cardiospermum kar-dee-ō-*sperm*-um *Sapindaceae*. From Gk. *kardia* (heart) and *spermum* (a seed) referring to the white, heart-shaped spot on the seed. Annual climber.
halicacabum hă-li-*kah*-ka-bum. The L. name for bladderwort (*Utricularia*). Love-in-a-Puff, Balloon Vine. S United States, Tropical America.

Cardoon see *Cynara cardunculus*

Carex kah-reks *Cyperaceae*. The L. name. Perennial, grass-like herbs. Sedge.
morrowii mo-*rō*-ee-ee. After Morrow. Japan.
pendula *pen*-dew-la. Pendulous (the flower spikes). Europe, Asia, N Africa.
pululifera pil-ew-*li*-fe-ra. Bearing little balls, referring to the globular female inflorescences. Europe.
pseudocyperus sood-ō-ki-*pe*-rus. False Cyperus. Widely distributed.
riparia ree-*pah*-ree-a. Of river banks. Great Pond Sedge. Europe, N Africa, W Asia.
scaposa ska-*pō*-sa. With a conspicuous scape. S China.
sylvatica sil-*vă*-ti-ka. Of woods. Europe.

Carlina kar-*lee*-na *Compositae*. A medieval name. Biennial and perennial herbs. Carline Thistle.

Carlina (continued)
acanthifolia a-kănth-i-*fo*-lee-a. With leaves
like *Acanthus*. S Europe, W Asia.
acaulis a-*kaw*-lis. without a stem. Europe.

Carmichaelia kar-mie-*keel*-ee-a *Leguminosae*,
After Capt. Dugald Carmichael
(1772–1827), Scottish army officer and plant
collector. New Zealand shrubs.
enysii e-*nis*-ee-ee. After John Davies Enys
(1837–1912), New Zealand magistrate
and naturalist who, with Kirk, collected
the type specimen.
petriei pet-*ree*-ee. After Petrie who
collected the type specimen in 1890, see
Coprosma petriei.

Carnation see *Dianthus caryophyllus*
Carolina Lupin see *Thermopsis caroliniana*

Carnegiea kar-*nee*-gee-a *Cactaceae*. After
Andrew Carnegie (1835–1919).
euphorbioides see *Neobuxbaumia
euphorbioides*.
gigantea gi-*găn*-tee-a. Very large. Saguaro.
SW United States.

Carpenteria kar-pen-*te*-ree-a *Hydrangeaceae*
[*Philadelphaceae*]. After Prof. William M.
Carpenter, Louisiana physician. Evergreen,
semi-hardy shrub.
californica kăl-i-*for*-ni-ka. Of California.

Carpinus kar-*peen*-us *Carpinaceae*. The L.
name. Deciduous trees. Hornbeam.
betulus bet-ew-lus. Like *Betula*. Common
Hornbeam. Europe, W Asia.
caroliniana kă-ro-lin-ee-*ah*-na. Of
Carolina. American Hornbeam. E N
America.
japonica ja-*pon*-i-ka. Of Japan.
turczaninowii tur-chah-ni-*nov*-ee-ee. After
Nicolai Stepanovich Turczaninow
(1796–1864) who discovered it in 1831.
N China, Japan.

Carpobrotus kar-po-*bro*-tus *Aizoaceae*. From
Gk. *karpos* (a fruit) and *brotus* (edible)
referring to the edible fruit. Succulent sub-
shrub.
edulis e-*dew*-lis. Edible (the fruit).
Hottentot Fig. S Africa.

Carrot see *Daucus carota sativus*
   Wild see *D. carota*
Cartwheel Flower see *Heracleum
mantegazzianum*

Carum kă-rum *Umbelliferae*. From *karon* the
Gk. name. Annual or biennial herb.

Carum (continued)
carvi kar-vee. The L. name. Caraway. W
Asia.

Carya kă-ree-a *Juglandaceae*. From *karya* Gk.
name of the walnut tree (*Juglans regia*).
Deciduous trees. Hickory. EN America.
cordiformis kor-di-*form*-is. Heart-shaped
(the nut). Bitternut Hickory.
glabra glă-bra. Glabrous (the shoots).
Pignut Hickory.
ovata ō-*vah*-ta. Ovate (the leaflets).
Shagbark Hickory.
tomentosa tō-men-*tō*-sa. Hairy (the young
shoots). Mockernut Hickory.

Caryopteris kă-ree-*op*-te-ris *Verbenaceae*.
From Gk. *karyon* (a nut) and *pteron* (a wing)
referring to the winged fruits. Deciduous
shrubs.
× *clandonensis* klăn-don-*en*-sis. *C. incana* ×
C. mongolica. From Clandon, Surrey
where it was raised.
incana in-*kah*-na. Grey (the leaves and
shoots). China, Japan.

Caryota kă-ree-*ō*-ta *Palmae*. From Gk.
karyota (a date-shaped nut). Tender Palms.
Fishtail Palm.
mitis mee-tis. Soft, not spiny. Burmese
Fishtail Palm. SE Asia.
urens ew-renz. Stinging, referring to the
stinging, needle-like crystals in the outer
covering of the fruit. Wine Palm, Toddy
Palm. Himalaya, SE Asia.

Cassia kă-see-a *Leguminosae*
(*Caesalpinoideae*). Gk. name for a species of
this or a related genus. Tender shrubs and
herbs.
alata see *Senna alata*
australis see *Senna australis*
corymbosa see *Senna corymbosa*
fistula fis-tew-la. Hollow, pipe-like.
Golden shower, Purging Cassia. SE Asia,
C & S America, N Australia.
grandis grănd-is. Large, showy (the flowers).
Pink Shower. C & S America.
marylandica see *Senna marylandica*
moschata mos-*kay*-ta. Musky. Bronze
Shower. C & S America, Cuba.

Cassinia ka-*seen*-ee-a *Compositae*. After
Count Henri de Cassini (1781–1832),
Italian botanist. Evergreen shrubs.
fulvida *ful*-vi-da. Slightly tawny (the
undersides of the leaves). New Zealand.

Cassiope ka-*see*-o-pay *Ericaceae*. After

Cassiope (continued)
Cassiope of Gk. mythology, the mother of Andromeda. Evergreen, heath-like shrubs.
fastigiata få-stig-ee-*ah*-ta. Erect (the shoots). Himalaya.
lycopodioides lik-ō-pō-dee-*oi*-deez. Like *Lycopodium*. Japan, Alaska.
tetragona tet-ra-*gō*-na. Four-angled (the shoots). Arctic N hemisphere.
wardii ward-ee-ee. After its discoverer Francis Kingdon Ward (1885–1958), plant collector, explorer and author who introduced numerous plants from China, Tibet and Burma. Himalaya.

Cast Iron Plant see *Aspidistra elatior*
Castor-oil Plant see *Ricinus communis*
  False see *Fatsia japonica*

Castanea ka-*stăn*-ee-a *Fagaceae*. The L. name from Castania, N Greece which was known for its trees. Deciduous tree.
sativa sa-*tee*-va. Cultivated. Sweet or Spanish Chestnut. S Europe, N Africa, W Asia.

Cat Thyme see *Teucrium marum*

Catalpa ka-*tăl*-pa *Bignoniaceae*. The N American Indian name. Deciduous trees.
bignonioides big-nō-nee-*oi*-deez. Like *Bignonia*. Indian Bean Tree. E United States.
× *erubescens* e-roo-*bes*-nez. *C. bignoniodes* × *C. ovata*. Blushing.
'Purpurea' pur-*pewr*-ree-a. Purple (the young shoots and leaves).
fargesii far-*geez*-ee-ee. After Farges who discovered it, see *Decaisnea fargesii*. China.
duclouxii dew-*cloo*-ee-ee. After Monsignor Fr Ducloux who collected in China.
speciosa spe-kee-ō-sa. Showy. Western Catalpa. United States.

Catananche kă-ta-*năn*-kee. *Compositae*. The Gk. name. Perennial herb.
caerulea kie-*ru*-lee-a. Deep blue (the flowers). Cupid's Dart. S Europe.

Catchfly, German see *Lychnis viscaria*
  Nodding see *Silene pendula*

Catharanthus kă-tha-*răn*-thus *Apocynaceae*. From Gk. *katharos* (pure) and *anthus* (a flower).
roseus ro-*see*-us. (= *Vinca rosea*). Rose-coloured (the flowers). Old Maid,

Catharanthus (continued)
Madagascar Periwinkle. Madagascar to India.

Cathedral Bells see *Cobaea scandens*
Cathedral Windows see *Calathea makoyana*
Catmint see *Nepeta cataria*
Catnip see *Nepeta cataria*
Cat's Foot see *Antennaria dioica*
Cat's Whiskers see *Tacca chantrieri*

Cattleya kăt-lee-a *Orchidaceae*. After William Cattley (died 1832), horticultural patron.
bicolor bi-ko-lor. Two-coloured (the flowers). Brazil.
bowringiana bow-ring-gee-*ah*-na. After John Charles Bowring (1821–93), orchid grower. C America.
citrina see *Encyclia citrina*
dowiana dow-ee-*ah*-na. After Captain J. M. Dow of the American Packet Service. C America.
intermedia in-ter-*me*-dee-a. Intermediate. Brazil.
labiata lă-bee-*ah*-ta. With a lip. Brazil.
loddigesii lod-ee-*jes*-ee-ee. After the Loddiges nursery of Hackney. Brazil.
mossiae mos-ee-ie. After Mrs Moss of Otterspool who grew the type specimen. Venezuela.
skinneri skin-a-ree. After Mr George Ure Skinner (1804–67) who collected orchids in South America. C America.
trianae tree-*ah*-nie. After J. J. Triana (1834–90). Colombia.
warscewiczii var-sha-*vich*-ee-ee. After Warscewicz, see *Alonsoa warscewiczii*. Colombia.

Cauliflower see *Brassica oleracea* Botrytis

Cautleya kawt-lee-a *Zingiberaceae*. After Sir Proby Thomas Cautley (1802–71), a military engineer who worked in India. Tender rhizomatous perennials.
gracilis gră-ki-lis. Graceful. Himalaya, W China.

Cayratia kay-ray-*tee*-a *Vitaceae*. From the native name, *cáy rat long*. Climbing or scrambling vines. Old World Tropics.
thomsonii tom-*son*-ee-ee (= *Parthenocissus thomsonii*). After Thomson, see *Aster thomsonii*. China, Assam.

Ceanothus kee-a-*nō*-thus *Rhamnaceae*. Gk. name of a spiny shrub. Evergreen and semi-evergreen shrubs.

Ceanothus (continued)
arboreus ar-*bo*-ree-us. Tree-like. Catalina
Ceanothus. California (islands).
'Burkwoodii' burk-*wud*-ee-ee. After
Burkwood and Skipwith who raised it.
dentatus den-*tah*-tus. Toothed (the leaves).
California.
gloriosus glō-ree-ō-sus. Glorious.
California.
papillosus pă-pil-*lō*-sus. With papillae,
referring to the wart-like glands on the
leaves. California.
  roweanus rō-ee-*ah*-nus. After E. Denys
  Rowe, Santa Barbara horticulturist
  who discovered it.
prostratus pros-*trah*-tus. Prostrate. W N
America.
thyrsiflorus thurs-i-*flō*-rus. With flowers in
a thyrse (a type of inflorescence).
California.
  repens ree-penz. Creeping.
× veitchianus veech-ee-*ah*-nus. After
Messrs Veitch, prominent 19th-century
nurserymen. California.

Cedar see *Cedrus*
  Atlas see *C. atlantica*
Cedar of Goa see *Cupressus lusitanica*
Cedar of Lebanon see *Cedrus libani*

Cedrela ked-rel-a *Meliaceae*. Diminutive of
Cedrus q.v. from the similar wood.
Deciduous tree.
  sinensis si-*nen*-sis. Of China. Chinese
  Cedar. N and W China.

Cedrus ke-drus *Pinaceae*. The L. name.
Evergreen conifers.
  deodara dee-ō-*dah*-ra. From the N Indian
  name. Deodar. Himalaya.
  libani li-ba-nee. Of Mount Lebanon.
  Cedar of Lebanon. Lebanon, Turkey.
    atlantica ăt-*lăn*-ti-ka. Of the Atlas
    Mountains.
    Glauca *glow*-ka. Glaucous (the leaves).

Celandine, Greater see *Chelidonium majus*
  Lesser see *Ranunculus ficaria*

Celastrus kel-ă-strus. *Celastraceae*. From
kelastros the Gk. name for an evergreen tree.
Deciduous climbers.
  hypoleucus hi-po-*loo*-kus. White beneath
  (the leaves). China.
  orbiculatus or-bik-ew-*lah*-tus. Orbicular
  (the leaves). NE Asia.

Celeriac see *Apium graveolens rapaceum*
Celery see *Apium graveolens dulce*

Celery (continued)
  Wild see *A. graveolens*
Celery Pine see *Phyllocladus alpinus*

Celmisia kel-*mis*-ee-a *Compositae*. After
Celmisios of Gk. mythology. Evergreen
herbaceous perennials. New Zealand Daisy.
New Zealand.
  coriacea ko-ree-*ah*-kee-a. Leathery (the
  leaves).
  hieraciifolia hee-e-rah-kee-i-*fo*-lee-a.
  Hieracium-leaved.
  spectabilis spek-*tah*-bi-lis. Spectacular.

Celosia ke-*lō*-see-a *Amaranthaceae*. From Gk.
keleos (burning) referring to the brilliantly
coloured flowers. Tender annual.
  argentea ah-jen-*tay*-a. Silvery (the flowers).
  Red Fox. Tropics.
    cristata kris-*tah*-ta. Crested (the
    inflorescence). Cockscomb. Cult.
    'Plumosa' ploo-*mō*-sa. Feathery (the
    inforescence).

Celsia arcturus see *Verbascum arcturus*
  cretica see *Verbascum creticum*

Celtis kel-tis *Ulmaceae*. Gk. name of a tree.
Deciduous trees. Nettle Tree.
  australis ow-*strah*-lis. Southern. S Europe,
  W Asia.
  laevigata lie-vi-*gah*-ta. Smooth (the leaves).
  Sugarberry. S United States.
  occidentalis ok-ki-den-*tah*-lis. Western.
  Hackberry. N America.

Centaurea kent-*ow*-ree-a *Compositae*. From
Gk. *kentaur* (a Centaur) which is said to
have used it medicinally. Annual and
perennial herbs.
  cineraria kin-e-*rah*-ree-a. (= *C.*
  gymnocarpa). From the resemblance to
  Cineraria maritima (= *Senecio cineraria*). W
  Italy, Sicily.
  cyanus see-*ah*-nus. Dark blue (the flowers).
  Cornflower. SE Europe, W Asia.
  dealbata dee-al-*bah*-ta. Whitened (the
  underside of the leaves). Caucasus.
  gymnocarpa see *C. cineraria*.
  hypoleuca hi-*pō*-loo-ka. White beneath (the
  leaves). W Asia.
  macrocephala măk-rō-*kef*-a-la. With a large
  head. Caucasus.
  montana mon-*tah*-na. Of mountains. C
  Europe.
  moschata see *Amberboa moschata*
  pulcherrima pul-*ke*-ri-ma. Very pretty.
  Caucasus, W Asia.

Centaurea (continued)
ragusina rah-goo-*see*-na. Of Dubrovnik (Ragusa). W Yugoslavia.
ruthenica roo-*then*-i-ka. Of Ruthenia (SW Russia). Caucasus, S Russia, Romania.
rutifolia roo-ti-*fo*-lee-a. *Ruta*-leaved. SE Europe.
simplicicaulis sim-plik-ee-*kaw*-lis. With an unbranched stem. Armenia.

Centranthus ken-*trăn*-thus *Valerianaceae*. From Gk. *kentron* (a spur) and *anthos* (a flower) referring to the spurred flowers. Perennial herb. Red Valerian.
ruber ru-ber. Red (the flowers). S Europe, N Africa.

Century Plant see *Agave americana*

Cephalaria kef-a-*lah*-ree-a *Dipsacaceae*. From Gk. *kephale* (a head), the flowers are borne in heads. Perennial herbs.
alpina ăl-*peen*-a. (= *Scabiosa alpina*). Alpine. Alps.
gigantea gi-*găn*-tee-a. (= *C. tatarica* hort.). Very large. Caucasus.

Cephalocereus kef-a-lō-*kay*-ree-us *Cactaceae*. From Gk. *kephale* (a head) and *Cereus* q.v., the flowers are borne from woolly heads.
palmeri see *Pilosocereus palmeri*
polylophus see *Neobuxbaumia polylopha*
senilis sen-*ee*-lis. Old, referring to the dense covering of long, white hairs. Old Man Cactus. C Mexico.

Cephalotaxus kef-a-lō-*tăks*-us *Cephalotaxaceae*. From Gk. *kephale* (a head) and *Taxus* q.v., from the resemblance to *Taxus*. Evergreen trees and shrubs. Plum Yew.
fortunei for-*tewn*-ee-ee. After its introducer, Robert Fortune, see *Fortunella*. N China.
harringtonii hă-ring-*ton*-ee-ee. After the Earl of Harrington. Cult.
drupacea droo-*pah*-kee-a. With a fleshy fruit. Japan, Korea.

Cerastium ke-*ră*-stee-um *Caryophyllaceae*. From Gk. *keras* (a horn), referring to the shape of the seed capsule. Perennial herbs.
alpinum ăl-*peen*-um. Alpine. N and alpine Europe.
tomentosum tō-men-*tō*-sum. Hairy. Snow in Summer. Italy, Sicily.
biebersteinii bee-ber-*stien*-ee-ee. After Friedrich August Marschall von

Cerastium (continued)
Bieberstein (1768–1826), German botanist. Crimea.

Ceratophyllum ke-ra-tō-*fil*-lum *Ceratophyllaceae*. From Gk. *keras* (a horn) and *phyllon* (a leaf) from the resemblance of the leaves to antlers. Aquatic herbs. Hornwort.
demersum day-*mer*-sum. Growing under water. Europe, N America.
submersum sub-*mer*-sum. Growing under water. Europe, N Africa, Asia.

Ceratopteris ke-ra-*top*-te-ris *Parkeriaceae*. From Gk. *keras* (a horn) and *pteris* (a fern) from the horn like appearance. Aquarium fern.
thalictroides tha-lik-*troi*-deez. Like *Thalictrum* (the foliage). Water Sprite. SE Asia.

Ceratostigma ke-ra-tō-*stig*-ma *Plumbaginaceae*. From Gk. *keras* (a horn) and *stigma* referring to horn-like growths on the stigma. Shrubs and herbs.
griffithii gri-*fith*-ee-ee. After William Griffith (1810–45), surgeon and botanist in India and SE Asia. E Himalaya, W. China.
plumbaginoides plum-bah-gi-*noi*-deez. Like *Plumbago*. China.
willmottianum wil-mot-ee-*ah*-num. After Miss Ellen Ann Willmott (1858–1934), a noted gardener who introduced many plants. She raised plants from the original introduction by Ernest Wilson, in her garden at Warley Place, Essex.

Cercidiphyllum ker-ki-di-*fil*-lum *Cercidiphyllaceae*. From *Cercis* q.v. and Gk. *phyllon* (a leaf), the leaves resemble those of *Cercis*. Deciduous tree.
japonicum ja-*pon*-i-kum. Of Japan.

Cercis ker-kis *Leguminosae*. From *kerkis* Gk. a weaver's shuttle, descriptive of the woody fruits. Deciduous trees.
canadensis kăn-a-*den*-sis. Of Canada or NE North America. Redbud. E and C United States.
siliquastrum si-li-*kwă*-strum. Like a siliqua, from the resemblance of the pod to a fruit found in *Cruciferae*. Judas Tree. E Mediterranean region, W Asia.

Cereus kay-ree-us *Cactaceae*. From L. *cereus* (a wax taper) referring to the shape.
aethiops ie-thee-ops. (= *C. caerulescens*). Of unusual appearance. N Argentina.

Cereus (continued)
jamacaru hǎ-ma-*kah*-roo. From the native name. Brazil.
saxicola saks-*ik*-ō-la (= *Monvillea cavendishii*). Of rocky places. S America.
spegazzinii speg-a-*zeen*-ee-ee (= *Monvillea spegazzinii*). After Carlos Spegazzini (1858–1926), Argentine botanist. Paraguay.
uruguayanus ū-roo-gwy-*ah*-nus (= *C. peruvianus*). Of Uruguay.
Peruvian Apple Cactus. SE South America

Ceropegia kay-rō-*pee*-gee-a *Asclepiadaceae*. From Gk. *keros* (wax) and *pege* (a fountain) referring to the waxy flowers. Tender succulents. S Africa.
barklyi bark-lee-ee. After Sir Henry Barkly (1815–98), Governor of the Cape Province and patron of botanists and plant collectors in the colonies.
linearis lin-ee-*ah*-ris. Narrow, with parallel sides (the corolla tube).
woodii wud-ee-ee. After John Medley Wood (1827–1915), farmer and botanist in South Africa. Rosary Vine, Hearts on a String.

Cestrum kes-trum *Solanaceae*. Gk. name of a plant. Tender and semi-hardy shrubs.
aurantiacum ow-rǎn-tee-*ah*-kum. Orange (the flowers). Guatemala.
elegans ay-le-gahnz. Elegant. Mexico.
fasciculatum fǎs-kik-ew-*lah*-tum. Clustered (the flowers). Mexico.
'Newellii' new-*el*-ee-ee. After Mr Newell, gardener at Ryston Hall, who raised it.
parqui par-kee. The Chilean name. Chile.

Ceterach *officinarum* see *Asplenium ceterach*

Chaenomeles kie-nō-*may*-leez *Rosaceae*. From Gk. *chaina* (to gape) and *melon* (an apple) referring to the belief that the fruit was split. Deciduous, spiny shrubs. Ornamental Quince.
cathayensis kǎ-thay-*en*-sis. Of China.
japonica ja-*pon*-i-ka. (= *Cydonia maulei, Cydonia japonica*). Of Japan.
speciosa spe-kee-ō-sa. (= *Cydonia lagenaria, Cydonia speciosa*). Showy. China.
× *superba* soo-*perb*-a. *C. japonica* × *C. speciosa*. Superb.

Chaenorrhinum kie-nō-*reen*-um *Scophulariaceae*. From Gk. *chaino* (to gape) and *rhis* (a snout) referring to the open mouth of the corolla tube. Perennial herb.
origanifolium o-ree-gahn-i-*fo*-lee-um. (=

Chaenorrhinum (continued)
Linaria origanifolia). *Origanum*-leaved. SW Europe.

Chain Plant see *Callisia navicularis*

Chamaecereus *sylvestrii* see *Echinopis chamacereus*

Chamaecyparis kǎ-mie-*kew*-pa-ris *Cupressaceae*. From Gk. *chamai* (low growng) and *kuparissos* (cypress). Evergreen confiers.
lawsoniana law-son-ee-*ah*-na. After Charles Lawson (1794–1874), Edinburgh nurseryman who raised it from the original introduction in 1854. Lawson Cypress. W N America.
'Ellwoodii' el-*wud*-ee-ee. After Ellwood, a gardener at Swanmore Park where it was raised.
'Fletcheri' *flech*-a-ree. After the Fletcher Bros. nursery who distributed it.
'Gimbornii' gim-*born*-ee-ee. After Von Gimborn, on whose estate in Holland it originated.
'Pottenii' po-*ten*-ee-ee. After the Potten nursery, Cranbrook, where it originated.
'Stewartii' stew-*art*-ee-ee. After the raisers, D. Stewart and son, Bournemouth.
'Wisselii' vis-*el*-ee-ee. After the raiser, F. van der Wissel, Dutch nurseryman.
nootkatensis nut-ka-*ten*-sis. Of Nootka sound, British Columbia, Nootka Cypress, W N America.
obtusa ob-*tew*-sa. Blunt (the leaves). Hinoki Cypress. Japan.
'Crippsii' *krips*-ee-ee. After the raisers, the Cripps nursery.
pisifera pee-*si*-fe-ra. Pea-bearing, referring to the small cones. Sawara Cypress. Japan.
'Filifera' fee-*le*-fe-ra. Thread-bearing, referring to the slender shoots.
'Plumosa' ploo-*mō*-sa. Feathery (the foliage).
'Squarrosa' skwah-*rō*-sa. With the leaves spreading at right angles.
thyoides thoo-*oi*-deez. Like *Thuja*. E N America.
'Andelyensis' ǎn-da-lee-*en*-sis. From Andelys, France.
'Ericoides' e-ri-*koi*-deez. *Erica*-like (the foliage).

Chamaecytisus kǎ-mie-si-*ti*-sus *Leguminosae*. From Gk. *chamai*, dwarf, on the ground, and *Cytisus* q.v., in which this genus was

Chamaecytisus (continued)
formerly included. Small trees, shrubs and
sub shrubs. Europe, Canary Islands.
 purpureus pur-*pewr*-ree-us (= *Cytisus
 purpureus*) Purple (the flowers). Alps, SE
 Europe.
 × *versicolor* ver-*si*-ko-lor (= *Cytisus* ×
 versicolor) *C. purpureus* × *C. hirsutus.*
 variously coloured (the flowers).
  'Hillieri' hil-ee-a-ree. After Hillier's
  nursery who raised it.

Chamadorea kă-mie-*do*-ree-a *Palmae.* From
Gk. *chamai* (on the ground) *dorea* (a gift).
Unlike many palms the fruits are borne
within reach. Tender palms.
 elegans ay-le-gahnz. (= *Collinia elegans,
 Neanthe bella*). Elegant. Parlour Palm.
 Mexico, Guatemala.
 erumpens ay-*rum*-penz. Breaking through,
 the inflorescence ruptures the sheath as it
 emerges. Bamboo Palm. C America.

Chamaemelum kă-mie-*may*-lum *Compositae.*
From Gk. *chamai* (on the ground) and *melon*
(an apple) referring to the apple-like scent
and the low habit. Perennial herb.
 nobile nŏ-bi-lee. (= *Anthemis nobilis*).
 Notable. Chamomile. W Europe, N
 Africa.

Chamaerops ka-*mie*-rops *Palmae.* From Gk.
chamai (low growing) and *rhops* (a bush)
referring to the dwarf habit compared to
most other palms. Semi-hardy palm.
 excelsa hort. see *Trachycarpus fortunei*
 humilis hum-i-lis. Low growing. Dwarf Fan
 Palm. Mediterranean region.

Chamomile see *Chamaemelum nobile*
 Dyer's see *Anthemis tinctoria*
 Yellow see *Anthemis tinctoria*
Chandelier Plant see *Kalanchoe delagonensis*
Chaste Tree see *Vitex agnus-castus*
Chatham Island Forget-me-not see
Myostidium hortensia

Cheilanthes kay-*lăn*-theez *Pteridaceae*
(*Adiantaceae*). From Gk. *cheilos* (a lip) and
anthos (a flower), the edge of the pinnules
forms a lip which covers the sporangia. Lip
Fern.
 distans dis-tanz. Widely spaced (the
 pinnae). Woolly Rock Fern. New
 Zealand, Australia.
 lanosa lah-*nŏ*-sa. Woolly (the undersides of
 the fronds). Hairy Lip Fern. E United
 States.

Cheilanthes (continued)
 pteridiodes ter-id-*oy*-deez (= *C. fragrans*).
 Resembling *Pteris.* S Europe.

Cheiranthus see *Erysimum*

Chelidonium ke-li-*dŏ*-nee-um *Papaveraceae.*
From Gk. *chelidon* (a swallow), it is said to
start flowering as the swallow arrives.
 majus mah-yus. Larger. Greater Celandine.
 Europe, W Asia.

Chelone ke-*lŏ*-nay *Scrophulariaceae.* From Gk.
chelone (a turtle), the corolla is shaped like a
turtle's head. Perennial herbs. Turtle Head.
 barbata see *Penstemon barbatus*
 glabra glă-bra. Glabrous. E United States.
 lyonii lie-*on*-ee-ee. After its discoverer John
 Lyon, see *Lyonia.* SE United States.
 obliqua o-*blee*-kwa. Oblique. E United
 States.

Chenille Plant see *Acalypha hispida*

Chenopodium kay-nŏ-*pŏ*-dee-um
Chenopodiaceae. From Gk. *chen* (a goose) and
podion (a foot) referring to the shape of the
leaves. Perennial herb.
 bonus-henricus bo-nus-hen-*ree*-kus. Good
 Henry, from 16th century German Güter
 Heinrich, a goblin with knowledge of
 healing plants. Good King Henry.
 Europe.

Cherry, Bird see *Prunus padus*
 Fuji see *P. incisa*
 Sour see *P. cerasus*
 Yoshino see *P.* × *yedoensis*
Cherry Laurel see *Prunus laurocerasus*
Cherry Pie see *Heliotropium arborescens*
Cherry Plum see *Prunus cerasifera*
Chervil see *Anthriscus cerafolium*
Chstnut, Sweet or Spanish see *Castanea
sativa*
Chestnut Vine see *Tetrastigma voinierianum*

Chiastophyllum kee-ă-stŏ-*fil*-lum
Crassulaceae. From Gk. *chiastos* (arranged
cross-wise) and *phyllon* (a leaf) referring to
the opposite leaves. Succulent perennial.
 oppositifolium o-po-si-ti-*fo*-lee-um. (=
 Cotyledon simplicifolia). With opposite
 leaves. Caucasus.

Chicory see *Cichorium intybus*
Chile Pine see *Araucaria araucana*
Chilean Bellflower see *Lapageria rosea,
Nolana*
Chilean Crocus see *Tecophilaea cyanocrocus*

Chilean Hazel see *Gevuina avellana*
Chilean Jasmine see *Mandevilla laxa*

Chimonanthus kee-mon-*ănth*-us
Calycanthaceae. From Gk. *cheima* (winter)
and *anthos* (a flower) referring to its winter-
flowering habit. Deciduous shrub.
 praecox prie-koks. Early (flowering).
 Winter Sweet. China.

China Aster see *Callistephus chinensis*
Chincherinchee see *Ornithogalum thyrsoides*
Chinese Cedar see *Toona sinensis*
Chinese Evergreen see *Aglaonema modestum*
Chinese Foxglove see *Rehmannia elata*
Chinese Gooseberry see *Actinidia deliciosa*
Chinese-hat Plant see *Holmskioldia
sanguinea*
Chinese Houses see *Collinsia bicolor*
Chinese Jade see *Crassula arborescens*
Chinese Lantern see *Physalis alkekengi,
Sandersonia aurantiaca*

Chionanthus kee-on-*ănth*-us *Oleaceae.* From
Gk. *chion*(snbow) and *anthos* (a flower)
referring to the white flowers. Deciduous
shrubs or trees.
 retusus re-*tew*-sus. Notched at the tip (the
 leaves). Chinese Fringe Tree. China.
 virginicus vir-*jin*-i-kus. Of Virginia. E
 United States.

Chionodoxa kee-on-o-*doks*-a *Liliaceae.* From
Gk. *chion* (snow) and *doxa* (glory). The
flowers are often borne with snow on the
ground. Bulbous perennials.
 albescens ăl-*bes*-enz. (= *C. nana*). White
 (the flowers). Crete.
 cretica see *C. nana*
 luciliae loo-*sil*-ee-ie. (= *C. gigantea*). After
 Lucile Boissier (1822–49) whose husband
 (see *Colchicum boissieri*) discovered and
 named it. W Turkey.
 nana nay-na. Dwarf. Crete.
 sardensis sar-*den*-sis. Of Sart, W Turkey.

Chives see *Allium schoenoprasum*
 Chinese see *A. tuberosum*

Chlidanthus lkid-*ănth*-us *Amaryllidaceae.*
From Gk. *chlide* (luxury) and *anthos* (a
flower). Bulbous perennial.
 fragrans frah-granz. Fragrant. Andes.

Chlorophytum klō-*ro*-fi-tum *Liliaceae.* From
Gk. *chloros* (green) and *phyton* (a plant).
Tender herbs. S Africa.
 capense ka-*pen*-see. Of the Cape of Good
 Hope.

Chlorophytum (continued)
 comosum ko-*mō*-sum. Tufted (the foliage).
 Spider Plant.
  'Vittatum' vi-*tah*-tum. Striped (the
  leaves).

Choisya shwūz-ee-a but usually *choy*-zee-a
Rutaceae. After Jacques Denis Choisy
(1799–1859), Swiss botanist. Evergreen
shrub.
 ternata ter-*nah*-ta. In threes (the leaflets).
 Mexican Orange Blossom. Mexico.

Chokeberry see *Aronia*
 Black see *A. melanocarpa*
 Red see *A. arbutifolia*
Christmas Box see *Sarcococca*
Christmas Cheer see *Sedum rubrotinctum*
Christmas Jewels see *Aechmea racineae*
Christmas Pride see *Ruellia macrantha*
Christmas Rose see *Helleborus niger*

Chrysalidocarpus kri-sah-li-dō-*kar*-pus
Palmae. From Gk. *chrysallis* (a chrysalis) and
karpos (a fruit), the fruit is said to resemble a
chrysalis. Tender palm.
 lutescens loo-*tes*-enz. Yellowish, the leaf
 sheaths and stalks. Yellow palm.
 Madagascar.

Chrysanthemum kris-*ănth*-e-mum *compositae.*
From Gk. *chrysos* (gold) and *anthos* (a
flower). Annual herbs.
 alpinum see *Leucanthemopsis alpina*
 carinatum kă-ri-*nah*-tum. Keeled (the
 involucral bracts). Painted Daisy.
 Morocco.
 coccineum see *Tanacetum coccineum*
 coronarium ko-rō-*nah*-um. Used in garlands
 Crown Daisy. Mediterranean region.
 corymbosum see *Tanacetum corymbosum*
 densum see *Tanacetum densum*
 frutescens see *Argyranthemum frutescens*
 haradjanii see *Tanacetum haradjanii*
 hosmariense see *Pyrethropsis hosmariensis*
 indicum in-di-kum see *Dendranthema
 indicum*
 maximum hort. see *Leucanthemum* ×
 superbum
 × *morifolium* see *Dendranthema* ×
 grandiflorum
 multicaule see *Coleostephus myconis*
 parthenium see *Tanacetum parthenium*
 segetum se-ge-tum. Of cornfields. Corn
 Marigold. Europe, W Asia.
 × *superbum* see *Leucanthemum x superbum*
 vulgare see *Tanacetum vulgare*
 weyrichii see *Dendranthema weyrichii*

Chrysogonum kris-*o*-go-num *Compositae.*
From Gk. *chrysos* (golden) and *gonu* (a knee)
referring to the yellow flowers and jointed
stem. Herbaceous perennial.
   virginianum vir-jin-ee-*ah*-num. Of
   Virginia. SE United States.

Cicerbita ki-*ker*-bi-ta *Compositae.* The Italian
name for a sow-thistle. Herbaceous
perennials.
   alpina ăl-*peen*-a. (= *Lactuca alpina*). Alpine.
   Europe.
   plumieri ploo-mee-*e*-ree. (= *Lactuca
   plumieri*). After Charles Plumier, see
   Plumeria. C Europe.

Cichorium ki-*ko*-ree-um *Compositae.*
Derived from the Arabic name. Annual,
biennial or perennial herbs.
   endivia en-*di*-vee-a. See below. Endive.
   India.
   intybus in-tew-bus. From *intubus* the L.
   name. Both this and the above were
   derived from Egyptian *tybi* (January) the
   month when it was eaten. Chicory.
   Europe, N Africa, W Asia.

Cider Gum see *Eucalyptus gunnii*
Cigar Plant see *Cuphea ignea*

Cimicifuga kee-mi-ki-*few*-ga *Ranunculaceae.*
From L. *cimex* (a bug) and *fugo* (to repel), *C.
foetida* has been used as an insect repellent.
Perennial herbs. Bugbane. Cohosh.
   americana a-me-ri-*kah*-na. American. E
   United States.
   dahurica da-*hewr*-ri-ka. Of Dahuria. SE
   Siberia. NE Asia.
   foetida foy-ti-da. Unpleasantly scented. SE
   Europe, N Asia.
   japonica ja-*pon*-i-ka. Of Japan.
   racemosa ră-kay-*mō*-sa. With flowers in
   racemes. Black Snakeroot. EN America.

Cinderella Slippers see *Sinningia regina*
Cineraria cruenta see *Pericallis cruenta*
   × **hybrida** see *Pericallis* × *hybrida*
   maritima see *Senecio cineraria*

Cionura kee-on-*ewr*-ra. *Asclepiadaceae.* From
Gk. *kion* (a column) and *oura* (a tail)
presumably referring to the stigma.
Deciduous climber.
   erecta e-*rek*-ta. (= *Marsdenia erecta*). Erect.
   SE Europe, W Asia.

Cissus kis-us *Vitaceae.* From Gk. *kissos* (ivy).
Tender climbers and succulents.

Cissus (continued)
   antarctica ăn-*tark*-ti-ka. Of Antarctic
   regions. Kangaro Vine. E Australia.
   bainesii baynz-ee-ee. After Thomas Baines
   (1820–75), artist and explorer in S Africa.
   SW Africa.
   capensis see *Rhoicissus capensis*
   discolor dis-ko-lor. Two-coloured (the
   leaves). Rex-begonia Vine. Indonesia.
   quadrangularis kwod-rang-gew-*lah*-ris.
   Four-angled (the stems). S Africa.
   rhombifolia rom-bi-*fo*-lee-a. (= *Rhoicissus
   rhomboidea* hort.). With diamond-shaped
   leaves. Grape Ivy. C and S America.
   sicyoides si-kee-*oi*-deez. Like *Sicyos
   (Cucurbitaceae).* Princess Fine. Tropical
   America.
   striata stree-*ah*-ta. Striped (the stems).
   Miniature Grape Ivy. Chile, S Brazil.

Cistus kis-tus. *Cistaceae.* From the Gk. name.
Evergreen shrubs. Rock Rose.
   × *aguilari* ă-gwi-*lah*-ree. *C. ladanifer* × *C.
   populifolius.* Of Aguilar, Spain. Spain,
   Portugal, Morocco.
      'Maculatus' măk-ew-*lah*-tus. Spotted
      (the petals).
   albidus ăl-bi-dus. Whitish (the leaves and
   shoots). SW Europe, N Africa.
   × *canescens* kah-*nes*-enz. *C. albidus* × *C.
   creticus.* Grey-hairy (the leaves and
   shoots). Algeria.
   × *corbariensis* kor-bah-ree-*en*-sis. (*C.
   populifolius* × *C. salviifolius*). From
   Corbières, S France.
   crispus kris-pus. Finely wavy (the leaves).
   SW Europe, N Africa.
   × *cyprius* kip-ree-us. *C. ladanifer* × *C.
   laurifolius.* Of Cyprus. W Mediterranean
   region.
   ladanifer la-*dah*-ni-fer. Bearing ladanum (a
   gum resin used in perfumery). S Europe,
   N Africa.
      latifolius lat-i-*fō*-lee-us (= *C. palhinhae*).
      Broad-leaved. Algarve.
   laurifolius low-ri-*fo*-lee-us. *Laurus*-leaved.
   Mediterranean region. SW Europe.
   × *lusitanicus* loo-si-*tah*-ni-kus. *C. hirsutus
   × C. ladanifer.* From Portugal.
      'Decumbens' day-*kum*-benz. Prostrate.
   populifolius pō-pul-i-*fol*-ee-us. *Populus*-
   leaved. SW Europe.
      lasiocalyx lă-see-ō-*kă*-liks. With a
      woolly calyx.
   × *pulverulentus* pul-ve-roo-*len*-tus. *C.
   albidus* × *C. crispus.* Dusted. W
   Mediterranean region.
   × *purpureus* pur-*pewr*-ree-us. *C. creticus* ×
   C. ladanifer. Purple (the flowers).

Cistus (continued)
× *skanbergii* skăn-*berg*-ee-ee. *C.
monspeliensis* × *C. parviflorus.* After
Skanberg. Greece.

Citrange see × *Citroncirus webberi*

× **Citrofortunella** kit-rō-for-tew-*nel*-a
Rutaceae. Intergeneric hybrid, from the
names of the parents. *Citrus* × *Fortunella.*
Tender evergreen shrub or tree.
mitis mee-tis. *Citrus reticulata* × *Fortunella*
sp. (= *Citrus mitis*). Not spiny.
Calamondin.

Citron see *Citrus medica*

× **Citroncirus** kit-ron-si-rus *Rutaceae.*
Intergeneric hybrid, from the names of the
parnets. *Citrus* × *Poncirus.* Tender evergreen
tree or shrub.
webberi web-a-ree. *Citrus sinensis* × *Poncirus
trifoliata.* After H. J. Webber. Citrange.

Citrullus kit-*rul*-us *Cucurbitaceae.* From *Citrus*
q.v., referring to the fruit. Tender, annual
herb.
lanatus lah-*nah*-tus. Woolly. Water Melon.
Tropical and S Africa.

Citrus kit-rus *Rutaceae.* The L. name for
citron (*C. medica*). Tender, evergreen trees.
aurantiifolia ow-răn-tee-i-*fo*-lee-a. With
leaves like *C. aurantium.* Lime. SE Asia.
aurantium ow-*răn*-tee-um. Orange (the
fruit). Seville Orange, Bitter Orange. S
Vietnam.
limon lee-mon. The L. name. Lemon. SE
Asia.
× *limonia* lee-*mō*-nee-a. *C. limon* × *C.
reticulata.* The L. name for the lemon tree.
Mandarin Lime.
maxima mahk-si-ma. Larger (the fruit).
Shaddock, Pummelo. Malay Peninsula.
Polynesia.
medica med-i-ka. Used in medicine.
Citron. India.
mitis see × *Citrofortunella mitis*
× *paradisi* pă-ra-*dee*-see. *C. maxima* × *C.
sinensis.* Of paradise. Grapefruit.
reticulata ray-tik-ew-*lah*-ta. Net-veined.
Mandarin Orange, Satsuma. Tangerine.
SE Asia.
sinensis si-*nen*-sis. Of China. Sweet
Orange. SE Asia.

Cladanthus kla-*dănth*-us *Compositae.* From
Gk. *klados* (a branch) and *anthos* (a flower),

Cladanthus (continued)
the flower heads are borne at the ends of the
shoots. Annual herb.
arabicus a-*ră*-bi-kus. (= *Anthemis arabica*).
Arabian. S Spain, N Africa.

Cladrastis kla-*drăs*-tis *Leguminosae.* From Gk.
klados (a branch) and *thraustos* (fragile)
referring to the brittle shoots. Deciduous
trees.
lutea loo-tee-a. Yellow (the wood). Yellow
Wood. SE United States.
sinensis si-*nen*-sis. Of China.

Clarkia klark-ee-a *Onagraceae.* (= *Godetia*).
After Captain William Clark (1770–1838).
Annual herbs.
amoena a-*mot*-na. Pleasant. Satin Flower.
California.
lindleyi lind-lee-ee (= *Oenothera
grandiflora*). After John Lindley
(1799–1865), botanist. Alabama.
concinna kon-*kin*-a. Elegant. Red Ribbons.
California.
pulchella pul-*kel*-la. Pretty. WN America.
unguiculata un-gwik-ew-*lah*-ta. (= *C.
elegans*). Clawed (the petals). California.

Clary see *Salvia sclarea*

Cleistocactus klay-stō-*kăk*-tus *Cactaceae.*
From Gk. *kleistos* (closed) and *Cactus* q.v.,
the flowers are tubular and nearly closed at
the mouth.
baumannii bow-*mahn*-ee-ee. After
Baumann, a cactus grower. S America.
jujuyensis hoo-hoo-ee-*en*-sis. Of Jujuy. N
Argentina.
smaragdiflorus sma-răg-di-*flō*-rus. With
emerald flowers, the inner perianth
segments are green. Firecracker Cactus. N
Argentina.
straussii strows-ee-ee. After Strauss. Silver
Torch. Bolivia, Argentina.
tominensis tō-mi-*nen*-sis. Of Tomina.
Bolivia.

Clematis klem-a-tis *Ranunculaceae.* Gk. name
for a climbing plant. Herbaceous perennials
and climbers.
alpina ăl-*peen*-a. Alpine. Europe, N Asia.
armandii ar-*mond*-ee-ee. After Armand
David, see *Davidia.* C and W China.
chrysocoma kris-o-ko-ma. Golden-haired
(the young growths). China.
cirrhosa ki-*rō*-sa. With tendrils, the leaf
stalks act as tendrils. Mediterranean
region.

Clematis (continued)
balearica bă-lee-*ah*-ri-ka. Of the Balearic Islands. Balearic Islands, Corsica.
× *eriostemon* e-ree-ō-*stay*-mon. *C. integrifolia* × *C. viticella*. With woolly stamens.
'Hendersonii' hen-der-*son*-ee-ee. After Henderson who raised it about 1830.
flammula flăm-ew-la. An old name for this plant from L. *flammula* (a little flame or small banner). S Europe.
florida flō-ri-da. Flowering. China.
'Sieboldii' see-*bōld*-ee-ee. After Siebold, from whose nursery it was introduced in 1836, see *Acanthopanax sieboldii*.
heracleifolia he-ra-klee-i-*fo*-lee-a. *Heracleum*-leaved. C and N China.
davidiana dă-vid-ee-*ah*-na. After David who introduced it to France in 1863, see *Davidia*.
integrifolia in-teg-ri-*fo*-lee-a. With entire leaves. E Europe to C Asia.
× *jackmanii* jăk-*măn*-ee-ee. *C. lanuginosa* × *C. viticella*. After Jackman's nursery where it was raised in 1860.
× *jouiniana* zhoo-ăn-ee-*ah*-na. *C. heracleifolia davidiana* × *C. vitalba*. After E. Join, manager of the Simon-Louis nursery, Metz, Franc.
macropetala măk-rō-*pe*-ta-la. With large petals, referring to the large, petal-like staminodes which distinguish it from *C. alpina*. China. Siberia.
montana mon-*tah*-na. Of mountains. Himalaya, China.
rubens ru-benz. Red (the flowers).
wilsonii wil-*son*-ee-ee. After Ernest Wilson who introduced it, see *Magnolia wilsonii*.
orientalis o-ree-en-*tah*-lis. Eastern. Caucasus to N China.
recta rek-ta. Erect. S Europe.
rehderiana ray-da-ree-*ah*-na. After Rehder, see *Rehderodendron*. Nepal to SW China.
serratifolia se-rah-ti-*fo*-lee-a. With toothed leaves (leaflets). Korea.
spooneri spoon-er-ee (= *C. chrysocoma* var. *sericea*). After Spooner. China.
tangutica tă-*gew*-ti-ka. Of Cansu (previously Kansu). NW China.
texensis teks-*en*-sis. Of Texas.
vitalba vee-*tăl*-bsa. Literally, white vine. Old Man's Beard. Traveller's Joy. Europe.
viticella vee-ti-*kel*-la. Diminutive of *Vitis*. S Europe.

Cleome klay-ō-mee *Capparidaceae*. Derivation uncertain, possibly from Gk. *kleos* (glory). Tender annual herb.

Cleome (continued)
hassleriana has-la-ree-*ah*-na. (= *C. spinosa* hort.). After Emile Hassler. Spider Flower. S America.

Clerodendrum kle-rō-*den*-drum *Verbenaceae*. From Gk. *kleros* (chance) and *dendrom* (a tree) referring to the variable medicinal properties. Hardy and tender, trees, shrubs and climbers.
bungei bung-gee-ee. After Alexander von Bunge (1803–90), Russian botanist. China.
speciosissimum spe-kee-ō-*sis*-i-mum. Most showy. Glory Bower. Java.
thomsoniae tom-*son*-ee-ie. After the wife of the Rev. W. C. Thomson, who was in Africa 1849–65. W Tropical Africa.
trichotomum tri-*ko*-to-mum. Branching into three. Japan, China.

Clethra kleth-ra *Clethraceae*. From Gk. *klethra* (alder). Deciduous trees and shrubs.
alnifolia ăl-ni-*fo*-lee-a. *Alnus*-leaved. Sweet Pepper Bush. E N America.
arborea ar-*bo*-ree-a. Tree-like. Lily of the Valley Tree. Madeira.
barbinervis bar-bi-*ner*-vis. With bearded veins. Japan.
delavayi del-a-*vay*-ee. After Delavay who discovered it in 1884, see *Abies delavayi*. China.
fargesii far-*geez*-ee-ee. After Farges who introduced it to France, see *Decaisnea fargessii*. China.

Cleyera klay-a-ra *Theaceae*. After Andreas Cleyer, 17th century doctor with the East India Co. Evergreen shrub.
japonica ja-*pon*-i-ka. Of Japan. Himalaya, E Asia.
'Tricolor' *tri*-ko-lor. (= *C. fortunei*). Three-coloured (the leaves).

Clianthus klee-*ănth*-us *Leguminosea (Papilionoidae)* From Gk. *kleos* (glory) and *anthos* (a flower). Tender sub-shrubs.
formosus for-*mō*-sus. (= *C. speciosus*). Beautiful. Sturt's Desert Pea. W Australia.
puniceus pew-*ni*-kee-us. Reddish-purple (the flowers). Glory Pea. New Zealand.

Cliff Brake, Green see *Pellaea viridis*
Purple see *P. atropurpurea*

Clivia klie-vee-a *Amaryllidaceae*. After Lady Charlotte Florentina Clive, Duchess of Northumberland (died 1868), granddaughter

Clivia (continued)
of Robert Clive. Tender evergreen perennial
herbs. S Africa.
   × *cyrtanthiflora* kur-tănth-i-*flō*-ra. *C.*
   miniata × C. nobilis. Cyrtanthus-
   flowered.
   miniata min-ee-*ah*-ta. Cinnabar-red (the
   flowers). Kaffir Lily.
   nobilis nō-bi-lis. Notable.

Clock Vine see *Thunbergia grandiflora*
Clover, White see *Trifolium repens.*

Cobaea kŏ-*bie*-a Polemoniaceae. After
Bernardo Coba (1572–1659), Spanish
missionary in Mexico and Peru. Annual
climber.
   scandens skăn-denz. Climbing. Cathedral
   Bells. Mexico.

Cobnut see *Corylus avellana*

Cochlioda kok-lee-ō-da Orchidaceae. From
Gk. *kochlos* (a snail-shell) referring to the
shell-like calluses on the lip of the type
species. Greenhouse orchid.
   densiflora dens-i-*flō*-ra. Densely flowered.
   Peru, Bolivia.
   rosea ro-see-a. Rose-coloured (the
   flowers). Peru.

Coconut see *Cocos nucifera*
Cockscomb see *Celosia cristata*
Cockspur Thorn see *Crataegus crus-galli*

Cocos kō-kos Palmae. From Portuguese *coco*
(a grinning face) from the appearance of the
fruit. Tender palms.
   nucifera new-*ki*-fe-ra. Nut-bearing.
   Coconut. Tropics.
   weddelliana see *Microcoelum weddellianum*

Codariocalyx ko-*dah*-ree-ō-kay-lix
Leguminosae (Papilionoideae). From Gk.
koidarion, a sheepskin, and *kalyx*, calyx; the
calyx is very hairy. Tropical shrubs.
   motorius mo-*to*-ree-us (= *Desmodium*
   gyrans, D. motorius). Moving (the leaflets).
   Telegraph Plant. Tropical Asia.

Codiaeum kō-dee-*ie*-um Euphorbiaceae. From
the native name. Tender evergreen shrub.
   variegatum vă-ree-a-*gah*-tum. Variegated.
   pictum pik-tum. Painted. Croton. SE Asia.

Codonopsis kō-dōn-*op*-sis Campanulaceae.
From Gk. *kodon* (a bell) and *-opsis* indicating
resemblance, referring to the bell-shaped
corolla. Annual and perennial climbers.

Codonopsis (continued)
   clematidea klem-a-*tid*-ee-a. Like *Clematis.*
   W Asia.
   convolvulacea kon-vol-vew-*lah*-kee-a. (=
   C. vinciflora). Like *Convolvulus*. Himalaya,
   W China.
   ovata ō-*vah*-ta. Ovate (the leaves).
   Himalaya, S China.

Coelogyne koy-*lo*-gin-ee but commonly see-
lō-*gie*-nee Orchidaceae. From Gk. *koilos*
(hollow) and *gyne* (female) referring to the
hollowed stigma. Greenhouse orchids.
   asperata ă-spe-*rah*-ta. Roughened. SE Asia.
   barbata bar-*bah*-ta. Bearded (the central
   lobe of the lip). Himalaya.
   Burfordiense bur-ford-ee-*en*-see. *C.*
   asperata × *C. pandurata*. Of Burford Lodge,
   Dorking.
   cristata kris-tah-ta. Crested (the lip).
   Himalaya.
   elata ay-*lah*-ta. Tall. Himalaya.
   flaccida flăk-ki-da. Drooping (the racemes).
   Himalaya.
   massangeana ma-son-zhee-*ah*-na. After M.
   de Massange, a 19th-century orchid
   grower.
   mooreana mor-ree-*ah*-na. After F. W.
   Moore who supplied the type specimen.
   S. Vietnam.
   nitida ni-ti-da (= *C. ochracea*) Shining (the
   flowers). Himalaya, Burma
   pandurata păn-dew-*rah*-ta. Fiddle-shaped
   (the lip). Black Orchid. Malaya, Borneo.
   speciosa spe-kee-ō-sa. Showy. Java.

Coffea kof-ee-a Rubiaceae. From *kahwah*, the
Arabic name. Tender, evergreen shrub.
   arabica a-*ră*-bi-ka. Arabian. Arabian Coffee
   Plant. Tropical Africa.

Coix kō-iks Gramineae. The Gk. name for a
similar plant. Annual grass.
   lacryma-jobi lă-kri-ma-*yō*-bee. Job's tears,
   from the tear-shaped, grey-white seeds.
   Job's Tears. SE Asia.

Colchicum kol-ki-kum Liliaceae
(*Colchidaceae*). Of Colchis, W Asia where
they were said to grow. Cormous perennials.
   aggripinum ăg-ri-*peen*-um. After
   Aggripina, Cult.
   autumnale ow-tum-*nah*-lee. Of autumn
   (flowering). Autumn Crocus. Europe.
   boissieri bwū-see-e-ree. After Pierre
   Edmund Boissier (1810–85), Genevese
   botanist. S Greece.
   bornmuelleri born-*moo*-la-ree. After Joseph

Colchicum (continued)
Bornmüller (1862–1942) who collected in the Balkans and W Asia. Turkey.
byzantinum bi-zan-*tee*-num. Of Istanbul (Byzantium). Cult.
cilicium ki-*li*-kee-um. Of Cilicia (S Turkey).
luteum loo-tee-um. Yellow (the flowers). C Asia, NW India.
speciosum spe-kee-ō-sum. Showy. Caucasus, N Turkey, Iran.

Coleus see *Solenostemon*

Colletia ko-*lay*-tee-a *Rhamnaceae*. After Philibert Collet (1643–1718). French botanist. More or less leafless, spiny shrubs.
armata ar-*mah*-ta. Spiny. S Chile.
paradoxa pă-ra-*doks*-a. (= *C. cruciata*). Unusual. Uruguay.

Collinia elegans see *Chamaedorea elegans*

Collinsia ko-*linz*-ee-a *Scrophulariaceae*. After Zaccheus Collins (1764–1831), Philadelphia botanist. Annual herbs.
bicolor bi-ko-lor (= *C. heterophylla*). Two-coloured (the flowers). Chinese Houses. California.
grandiflora grăn-di-*flō*-ra. Large-flowered. Blue Lips. W N America.
heterophylla see *C. bicolor*
verna ver-na. Of spring (flowering). Blue-eyed Mary. E United States.

Colombia Buttercup see *Oncidium cheirophorum*

Colquhounia ka-*hoon*-ee-a *Labiatae*. After Sir Robert Colquhoun (died 1838), who collected in the Himalaya. Semi-hardy shrub.
coccinea kok-*kin*-ee-a. Scarlet (the flowers). Himalaya, SW China.

Columbine see *Aquilegia*

Columnea ko-*lum*-nee-a *Gesneriaceae*. After Fabius Columna (1567–1640). Tender, evergreen climbers. Costa Rica.
gloriosa glo-ree-ō-sa. Glorious.
linearis line-ee-*ah*-ris. Linear (the leaves).
microphylla mik-rō-*fil*-la. With small leaves.

Colutea ko-*loo*-tee-a *Leguminosae*. From *kolutea* the Gk. name. Deciduous shrubs and trees.
arborescens ar-bo-*res*-enz. Tree-like. Bladder Senna. Mediterranean region, SE Europe.

Colutea (continued)
× *media* me-dee-a. *C. arborescens* × *C. orientalis*. Intermediate (between the parents).
orientalis o-ree-en-*tah*-lis. Eastern. W Asia.

Comfrey see *Symphytum*
Russian see *S.* × *uplandicum*

Commelina kom-e-*leem*-a *Commelinaceae*. After two Dutch botanists, Johan (1629–92) and Caspar (1667–1731) Commelin. Semi-hardy and tender perennials. Day Flower.
coelestis koy-*les*-tis. Sky-blue (the flowers). Mexico.
erecta e-*rek*-ta. Erect. E and S United States.
tuberosa tew-be-*rō*-sa. Tuberous. Mexico.

Comptonia komp-*ton*-ee-a *Myricaceae*. After Henry Compton, Bishop of London (1632–1713). Deciduous, suckering shrub.
peregrina pe-re-*gree*-na. Foreign. Sweet Fern. E N America.

Cone Flower see *Rudbeckia*

Conoclinium kon-ō-*klin*-ee-um *Compositae*. From Gk. *konos*, a cone, and *klino*, to lean; the receptacle is conical. Perennial herbs, once included in *Eupatorium*.
coelestinum koy-les-*teen*-um (= *Eupatorium coelestinum*). Sky-blue (the flowers). Mistflower. S and E N America, West Indies.

Conophytum kon-o-*fit*-um *Aizoaceae*. From Gk. *konos* (a cone) and *phyton* (a plant) referring to the conical shape of the plant body. S Africa.
albescens see *C. bilobum*
bilobum bi-*lō*-bum. Two lobed.
frutescens froo-*tes*-enz. Shrubby.
gratum grah-tum. Pleasing.
minutum mi-*new*-tum. Very small.
obcordellum ob-kor-*del*-lum. Shaped like a small heart.
pearsonii peer-*son*-ee-ee. After Prof. Pearson.

Consolida kon-*so*-li-da *Ranunculaceae*. From L. *consolida* (to make whole) referring to medicinal properties. Annual herbs. Larkspur.
ambigua ăm-*big*-ew-a. (= *Delphinium ajacis*). Doubtful. Mediterranean region.
regalis ray-*gah*-lis. (= *Delphinium consolida*). Royal. SE Europe, Turkey.

Convallaria kon-va-*lah*-ree-a *Liliaceae*. From L. *convallis* (a valley). Herbaceous perennial.
majalis mah-*yah*-lis. Flowering in May. Lily of the Valley. Europe.

Convolvulus kon-*vol*-vew-lus *Convolvulaceae*. From L. *convolva* (to twine around). Annual and perennial herbs and shrubs.
althaeoides ahl-thie-*oi*-dezz. Like *Althaea*. S Europe.
cneorum nee-o-rum. From *kneorum*, Gk. name for a dwarf, olive-like shrub. Silver Bush. S Europe.
sabatius sa-*bah*-tee-us. (= *C. mauritanicus*). Of Savona (Sabbatia), N Italy. Italy, Sicily, N Africa.
tricolor tri-ko-lor. Three-coloured (the flowers). S Europe.

Coprosma ko-*pros*-ma *Rubiaceae*. From Gk. *kopros* (dung) and *osme* (smell) referring to the odour of the foliage. Evergreen, tender and hardy shrubs. New Zealand.
lucida loo-ki-da. Glossy (the leaves).
petriei pet-ree-ee. After Donald Petrie (1846–1925), Scottish inspector of schools and amateur naturalist in New Zealand.
repens ree-penz. Creeping.

Coral Berry see *Ardisia crispa*.
Coralberry see *Aechmea fulgens*.
Coral Drops see *Bessera elegans*.
Coral Plant see *Berberidopsis corallina*, *Rusellia equisetiformis*.
Coral Tree see *Erythrina crista-galli*.
Coral Vine see *Antigonon leptopus*.

Cordyline kor-*di*-li-nee *Agavaceae*. From Gk. *kordyle* (a club) referring to the large, fleshy roots of some species. Tender and semi-hardy, evergreen trees and shrubs.
australis ow-*strah*-lis. Southern. Cabbage Tree. New Zealand.
indivisa in-dee-*vee*-sa. Undivided, it is usually single-stemmed. New Zealand.
rubra rub-ra. Red (? the flowers). Cult.
stricta strik-ta. Upright. New South Wales.
terminalis ter-min-*ah*-lis. Terminal (the inflorescence). Good Luck Plant. E Asia.

Coreopsis ko-ree-*op*-sis *Compositae*. From Gk. *koris* (a bug) and -*opsis* indicating resemblance, the seeds look like ticks. Annual and perennial herbs. Tickseed.
auriculata ow-rik-ew-*lah*-ta. With basal lobes (the leaves). SE United States.
basilis ba-*sah*-lis. (= *C. drummondii*). Basal,

Coreopsis (continued)
perhaps referring to the red bases of the ray florets. S United States.
grandiflora grăn-di-*flō*-ra. Large-flowered. SE United States.
lanceolata lăn-kee-ō-*lah*-ta. Lanceolate (the leaves). SE United States.
tinctoria tink-*to*-ree-a. Used in dyeing. C and W United States.
verticillata ver-ti-ki-*lah*-ta. Whorled, the leaves are finely cut, appearing whorled. SE United States.

Coriandrum ko-ree-*ăn*-drum *Umbelliferae*. From *koriandrum* the Gk. name. Annual herb.
sativum sa-*tee*-vum. Cultivated. Coriander. N Africa, W Asia.

Coriaria ko-ree-*ah*-ree-a *Coriariaceae*. From L. *corium* (leather), some species are used in tanning. Deciduous shrubs.
japonica ja-*pon*-i-ka. Of Japan.
terminalis ter-mi-*nah*-lis. Terminal (the racemes). Himalaya, China.
  xanthocarpa zănth-ō-*kar*-pa. With yellow fruits.

Corkscrew Rush see *Juncus effusus* 'Spiralis'
Corn see *Zea mays*
Corn Cockle see *Agrostemma githago*
Corn Lily see *Ixia*
Corn Marigold see *Chrysanthemum segetum*
Corn Salad see *Valerianella locusta*
Cornelian Cherry see *Cornus mas*
Cornflower see *Centurea cyanus*

Cornus kor-nus *Cornaceae*. The L. name for *C. mas*. Deciduous shrubs and trees. Dogwood.
alba ăl-ba. White (the fruit). Siberia, China.
alternifolia ăl-ter-ni-*fo*-lee-a. With alternate leaves. EN America.
amomum a-*mō*-mum. Gk. name of a spice plant.
canadensis kăn-a-*den*-sis. Of Canada or NE North America. N America. E Asia.
capitata kăp-i-*tah*-ta. In a dense head (the flowers). Himalaya, China.
florida flō-ri-da. Flowering, Flowering Dogwood. E United States.
kousa koo-sa. The Japenese name. Japan, China.
  chinensis chin-*en*-sis. Of China.
macrophylla măk-ro-*fil*-la. With large leaves. Himalaya.
mas mahs. Male, an epithet used to distinguish a robust specimen from a

Cornus (continued)
more delicate one which was regarded as
female. Cornelian Cherry. Europe.
nuttallii nū-*tahl*-ee-ee. After Thomas
Nuttall (1786–1839). Pacific Dogwood.
W N America.
sanguinea săng-*gwin*-ee-a. Red (autumn
colour). Common Dogwood. Europe.
stolonifera stol-lō-*ni*-fe-ra. Bearing stolons.
N America.
'Flaviramea' flah-vi-*rahm*-ee-a. With
yellow shoots.

Corokia ko-*rō*-kee-a *Cornaceae*. From the
Maori name *korokia*. Evergreen shrubs.
New Zealand.
cotoneaster ko-tōn-ee-*ă*-ster. The
intricately branched habit is reminiscent
of some cotoneasters.
× *virgata* vir-*gah*-ta. *C. buddleioides* × *C.
cotoneaster*. Twiggy.

Coronilla ko-rō-*nil*-la *Leguminosae*
(*Papilionoideae*). Diminutive of L. *corona* (a
crown) referring to the arrangement of the
flowers. Hardy and semi-hardy shrubs.
emerus e-*me*-rus. From the Italian name
emero. Scorpion Senna. C and S Europe.
emeroides e-me-*roi*-des. Like *C. emerus*.
SE Europe. Syria.
valentina văl-en-*teen*-a. Of Valencia, Spain.
S Europe.
glauca glow-ka. Glaucous (the leaves). S
Europe.

Correa ko-ree-a *Rutaceae*. After José
Francesco Correa de Serra (1751–1823),
Portuguese botanist. Tender and semi-hardy
evergreen shrubs. SE Australia.
alba ăl-ba. White (the flowers).
reflexa re-*fleks*-a. (= *C. speciosa*). Reflexed (the
corolla lobes).

Cortaderia kor-ta-*de*-ree-a *Gramineae*. From
the Argentinian name. Perennial grasses.
jubata yoo-*bah*-ta. Like a mane (the
inflorescence). S America.
richardii ree-*shard*-ee-ee. After Achille
Richard (1794–1852), French botanist.
New Zealand.
selloana sel-ō-*ah*-na. (= *C. argentea*). After
Sellow, see *Feijoa sellowiana*. Pampas Grass.
S America.

Corydalis ko-*ri*-da-lis *Fumariaceae*. The Gk.
name for a lark, the spur of the flower
resembles that of a lark. Perennial herbs.
bulbosa see *C. solida*

Corydalis (continued)
cashmeriana kăsh-me-ree-*ah*-na. Of Kashmir.
Himalaya.
cheilanthifolia kay-lănth-i-*fo*-lee-a. With
leaves like *Cheilanthus*. China.
lutea loo-tee-a. Yellow (the flowers).
Europe.
nobilis nō-bi-lis. Notable. C Asia.
solida so-li-da. Solid (the tuber). Europe.

Corylopsis ko-ril-*op*-sis *Hamamelidaceae*.
From *Corylus* q.v. and Gk. -*opsis* indicating
resemblance, the habit and leaves are similar
to *Corylus*. Deciduous trees and shrubs.
glabrescens gla-*bres*-enz. Nearly glabrous
(the leaves). Japan, Korea.
pauciflora paw-si-*flō*-ra. Few-flowered (the
spikes). Japan, Taiwan.
sinensis si-*nen*-sis. Of China. China, Tibet.
spicata spee-*kah*-ta. With flowers in spikes.
Japan.

Corylus ko-ril-us *Corylaceae*. From *korylos* the
Gk. name. Deciduous shrubs and trees.
avellana ă-ve-*lah*-na. Of Avella Vecchia, S
Italy. Hazel, Cobnut. Europe, N Africa, W
Asia.
'Contorta' kon-*tor*-ta. Twisted (the
shoots). Harry Lauder's Walking Stick.
colurna ko-*lurn*-a. The classical name.
Turkish Hazel. SE Europe, W Asia.
maxima mahk-si-ma. Larger. Filbert. S
Europe.

Cosmos kos-mos *Compositae. From Gk.
kosmos* (beautiful). Annual herbs.
bipinnatus bi-pin-*ah*-tus. With bi-pinnate
leaves. Mexico.
sulphureus sul-*fewr*-ree-us. Sulphur-yellow
(the flowers). Mexico.

Costmary see *Tanacetum balsamita*

Cotinus ko-ti-nus *Anacardiaceae*. From *kotinos*
Gk. name for the olive. Deciduous shrubs
or trees.
coggygria ko-*gig*-ree-a. (= *Rhus cotinus*).
From *kokkugia* the Gk. name. Smoke Tree.
C and S Europe, Himalaya, China.
'Foliis Purpureis' *fo*-lee-is pur-*pewr*-ree-
is. With purple leaves.
obovatus ob-ō-*vah*-tus. (= *Rhus cotinoides*).
Obovate (the leaves). SE United States.

Cotoneaster ko-tōn-ee-*ă*-ster *Rosaceae*. From
L. *cotoneum* (quince) and -*aster* (resembling
somewhat) from the similiarity of the leaves
of some species. Deciduous and evergreen
shrubs.

Cotoneaster (continued)
adpressus ăd-*pres*-us. Pressed against,
referring to its low habit. Himalaya,
China.
   praecox prie-koks. Early (fruit ripening).
affinis a-*fee*-nis. Related to (another
species). Himalaya, W China.
bacillaris bă-ki-*lah*-ris. Staff-like. The
branches are used to make walking-sticks
in the Himalaya.
bullatus bul-*lah*-tus. With impressed veins
on the leaves. W China, Tibet.
   floribundus flŏ-ri-*bun*-dus. Profusely
flowering.
congestus con-*ges*-tus. Congested (the
habit). Himalaya, W China.
conspicuus kon-*spik*-ew-us. Conspicuous
(the fruit). Tibet.
'Cornubia' kor-*new*-bee-a. L. name for
Cornwall.
dammeri dăm-a-ree. After Dammer. C
China.
distichus see *C. nitidus*
divaricatus di-vah-ri-*kah*-tus. With
spreading branches. China.
'Exburiensis' eks-ba-ree-*en*-sis. *C. frigidus*
× *C. salicifolius*. From Exbury, where it
was raised.
franchetti fron-*shay*-tee-ee. After Adrien
Franchet (1834–1900), French botanist.
Tibet, W China.
   sternianus stern-ee-*ah*-nus. After Sir
Frederick Stern. Burma.
frigidus fri-gi-dus. Growing in cold regions.
Himalaya, W China.
horizontalis ho-ri-zon-*tah*-lis. Horizontal
(the habit). China.
'Hybridus Pendulus' *hib*-ri-dus *pen*-dew-
lus. A hybrid of pendulous habit.
lacteus lăl-tee-us. Milky (the flowers).
China.
microphyllus mik-rŏ-*fil*-lus. Small-leaved.
Himalaya, SW China.
   cochleatus kok-lee-*ah*-tus. Shell-like (the
leaves).
   thymifolius tiem-i-*fo*-lee-us. *Thymus*-
leaved.
multiflorus mul-tee-*flŏ*-rus. Many-
flowered. Caucasus to China.
nitidus ni-ti-dus (= *C. distichus*). Shining
(the leaves). Himalaya.
pannosus pă-*nŏ*-sus. Felt-like (the young
shoots and the undersides of the leaves).
China.
prostratus pros-*trah*-tus. Prostrate.
Himalaya, SW China.
'Rothschildianus' roths-chield-ee-*ah*-nus.
After Baron Rothschild.

Cotoneaster (continued)
salicifolius să-lik-i-*fo*-lee-us. *Salix*-leaved.
China.
   floccosus flok-ŏ-sus. Woolly (the
underside of the young leaves).
   rugosus roog-ŏ-sus. Wrinkled (the
leaves).
simonsii sie-*monz*-ee-ee. After Mr Simons
who introduced it. Himalaya.
× *watereri waw*-ta-ra-ree. *C. frigidus* × *C.
henryanus*. After Waterer's who raised it.

Cotton, Levant see *Gossypium herbaceum*
  Tree see *G. arboreum*
Cotton Rose see *Hibiscus mutabilis*

Cotula *kot*-ew-la Compositae. From Gk.
kotula (a small cup), the base of the leaves
often clasp the stem making a small cup.
Annual and perennial herbs.
   atrata see *Leptinella atrata*
   barbata bar-*bah*-ta. Bearded (the stems). S
Africa.
   coronopifolia ko-rŏ-no-pi-*fo*-lee-a.
Coronopus-leaved. Brass Buttons. S Africa.
   potentillina see *Leptinella potentillina*
   squalida see *Leptinella squalida*

Cotyledon ko-ti-*lay*-don Crassulaceae. From
Gk. *kotlye* (a small cup) referring to the cup-
shaped leaves of some species. Tender
succulents.
   barbeyi bar-bee-ee. After William Barbey
(1842–1914), Geneva botanist. E. Africa,
Arabia.
   campanulata kam-pan-yew-*lah*-ta. Bell-
shaped (the flowers). South Africa.
   ladysmithensis lay-dee-smith-*en*-sis. From
Ladysmith. S Africa.
   orbiculata or-bik-ew-*lah*-ta. Orbicular (the
leaves). S Africa.
   paniculata see *Tylocodon paniculatus*
   reticulata see *Tylocodon reticulatus*
   simplicifolia see *Chiastophyllum
oppositifolium*
   teretiflora see *C. campanulata*
   undulata un-dew-*lah*-ta. Wavy-edged (the
leaves). Silver Crown. S. Africa.

Couve Tronchuda see *Brassica oleracea*
Tronchuda.
Cow Herb see *Vaccaria hispanica*
Cowberry see *Vaccinium vitis-idaea*
Cowslip see *Primula veris*
Crab see *Malus*
  Siberian see *M. baccata*

Crambe kram-bay Cruciferae. Gk. name for

Crambe (continued)
cabbage from the similar leaves. Perennial
herbs.
   cordifolia kor-die-*fo*-lee-a. With heart-
   shaped leaves. Caucasus.
   maritima ma-*ri*-ti-ma. Growing near the
   sea. Sea Kale. W Europe to W Asia.

Cranberry see *Vaccinium*
   American see *V. macrocarpon*
   Small see *V. oxycoccus*
Crane Lily see *Strelitzia reginae*
Cranesbill see *Geranium*
   Bloody see *G. sanguineum*
   Dusky see *G. phaeum*
   Meadow see *G. pratense*
   Wood see *G. sylvaticum*
Crape Myrtle see *Lagerstroemia indica*

Crassula krăs-ew-la *Crassulaceae*. From L.
crassus (thick) referring to the fleshy leaves.
Tender succulents. S Africa.
   arborescens ar-bo-*res*-enz. Tree-like.
   Chinese Jade.
   argentea see *C. ovata*
   coccinea kok-*kin*-ee-a (× *Rochea coccinea*).
   Scarlet (the flowers).
   excelsis ex-*sell*-sis. Tall. Cape Province
      cooperi koo-pa-ree. After Thomas
      Cooper (1815–1913).
   lactea lăk-tee-a. Milky (the flowers).
   Tailor's Patch.
   lycopodioides see *C. mucosa*
   milfordiae mil-*ford*-ee-ie. After Mrs Helen
   A. Milford (died 1940) who collected in
   S Africa.
   muscosa mus-*kō*-sa (= *C. lycopodioides*).
   Moss-like.
   ovata ō-*vay*-ta (= *C. argentea, C. portulaca*).
   Ovate, egg-shaped with the broad end
   lowermost. Jade Plant.
   perfoliata per-fo-li-*ay*-ta. With the leaf
   embracing the stem (perfoliate).
      falcata făl-*kah*-ta. Sickle-shaped (the
      leaves). Propeller Plant.
   perforata per-fo-*rah*-ta. Perforated, the
   paired leaves are joined at the base and thus
   perforated by the stem.
   portulaca see *C. ovata*
   rupestris roo-pes-tris. Growing on rocks.
   Buttons on a String.
   sarcocaulis sar-kō-*kaw*-lis. Fleshy-stemmed.
   schmidtii shmit-ee-ee. After E. Schmidt
   socialis so-kee-*ah*-lis. Growing in colonies.
   teres te-res. Cylindrical.

Crataegus kra-*tie*-gus *Roseceae*. The Gk.
name from *kratos* (strength) referring to the
hard wood. Deciduous trees. Hawthorn.

Crataegus (continued)
   crus-galli kroos-*gă*-lee. A cock's spur,
   referring to the long thorns. Cockspur
   Thorn. E N America.
   laciniata La-kin-ee-*ah*-ta. Deeply cut (the
   leaves). W Asia.
   laevigata lie-vi-*gah*-ta. (= *C. oxyacantha*).
   Smooth (the leaves). Midland Hawthorn.
   Europe.
   × *lavallei* la-*vahl*-ee-ee. After Pierre
   Alphonse Martin Lavellée (1836–84).
      'Carrierei' kă-ree-*e*-ree-ee. After Elie
      Abel Carrière (1816–96), French
      botanist and horticulturist.
   monogyna mon-*o*-gi-na. With one pistil.
   Europe, N Africa, W Asia.
   × *prunifolia* proon-if-*fo*-lee-a. *C. crus-galli*
   × *C. macrantha*. *Prunus*-leaved. Origin
   unknown.
   tanacetifolia tă-na-set-i-*fo*-lee-a. *Tanacetum*-
   leaved. W Asia.

Cream Cups see *Platystemon californicus*
Creeping Charlie see *Pilea nummularifolia*
Creeping Jenny see *Lysimachia nummularia*
Creeping Wintergreen see *Gaultheria*
procumbens
Creeping Zinnia see *Sanvitalia procumbens*

Crepis kre-pis *Compositae*. From Gk. *krepis* (a
boot). Annual and perennial herbs.
   aurea ow-ree-a. Golden (the flowers). SE
   Europe.
   incana in-*kah*-na. Grey-hairy (the leaves).
   S Greece.
   rubra rub-ra. Red (the flowers). SE Europe.

Crinodendron krin-ō-*den*-dron
Elaeocarpaceae. From Gk. *krinon* (a lily) and
dendron (a tree). Evergreen, semi-hardy
shrub.
   hookerianum huk-a-ree-*ah*-num. After
   Hooker. Chile.

Crinum kree-num *Amaryllidaceae*. From Gk.
krinon (a lily). Semi-hardy, bulbous
perennials.
   bulbispermum bul-bee-*sperm*-um. (= *C.
   capense*). With bulbous seeds. S Africa.
   moorei mor-ree-ee. After Moore. Natal.
   natans nă-tănz. Floating. W Africa.
   × *powellii* powl-ee-ee. *C. bulbispermum* × *C.
   moorei*. After C. Baden-Powell who raised
   it about 1885.

Crocosmia krō-*kos*-mee-a *Iridaceae*. From Gk.
krokos (saffron) and *osme* (smell), the dried
flowers smell of saffron. Cormous perennials.
   × *crocosmiiflora* krō-kos-mee-i-*flō*-ra. *C.*

Crocosmia (continued)

   aurea × *C. pottsii*. With flowers like *Crocosmia*, it was originally described in another genus. Montbretia.

   masonorum may-son-ee-*or*-rum. After Canon G. E. and Miss Mason, collectors of S African plants. S Africa.

   paniculata pa-nik-ew-*lah*-ta. (= *Antholyza paniculata*. *Curtonus paniculatus*). With flowers in panicles. S Africa.

   pottsii pots-ee-ee. After George Harrington Potts (1830–1907), who introduced it in 1877. S. Africa.

Crocus *krō*-kus *Iridaceae*. From Gk. *krokos* (saffron). Cormus herbs.

   asturicus see *C. serotinus salzmannii*

   ancyrensis ăn-ki-*ren*-sis. Of Ankara. W Turkey.

   angustifolius ăn-gus-ti-*fo*-lee-us. (= *C. susianus*). Narrow-leaved. Crimea.

   aureus see *C. flavus*.

   banaticus ba-*nah*-ti-kus. Of Banat (N Romania). E Europe.

   biflorus bi-*flō*-rus. Two-flowered. Scotch Crocus. S Europe, W Asia.

   cancellatus kăn-ke-*lah*-tus. Latticed (the corm tunic). Greece, Turkey.

   candidus *kăn*-di-dus. White (the flowers). W Turkey.

   chrysanthus kris-*ănth*-us. Golden-flowered. SE Europe, S Turkey.

   clusii see *C. serotinus clusii*

   corsicus *kov*-si-kus. Of Corsica.

   dalmaticus dăl-*mă*-ti-kus. Of Dalmatia. Yugoslavia. Albania.

   etruscus e-*troos*-kus. Of Tuscany. W Italy.

   flavus flah-vus. (= *C. aureus*). Yellow (the flowers). SE Europe, W Asia.

   fleischeri flie-sha-ree. After M. Fleischer (1861–1930). W and S Turkey.

   imperati im-pe-*rah*-tee. After Ferrante Imperato (1550–1625), Naples apothecary. W Italy.

     suaveolens swah-vee-*ō*-lenz. (= *C. suaveolens*). Sweet-scented.

   korolkowii ko-rol-*kov*-ee-ee. After General Nikolai Iwanovitsch Korolkow (born 1837). C Asia.

   kotschyanus kot-shee-*ah*-nus. After Theodor Kotschy (1813–66), Austrian botanist. W Asia.

   laevigatus lie-vi-*gah*-tus. Smooth. (the corm tunic). S Greece.

   longiflorus long-gi-*flō*-rus. Long-flowered. SW Italy, Sicily, Malta.

   medius me-dee-us. Intermediate. SE France, NW Italy.

Crocus (continued)

   minimus min-i-mus. Smaller. Corsica, Sardinia.

   niveus niv-ee-us. Snow white. S Greece.

   nudiflorus new-di-*flō*-rus. Flowering before the leaves emerge. SW Europe.

   ochroleucus ok-roo-*loo*-kus. Yellowish-white. W Asia.

   olivieri o-liv-ee-*e*-ree. After Guillaume Antoine Olivier (1756–1814), French naturalist. SE Europe, Turkey.

     balansae ba-*lahn*-zie. After Benedict Balansa (1825–91), French botanist who collected in W Asia. W Asia.

   pulchellus pul-*kel*-lus. Pretty. Balkans, W Turkey.

   reticulatus ray-tik-ew-*lah*-tus. Net-veined (the corm tunic). SE Europe.

   sativus sa-*tee*-vus. Cultivated. Saffron. Cult.

   serotinus se-*ro*-ti-nus. Late flowering. Portugal.

     clusii clooz-ee-ee. (= *C. clusii*). After Clusius, see *Gentiana clusii*. Portugal, Spain.

     salzmannii sahlts-*mahn*-ee-ee. (= *C. asturicus*). After Philip Salzmann (1781–1851), French botanist. Spain.

   sieberi see-ba-ree. After France William Sieber (1789–1844), a plant collector. SE Europe.

   speciosus spe-kee-*ō*-sus. Showy. W Asia.

   suaveolens see *C. imperati suaveolens*

   susianus see *C. angustifolius*

   tommasinianus tom-a-see-nee-*ah*-nus. After Muzio Giuseppe Spirito de Tommasini (1794–1879), Italian botanist. E Europe.

   vernus ver-nus. Of spring (flowering). Dutch Crocus. Europe.

   versicolor ver-*si*-ko-lor. Variously coloured. Maritime Alps.

Crocus, Autumn see *Colchicum autumnale*
Cross Vine see *Bignonia capreolata*

Crossandra kros-*ăn*-dra *Acanthaceae*. From Gk. *krossos* (a fringe) and *aner* (male) referring to the fringed anthers. Tender, evergreen shrubs.

   infundibuliformis in-fun-dib-ew-lee-*form*-is. (= *C. undulifolia*). Trumpet-shaped (the flowers). Firecracker Flower. S India, Sri Lanka.

   nilotica ni-*lo*-ti-ka. Of the Nile Valley. Tropical Africa.

Croton see *Codiaeum variegatum pictum*
Crowberry see *Empetrum*
Crown Daisy see *Chrysanthemum coronarium*

Crown of Thorns see *Euphorbia milii*
Cruel Plant see *Araujia sericofera*

Cryophytum crystallinum see
Mesembryanthemum crystallinum

Cryptanthus krip-*tǎnth*-us *Bromeliaceae*. From
Gk. *krypto* (to hide) and *anthos* (a flower)
referring to hidden flowers. Tender
perennials. Earth Star. Brazil.
  acaulis a-*kaw*-lis. Stemless. Starfish Plant.
  bivittatus bi-vi-*tah*-tus. With two stripes
  (the leaves).
  bromeliodes brom-ee-lee-*oi*-deez. Like
  Bromelia. Rainbow Star.
  fosterianus fos-ta-ree-*ah*-nus. After Milford
  Bateman Foster (born 1888) a collector
  and grower of bromeliads.
  zonatus zō-*nah*-tus. Banded (the leaves).
  Zebra Plant.

Cryptocoryne krip-tō-*ko*-ri-nee *Araceae*.
From Gk. *krypto* (to hide) and *coryne* (a
club). The spadix (i.e. the club) is hidden by
the spathe. Aquarium plants. Water
Trumpet.
  aponogetifolia see *C. usteriana*
  balansae see *C. crispatula*
  beckettii be-*ket*-ee-ee. After Thomas W.
  Naylor Becket (1839–1906), a coffee
  planter who collected in Sri Lanka. Sri
  Lanka.
  blassii see *C. cordata*
  ciliata ki-lee-*ah*-ta. Fringed with hairs (the
  spathe). SE Asia.
  cordata caw-*day*-ta. (= *C. blassii*). Heart-
  shaped, cordate. Malay Peninsular.
  crispatula kris-*pat*-ew-la (= *C. balansae*).
  Closely curled. India to S China.
  griffithii gri-*fith*-ee-ee. After Griffith, see
  Ceratostigma griffithii. Malay Peninsula.
  nevillii ne-*vil*-ee-ee. After H. Nevill who
  collected the type specimen. Sri Lanka.
  petchii see *C. beckettii*
  undulata un-dew-*lah*-ta. Wavy-edged (the
  leaves).
  usteriana ew-ster-ee-ah-na (= *C.
  aponogetonifolia*). After Alfred Usteri
  (1869–1948), botanist. Philippines.
  willisii wil-*is*-ee-ee. After Willis. Sri Lanka.

Cryptogramma krip-tō-*grǎm*-ma
Cryptogrammaceae. From Gk. *krypto* (to hide)
and *gramma* (a line), the sori which form a
line around the margins of the pinnules are
hidden by the rolled leaf margins. Fern.
  crispa kris-pa. Crisped (the fronds). Parsley
  Fern. Europe, W Asia.

Cryptomeria krip-to-*me*-ree-a *Cupressaceae*.
From Gk. *krypto* (to hide) and *meris* (a part)
referring to the concealed parts of the
flowers.
  japonica ja-*pon*-i-ka. Of Japan. Japan,
  China.
    'Elegans' *ay*-le-gahnz. Elegant.
    'Lobbii' *lob*-ee-ee. After its introducer
    Thomas Lobb (1820–94), who collected
    for Veitch in SE Asia.
    'Vilmoriniana' vil-mo-rin-ee-*ah*-na. After
    M. de Vilmorin, French nurseryman, in
    whose garden it was found.

Ctenanthe ten-*ǎnth*-ee *Marantaceae*. From
Gk. *kteinos* (a comb) and *anthos* (a flower)
from the arrangement of the bracts. Tender
perennials. Brazil.
  amabilis a-*mah*-bi-lis (= *Stromanthe
  amabilis*). Beautiful.
  lubbersiana lub-erz-ee-*ah*-na. After C.
  Lubbers, head gardener at the Brussels'
  Botanic Garden at the end of the 19th
  century.
  oppenheimiana o-pan-hiem-ee-*ah*-na. After
  Edouard Oppenheim. Never-never Plant.
  Brazil.

Cuckoo Flower see *Cardamine pratensis*
Cuckoo Pint see *Arum maculatum*
Cucumber see *Cucumis sativus*
Cucumber Tree see *Magnolia acuminata*

Cucumis kew-kew-mis *Cucurbitaceae*. The L.
name for cucumber. Tender annuals.
  melo may-lo. L. Name for an apple-shaped
  melon. Melon. W. Africa.
  sativus sa-*tee*-vus. Cultivated. Cucumber,
  Gherkin. S Asia.

Cucurbita kew-*kur*-bi-ta *Cucurbitaceae*. L.
name for a gourd. Annual herbs.
  maxima mahk-si-ma. Largest. Pumpkin.
  Cult.
  pepo pe-pō. L. name for a large pumpkin
  or marrow. Marrow. Cult.

Cuitlauzina coot-low-*zeen*-a *Orchidaceae*.
After King Cuitlahuatzin of the Iztapalapae
people of W Mexico, a noted collector of
rare plants. A single species of ephiphytic or
terrestrial orchid previously included in
Odontoglossum.
  pendula pen-dew-la (= *Odontoglossum
  pendulum*).
  Pendulous (the infloresence). Mexico,
  Guatemala.

Cunninghamia kūn-ing-*hǎ*-ee-a.

Cunninghamia (continued)
Cupressaceae. After James Cunningham, a
surgeon with the East India Co. who found
the following in 1701. Evergreen conifer.
   lanceolata lăn-kee-ō-*lah*-ta. Lanceolate (the
   leaves). China.

Cup Flower see *Nierembergia*

Cuphea kew-fee-a *Lythraceae.* From Gk.
kyphos (curved) referring to the curved seed
capsule. Tender herbs and shrubs.
   cyanea see-*ah*-nee-a. Blue (the petals).
   Mexico.
   hyssopifolia hi-sōp-i-*fo*-lee-a. *Hyssopus*-
   leaved. Mexico, Guatemala.
   ignea ig-nee-a. Glowing (the red calyx) or
   on fire (the flowers resemble a lighted
   cigar). Cigar Plant. Mexico, Jamaica.

Cupid's Dart see *Catananche caerulea*

× **Cupressocyparis** kew-pres-ō-*kew*-pa-ris
Cupressaceae. Intergeneric hybrid, from the
names of the parents. *Chamaecyparis* ×
Cupressus. Evergreen conifer.
   leylandii lay-*lǎnd*-ee-ee. *Chamaecyparis
   nootkatensis* × *Cupressus macrocarpa.* After
   C. J. Leyland who grew some of the first-
   raised trees at Haggerston Hall. Leyland
   Cypress.

Cupressus kew-*pres*-us *Cupressaceae.* The L.
name for *C. sempervirens.* Evergreen
conifers. Cypress.
   arizonica a-ri-*zon*-ik-a. From Arizona.
   Arizona Cypress.
      glabra glǎ-bra. Smooth (the bark).
      Arizona.
   cashmerinana see *C. himalaica* var.
   darjeelingensis
   himalaica him-ah-*lay*-ik-a. From the
   Himalayas. Bhutan Cypress.
      darjeelingensis dar-gee-lin-*jen*-sis. From
      Darjeeling. Kashmir Cypress.
   lusitanica loo-si-*tah*-ni-ka. Of Portugal,
   from where it was introduced to England.
   Cedar of Goa. Mexico.
   macrocarpa măk-rō-*kar*-pa. Large-fruited.
   Monterey Cypress. California.
   sempervirens sem-per-*vi*-renz. Evergreen.
   Italian Cypress. Cult.

Currant, Black see *Ribes nigrum*
   Buffalo see *R. odoratum*
   Flowering see *R. sanguineum*
   Mountain see *R. alpinum*
   Red see *R. silvestre*

Curry Plant see *Helichrysum italicum
serotinum*
Curtonus paniculatus see *Crocosmia paniculata*
Custard Apple see *Annona cherimola, A.
reticulata*
Cut-leaved Bramble see *Rubus laciniatus*

Cyananthus kee-a-*nǎnth*-us *Campanulaceae.*
From Gk. *kyanos* (blue) and *anthos* (a flower).
Perennial herbs.
   lobatus lō-*bah*-tus. Lobed (the leaves).
   Himalaya, Assam. W. China.
   microphyllus mik-rō-*fil*-lus. Small-leaved.
   Himalaya, Nepal.
   sherriffi she-*rif*-ee-ee. After George Sherriff
   (1896–1967), who collected in the
   Himalaya. Himalaya.

Cyanotis kee-a-*nō*tis *Commelinaeceae.* From
Gk. *kyanos* (blue) and *ous* (an ear) referring
to the ear-like, blue petals. Tender perennial
herbs.
   kewensis kew-*en*-sis. Of Kew. Teddy Bear
   Plant. India.
   somaliensis so-mah-lee-*en*-sis. Of Somalia.
   Pussy Ears. Tropical Africa.

Cyathodes kee-a-*thō*-deez *Epacridaceae.* From
Gk. *kyathodes* (a small cup) referring to the
cup-shaped disc under the ovary. Evergreen,
heath-like shrub.
   colensoi ko-*len*-zo-ee. After the Rev.
   William Colenso (1811–99), printer,
   missionary and botanist in New Zealand,
   who collected the type specimen. New
   Zealand.

Cycas see-kas *Cycadaceae.* Gk. name for a
palm, which this resembles in habit and leaf.
Evergreen, tender, palm-like plant.
   revoluta re-vo-*loo*-ta. Revolute (the
   margins of the leaflets). Sago Palm.
   Ryuku Islands, S Japan.

Cyclamen sik-la-men, classically *koo*-kla-men
Primulaceae. The Gk. name. Cormous herbs.
Sowbread.
   cilicium ki-*li*-kee-um. Of Cilicia, S Turkey.
   Turkey.
   coum kō-um (= *C. atkinsii, C. orbiculatum*).
   Of Kos. SE Europe, W Asia.
   creticum kray-ti-kum. Of Crete.
   cyprium kip-ree-um. Of Cyprus.
   graecum grie-kum. of Greece. Greece,
   Turkey, Cyprus.
   hederifolium he-di-ri-*fo*-lee-um. (= *C.
   neapolitanum*). *Hedera*-leaved. SE Europe,
   Turkey.
   persicum per-si-kum. Of Iran (Persia).

Cyclamen (continued)
Florist's Cyclamen. E Mediterranean region, W Asia.
purpurascens pur-pew-*răs*-enz. (= *C. europaeum*). Purplish (the flowers). Europe, Causcasus.
repandum re-*pǎn*dum. Wavy-margined (the leaves). Mediterranean region.
rohlfsianum rolf-ee-*ah*-num. After Rohlfs who first collected it. Libya.

Cydonia si-*dō*-nee-a or classically koo-*dō*-nee-a *Rosaceae*. L. name, from Cydonia, NE Crete, now Khania. Deciduous tree.
japonica see *Chaenomeles japonica*
lagenaria see *Chaenomeles speciosa*
maulei see *Chaenomeles japonica*
oblonga ob-*long*-ga. Oblong (the leaves). Quince. W and C Asia.
speciosa see *Chaenomeles speciosa*.

Cymbalaria sim-ba-*lah*-ree-a *Scrophulariaceae*. From Gk. *kymbalon* (a cymbal) referring to the shape of the leaves. Herbaceous perennials.
aequitriloba ie-kwee-tri-*lō*-ba. With three equal lobes (the leaves). S Europe.
hepaticifolia he-pǎ-ti-ki-*fo*-lee-a. *Hepatica*-leaved. Corsica.
muralis mew-*rah*-lis. (= *Linaria cymbalaria*). Growing on walls. Ivy-leaved Toadflax. Europe.
pallida pǎ-li-da. Pale (the flowers). Italy.

Cymbidium kim-*bid*-ee-um *Orchidaceae*. From Gk. *kymbe* (a boat), referring to the hollowed lip. Greenhouse orchids.
Alexanderi ǎ-leks-*ahn*-da-ree. After H. G. Alexander (1875–1972), orchid hybridist.
aloifolium a-lō-ee-*fo*-lee-um. *Aloe*-leaved. Himalaya, SE Asia.
canaliculatum kǎ-na-lik-ew-*lah*-tum. Channelled (the leaves). Queensland.
Coningsbyanum kon-ingz-bee-*ah*-num. *C. grandiflorum* × *C. insigne*. After the raiser. Arthur Coningsby (c. 1888–1966), an orchid grower with Sander's.
dayanum day-*ah*-num. After John Day (1824–88), amateur orchid grower.
devonianum de-vō-nee-*ah*-num. Of Devon. Himalaya.
eburneum e-*burn*-ee um. Ivory white. Himalaya, SE Asia.
elegans -*ay*-le-gahnz. Elegant. N India.
ensifolium ayn-si-*fo*-lee-um. With sword-shaped leaves. E Asia.
erythrostylum e-rith-ro-*stil*-um. With a red column. Vietnam.
findlaysonianum fin-lay-son-ee-*ah*-num.

Cymbidium (continued)
After George Finlayson (died 1823), a naturalist with the East India Co. SE Asia.
floribundum flor-i-*bund*-um (= *C. pumilum*). Floriferous. China.
giganteum see *C. iridioides*
grandiflorum see *C. hookerianum*
hookerianum huk-a-ree-*ah*-num (= *C. grandiflorum*). After Hooker. Himalaya, China.
insigne in-*sig*-nee. Remarkable. Indochina.
iridioides eye-rid-i-oy-deez (= *C. giganteum*). Resembling *Iris*. Himalaya to SE Asia.
lowianum lō-ee-*ah*-num. After Sir Hugh Low (1824–1905), a diplomat in SE Asia and an authority on orchids. Burma.
Pauwelsii pow-*elz*-ee-ee. *C. insigne* × *C. lowianum concolor*. After Pauwels who collected orchids for Sander's.
pendulum see *C. aloifolium*
pumilum see *C. floribundum*
tigrinum ti-*gree*-num. Striped like a tiger. SE Asia.
tracyanum tray-see-*ah*-num. After Henry Amos Tracy (c. 1850–1910), orchid and bulb nurseryman. SE Asia.

Cynara si-*nah*-ra or classically koo-*nah*-ra *Compositae*. The L. name. Perennial herbs.
cardunculus kar-dun-*kew*-lus. Diminutive of *Carduus*, a genus of thistles. Cardoon. SW Europe.
scolymus *sko*-li-mus. The L. name for *Scolymus hispanicus*. Globe Artichoke. Cult.

Cynoglossum si-no-*glos*-um *Boraginaceae*. From Gk. *kyon* (a dog) and *glossum* (a tongue) referring to the leaves.
amabile a-*mah*-bi-lee. Beautiful. E Himalaya, W China.
nervosum ner-*vō*-sum. Distinctly veined (the leaves). Himalaya.
officinale o-fi-ki-*nah*-lee. Sold as a herb. Hound's Tongue. Europe.

Cyperus si-*pe*-rus or classically koo-*pe*-rus *Cyperaceae*. Gk. name for a sedge. Tender and hardy, grass-like plants. Sedge.
albostriatus ǎl-bō-stree-*ah*-tus. White-striped, referring to the pale leaf veins. S Africa.
involucratus in-vo-loo-*krah*-tus. With an involucre, referring to the large, leaf-like bracts borne under the inflorescence. Africa.
longus long-gus. Long, the tall stems. Galingale. Europe to India.

Cyperus (continued)
  papyrus pa-*pi*-rus. Gk. name for the paper
  made from this plant. Papyrus. C Africa,
  Nile Valley.

Cypress see *Cupressus*
  Hinoki see *Chamaecyparis obtusa*
  Italian see *Cupressus sempervirens*
  Lawson see *Chamaecyparis lawsoniana*
  Leyland see × *Cupressocyparis leylandii*
  Monterey see *Cupressus macrocarpa*
  Nootka see *Chamaecyparis nootkatensis*
  Sawara see *Chamaecyparis pisifera*
Cypress Spurge see *Euphorbia cyparissias*
Cypress Vine see *Ipomaea quamoclit*

Cypripedium kip-ree-*pee*-dee-um
Orchidaceae. From Gk. *kypris* (Venus) and
pedilon (a slipper) referring to the shape of
the flowers. Hardy orchids. Lady's Slipper
Orchid. For the tender orchids sometimes
listed here see *Paphiopedilum*.
  acaule a-*kaw*-lee. Stemless. E United States.
  arietinum ă-ree-ee-*tee*-num. Like a ram's-
  head. Ram's Head Lady's Slipper Orchid.
  EN America.
  calceolus kăl-*kee*-o-lus. A small shoe.
  Europe, Asia.
  candidum kăn-di-dum. White. E United
  States.
  reginae ray-*geen*-ie. Of the Queen. E N
  America.

Cyrtanthus kur-*tănth*-us *Amaryllidaceae*.
From Gk. *kyrtos* (arched) and *anthos* (a flower)
referring to the curved perianth tube. Tender
and semi-hardy bulbous herbs. S Africa.
  angustifolius ăn-gus-ti-*fo*-lee-us. Narrow-
  leaved.
  elatus el-*ay*-tus (= *Vallota speciosa*). Tall.
  Scarborough Lily. S Africa.
  mackenii ma-*ken*-ee-ee. After Mark John
  M'ken (1823–72), a Scotsman who became
  curator of the Natal Botanic Garden and
  collected in S Africa.
  obrienii ō-*brie*-an-ee-ee. After Mr Jas.
  O'Brien who imported it.
  ochroleucus ok-rō-*loo*-kus. Yellowish-white.
  parviflorus par-vi-*flō*-rus. Small flowered.
  sanguineus san-*win*-ee-us. Blood-red (the
  flowers).

Cyrtomium ser-tō-mee-um *Dryopteridaceae*.
From Gk *kyrtos*, meaning arched, referring
to the habit of growth. Terrestrial Ferns.
  falcatum fal-*kay*-tum (= *Polystichum
  falcatum*). Sickle-shaped (the pinnae).
  Japanese Holly Fern. China, Malaysia,
  India to S & E Africa, Hawaii.

Cystopteris kis-*top*-te-ris (*Athyriaceae*)
Dryopteridaceae. From Gk. *kystos* (a bladder)
and *pteris* (a fern) referring to the bladder-
like indusium. Bladder Fern.
  bulbifera bul-*bi*-fe-ra. Bulb-bearing, the
  small, bulb-like growths on the fronds
  from which new plants grow. Berry
  Bladder Fern. N America.
  dickieana dik-ee-*ah*-na. After Prof. George
  Dickie (1812–82), of Aberdeen. Arctic
  Bladder Fern. N hemisphere.
  fragilis *fră*-gi-lis. (= *C. alpina*). Brittle (the
  stalks). Brittle Bladder Fern. Widely
  distributed.
  montana mon-*tah*-na. Of mountains.
  Mountain Bladder Fern. N hemisphere.

Cytisus si-ti-sus *Leguminosae*. From *kytisos*,
Gk. name for these or similar shrubs.
Deciduous and evergreen shrubs. Broom.
  battandieri see *Argyrocytisus battandieri*
  × *beanii* been-ee-ee. *C. ardoinii* × *C.
  purgans*. After William Jackson Bean
  (1863–1947), Kew botanist and authority
  on hardy, woody plants.
  'Burkwoodii' burk-*wud*-ee-ee. After
  Albert Burkwood of Burkwood and
  Skipwith.
  decumbens day-*kum*-benz. Prostrate. S
  Europe.
  × *kewensis* kew-en-sis. *C. ardoinii* × *C.
  multiflorus*. Of Kew where it was raised in
  1891.
  monspessulanus see *Genista monspessulanus*
  multiflorus mul-tee-*flō*-rus. Many-
  flowered. White Spanish Broom. Spain,
  Portugal.
  nigricans nig-ri-kănz. Blackish, the flowers
  turn black when dried. C and S Europe.
  × *praecox* prie-koks. *C. multiflorus* × *C.
  purgans*. Early (flowering).
  procumbens prō-*kum*-benz. Prostrate. SE
  Europe.
  purgans pur-ganz. Purging. SW Europe.
  purpureus see *Chamaecytisus purpureus*.
  racemosus hort. see *Genista* × *spachianus*
  scoparius skō-*pah*-ree-us. Broom-like.
  'Andreanus' on-dray-*ah*-nus. After
  André who discovered it in Normandy,
  see *Anthurium andreanum*.
  × *spachianus* see *Genista* × *spachianus*
  supranubia soo-pra-*new*-bee-as. (=
  Spartocytisus nubigenus). From above the
  clouds. Canary Islands.
  × *versicolor* See *Chamaecytisus* × *versicolor*

D

Daboecia dă-bō-*ee*-kee-a *Ericaceae*. After St. Dabeoc. Evergreen, heath-like shrubs.
  azorica a-*zo*-ri-ka. Of the Azores.
  cantabrica kăn-*tá*-bri-ka. Of Cantabria, N Spain. W Europe.
    'Atropurpurea' ah-trō-pur-*pewr*-ree-a. Deep purple.
    'Bicolor' *bi*-ko-lor. Two coloured.
    'Praegerae' *pray*-ga-rie. After Mrs Prager who found it.
  × *scotica sko*-ti-ka. *D. azorica* × *D. cantabrica*. Of Scotland where it was raised.

Dacrydium da-*krid*-ee-um *Podocarpaceae*. From Gk. *dacrydion* (a small tear), referring to the exuded resin drops. Evergreen conifer.
  franklinii see *Lagarostrobus franklinii*

Dactylorhiza dăk-til-ō-*ree*-za *Orchidaceae*. From Gk. *dactylos* (a finger) and *rhiza* (a root) referring to the finger-like tubers. Hardy orchids.
  elata ay-*lah*-ta. (= *Orchis elata*). Tall. N Africa.
  foliosa fo-lee-ō-sa. (= *Orchis foliosa*, *Orchis maderensis*). Leafy. Madeira.
  fuchsii fuks-ee-ee. After Fuchs. Common Spotted Orchid. Europe.
  incarnata in-kar-*nah*-ta. (= *Orchis incarnata*). Pink (the flowers). Meadow Orchid. Europe.
  majalis mah-*yah*-lis. Flowering in May. Europe.
  praetermissa prie-ter-*mis*-a. (= *Orchis praetermissa*). Overlooked. NW Europe.

Daffodil, Hoop-petticoat see *Narcissus bulbocodium*
  Pheasant's Eye see *N. poeticus recurvus*
  Wild see *N. pseudonarcissus*

Dahlia dah-lee-a *Compositae*. After Dr Anders Dahl (1751–89), Swedish botanist. Tender and semi-hardy tuberous perenials. Mexico.
  coccinea kok-*kin*-ee-a. Scarlet (the ray flowers).
  pinnata pin-*ah*-ta. Pinnate (the leaves).

Daisy see *Bellis perennis*
Daisy Bush see *Olearia*
Dame's Violet see *Hesperis matronalis*

Danae dă-na-ay *Liliaceae* (*Ruscaceae*). After Danae of Gk. mythology. Evergreen shrub.
  racemosa ră-kay-*mō*-sa (= *Ruscus racemosus*).

Danae (continued)
  With flowers in racemes. Alexandrian Laurel. SW Asia.

Daphne dăf-nay *Thymelaeaceae*. The Gk. name for *Laurus nobilis*. Deciduous and evergreen shrubs.
  arbuscula ar-*bus*-kew-la. Like a dwarf tree. E Czechoslovakia.
  bholua bo-loo-a. From the native name, Bholu Swa. Himalaya.
  blagayana blă-gay-*ah*-na. After Count Blagay who discovered it in 1837. SE Europe.
  × *burkwoodii* burk-*wud*-ee-ee. *D. caucasica* × *cneorum*. After Albert Burkwood who raised it.
  cneorum nee-*o*-rum. Gk. name for an olive-like shrub. Garland Flower. Europe.
  genkwa genk-wa. The Japanese version of the Chinese name, Genk' wa. China.
  laureola low-*ree*-o-la. L. name for a little laurel crown. Spurge Laurel. S and W Europe, N Africa.
  × *mantensiana* măn-tenz-ee-*ah*-na. *D.* × *burkwoodii* × *D. retusa*. After Manten's nursery. British Columbia.
  mezereum me-*ze*-ree-um. From *mezereon* the L. name. Mezereon. Europe, Siberia.
  odora o-*dō*-ra. Fragrant. China.
    'Aureo-marginata' *ow*-ree-ō-mar-gi-*nah*-ta. Gold-margined (the leaves).
  petraea pe-*trie*-a. Growing on rocks. N Italy.
  retusa see *D. tangutica*
  tangutica tăn-*gew*-ti-ka. Of Gansu (Kansu). W China.

Daphniphyllum dăf-nee-*fil*-lum *Daphniphyllaceae*. From Gk. *daphne* (laurel) and *phyllon* (a leaf) from the resemblance of the leaves to those of *Laurus nobilis*. Evergreen shrub.
  macropodum ma-*kro*-po-dum. With a large stalk (the leaves). Japan.

Darlingtonia dar-ling-*ton*-ee-a *Sarraceniaceae*. After Dr William Darlington (1782–1863). Carnivorous herb.
  californica kăl-i-*forn*-i-ka. Californian. California Pitcher Plant. California, Oregon.

Darmera dah-*mer*-a *Saxifragaceae*. Derivation not found. A single species of rhizomatous perennial. NW California to Oregon.
  peltata pel-*tay*-ta (= *Peltiphyllum peltatum*). Peltate (the leaves). Umbrella Plant.

Date Plum see *Diospyros lotus*

Datura de-*tewr*-ra *Solanaceae*. From a native name. Tender herbs and shrubs.
arborea see *Brugmansia arborea*
ceratocaula ke-ra-tō-*kaw*-la. With a horn-like stem. Mexico.
inoxia in-*oks*-ee-a. (= *D. meteloides*). Not spiny. SW United States, Mexico.
metel me-tel. The native name. SW China.
sanguinea see *Brugmansia sanguinea*
suaveolens see *Brugmansia suaveolens*

Daucus dow-kus *Umbelliferae*. The L. name. Biennial herb.
carota ka-*rot*-a. From *karoton* the Gk. name. Wild Carrot. Europe, Asia.
sativus sa-*tee*-vus. Cultivated. Carrot.

Davallia da-*vahl*-ee-a *Davalliaceae*. After Edmond Davall (1763–1798), Swiss botanist. Evergreen, tender ferns.
canariensis ka-nah-ree-*en*-sis. Of the Canary Islands. Deer's-foot Fern. Canary Islands, Portugal, Spain, Madeira.
fejeensis fee-jee-*en*-sis. Of Fiji. Rabbit's-foot Fern.
mariesii ma-*reez*-ee-ee. After Charles Maries (c. 1851–1902), who collected for Veitch in China and Japan. Squirrel's-foot Fern, Ball Fern. E Asia.
trichomanoides trik-ō-mahn-*oi*-deez. Like *Trichomanes*. Squirrel's-foot Fern. Malaysia.

Davidia da-*vid*-ee-a *Davidiaceae*. After Abbé Armand David (1826–1900), French missionary and plant collector in China who discovered the following in 1869. Deciduous tree.
involucrata in-vo-loo-*krah*-ta. With an involucre, referring to the showy bracts. Dove Tree, Handkerchief Tree. China.
vilmoriniana vil-mo-rin-ee-*ah*-na. After Maurice Vilmorin, French nurseryman who first raised it.

Dawn Redwood see *Metasequoia glyptostroboides*
Day Flower see *Commelina*
Day Lily see *Hemerocallis*
Dead Nettle see *Lamium*
   Giant see *L. orvala*
   Spotted see *L. maculatum*

Decaisnea de-*kayz*-nee-a *Lardizabalaceae*. After Joseph Decaisne (1807–82), French botanist and director of the Jardin des Plantes, Paris. Deciduous shrub.
fargesii far-*geez*-ee-ee. After Père Paul

Decaisnea (continued)
Guillaume Farges (1844–1912), French missionary who introduced it to France in 1895. W China.

Decumaria dek-ew-*mah*-ree-a *Hydrangeaceae*. From L. *decimus* (ten), the parts of the flower are in tens. Woody climbers.
barbara bar-ba-ra. Foreign, it was originally thought to be introduced. SE United States.
sinensis sin-*en*-sis. Of China.

Delairea del-*aer*-ee-a *Compositae*. After M. Delaire, 19th century French botanist. German Ivy, Parlour Ivy. South Africa.
odorata ō-do-*ray*-ta (= *Senecio mikanioides*) fragrant.

Delphinium del-*fin*-ee-um *Ranunculaceae*. From the Gk. name, from *delphis* (a dolphin) referring to the shape of the flowers. Perennial herbs.
ajacis see *Consolida ambigua*
cardinale kar-di-*nah*-lee. Scarlet (the flowers). California.
consolida see *Consolida regalis*
elatum ay-*lah*-tum. Tall. Europe, Siberia.
exaltatum eks-al-*tah*-tum. Very tall. SE United States.
grandiflorum grăn-di-*flō*-rum. Large-flowered. Nepal to China.
× *magnificum* mahg-ni-fi-kum. *D. cheilanthifolium* × *D. grandiflorum*. (*D. formosum* hort.). Magnificent.
nudicaule new-di-*kaw*-lee. With a bare stem. California, Oregon.
× *ruysii* ries-ee-ee. *D. elatum* × *D. nudicaule*. After B. Ruys, Dutch nurseryman.
tatsienense tăt-see-en-*en*-see. Of Tatsienlu (now K'ang-ling), Sichuan, W China.
zalil zah-lil. The native name. Afghanistan.

Dendranthema den-dran-*theem*-a *Compositae*. From Gk. *dendron*, tree, and *anthemon*, flower. Perennial herbs, sometimes woody at base; formerly included in *Chrysanthemum*. Europe, C and E Asia.
× *grandiflorum* grăn-di-*flō*-rum (= *Chrysanthemum* × *morifolium*). Large flowered. Florists' Chrysanthemum.
indicum in-di-kum (= *Chrysanthemum indicum*). Indian. A parent of the florists' Chrysanthemum. China, Japan.
weyrichii way-rik-ee-ee (= *Chrysanthemum weyrichii*). After Dr. Weyrich, a Russian naval surgeon. Japan.

Dendrobium den-*drō*-bee-um *Orchidaceae*.

Dendrobium (continued)
From Gk. *dendron* (a tree) and *bios* (life)
referring to their epiphytic habit.
Greenhouse orchids.
   aphyllum a-*fil*-lum. (= *D. pierardii*).
   Leafless. Himalaya, SE Asia.
   aureum see *D. heterocarpum*
   bigibbum bi-*gib*-um. With two swellings (at
   the base of the lip). N Australia.
   brymerianum brie-ma-ree-*ah*-num. After
   W. E. Brymer. SE Asia.
   densiflorum dens-i-*flō*-rum. Densely-
   flowered. Himalaya, Assam, Burma.
   fimbriatum fim-bree-*ah*-tum. Fringed (the
   lip). Himalaya, India, Burma.
   heterocarpum he-te-ro-*kar*-pum. (= *D.
   aureum*). With variable fruit. Nepal,
   Sikkim, Burma.
   infundibulum in-fun-*dib*-ew-lum. Funnel-
   shaped (the flowers). Burma.
   kingianum king-ee-*ah*-num. After Captain
   King. Australia.
   longicornu long-gi-*kor*-noo. With a long
   horn. Himalaya, Assam, Burma.
   moschatum mos-*kah*-tum. Musk-scented.
   Himalaya, India, Burma.
   nobile nō-bi-lee. Notable. Himalaya, SE
   Asia.
   pierardii see *D. aphyllum*
   primulinum preem-ew-*leen*-um. Primrose-
   coloured. Himalaya, N Burma.
   victoriae-reginae vik-*tor*-ree-ie-ray-*geen*-ie.
   After Queen Victoria. Philippines.
   williamsonii wil-yam-*son*-ee-ee. After Mr
   W. J. Williamson who discovered it in
   1868. Himalaya.

Dendrochilum den-drō-*keel*-um *Orchidaceae*.
From Gk. *dendron* (a tree) and *cheilos* (a lip),
referring to the epiphytic habit and
conspicuous lip. Greenhouse orchids.
Philippines.
   cobbianum cob-ee-*ah*-num. After Walter
   Cobb (c. 1836–1922) of Sydenham who
   first flowered it.
   filiforme fee-lee-*form*-ee. Thread-like.
   glumaceum gloo-*mah*-kee-um. With dry,
   chaffy bracts.

Dennstaedtia den-*stet*-ee-a *Dennstaedtiaceae*.
After August Wilhelm Dennstedt
(1776–1826). German botanist. Fern.
   punctilobula punk-tee-*lob*-ew-la. With
   dotted lobules. Hay scented Fern. E N
   America.

Dentaria see *Cardamine*

Dendromecon den-drō-*may*-kon *Papaveraceae*.

Dendromecon (continued)
From Gk. *dendron* (a tree) and *mecon* (a
poppy). Semi-hardy evergreen shrub.
   rigida ri-gi-da. Rigid (the leaves).
   Calfornia.

Deodar see *Cedrus deodara*
Desert Privet see *Peperomia magnoliaefolia*

Desfontainea des-(or day-)fon-*tay*-nee-a
Loganiaceae After René Louiche
Desfontaines (1753–1833), French botanist.
Semi-hardy, evergreen shrub.
   spinosa spee-*nō*-sa. Spiny (the leaves).
   Andes.

Desmodium des-*mō*-dee-um *Leguminosae*
(*Papilionoideae*) From Gk. *desmos* (a chain)
referring to the jointed stamen. Annual herb.
   motorium see *Codariocalyx motorius*

Deutzia doytz-ee-a *Philadelphaceae*. After
Johann van der Deutz (1743–88), a patron
of Thunberg. Deciduous shrubs.
compacta com-*păk*-ta. Compact (the
inflorescence). China.
× *elegantissima* ay-le-gan-*tis*-i-ma. *D.
purpurascens* × *D. sieboldiana*. Most elegant.
gracilis *grā*-ki-lis. Graceful. Japan.
longifolia long-gi-*fo*-lee-a. With long leaves.
W China.
× *magnifica* magh-*ni*-fi-ka. *D. scabra* × *D.
vilmoriniae*. Magnificent.
pulchra pul-kra. Pretty. Philippines, Taiwan.
× *rosea* ros-ee-a. *D. gracilis* × *D. purpurascens*.
Rose-coloured (the flowers).
scabra skā-bra. Rough (the leaves). Japan,
China.
setchuanensis sech-wahn-*en*-sis. Of Sichuan
(Setchwan). China.
   corymbiflora ko-rim-bee-*flō*-ra. With
   flowers in corymbs.

Devil Flower see *Tacca chantrieri*
Devil's Backbone see *Kalanchoe
daigremontiana*
Devil's Claw see *Physoplexis comosa*
Devil's Fig see *Argemone mexicana*
Devil's Ivy see *Epipremnum aureum*
Devil's Paintbrush see *Pilosella aurantiacum*
Devil's Tongue see *Amorphophallus rivieri*

Dianella dee-a-*nel*-la *Liliacae* (*Phormiaceae*).
Diminutive of Diana, goddess of the chase.
Rhizomatous perennials. Flax Lily.
   caerulea kie-*ru*-lee-a. Deep blue. New
   South Wales.
   intermedia in-ter-*me*-dee-a. Intermediate.
   New Zealand.

Dianella (continued)
tasmanica tăz-*măn*-i-ka. Of Tasmania.

Dianthus dee-*ănth*-us *Caryophyllaceae*. From
Gk. *Di* (of Zeus or Jove) and *anthos* (a
flower). Annual, biennial and perennial
herbs. Pink.
alpinus ăl-*peen*-us. Alpine. Alps.
arenarius ă-ray-*nah*-ree-us. Growing in
sandy places. S Sweden.
× *arvenensis* ar-ven-*en*-sis. *D.*
monspessulanus × *D. seguieri*. Of the
Auvergne. France.
barbatus bar-*bah*-tus. Bearded (the petals).
Sweet William. S and E Europe.
caesius see *D. gratianopolitanus*
× *calalpinus* kăl-al-*peen*-us. *D. alpinus* × *D.*
callizonus. From the names of the parents.
callizonus kă-lee-*zōn*-us. Beautifully zoned
(the petals). S Carpathians.
carthusianorum kar-thew-zee-a-*nor*-rum.
Of the monks of the Carthusian Monastery
nr. Grenoble. SW and C Europe.
caryophyllus kă-ree-ō-*fil*-lus. Smelling of
walnut leaves. Carnation, Clove Pink.
Mediterranean region.
chinensis chin-*en*-sis. Of China. Indian
Pink.
deltoides del-*toi*-deez. Shaped like the Gk.
letter delta (△) (the petals). Europe.
glacialis glă-kee-*ah*-lis. Growing near
glaciers. Glacier Pink. E Alps,
Carpathians.
gratianopolitanus grah-tee-ah-nō-po-li-*tah*-
nus. Of Grenoble. Cheddar Pink.
Europe.
haematocalyx hie-mah-tō-*kă*-liks. With a
blood-red calyx. Yugoslavia, Greece.
knappii nahp-ee-ee. After Joseph Armin
Knapp (1843–99), botanist of Vienna. W.
Yugoslavia.
myrtinervius mur-tee-*ner*-vee-us. Veined
like *Myrtus*. Greece, Yugoslavia.
neglectus see *D. pavonius*
noeanus see *D. petraeus noeanus*
pavonius pah-*vō*-nee-us. (= *D. neglectus*).
From L. *pavo* (a peacock) referring to the
brilliantly coloured flowers. Alps.
petraeus pe-*trie*-us. Growing on rocks. SE
Europe.
  noeanus nō-ee-*ah*-nus. (= *D. noeanus*).
  After Friedrich Wilhelm Noë. Bulgaria.
plumarius ploo-*mah*-ree-us. Plumed (the
fringed petals). E and C Europe.
'Allwoodii' awl-*wud*-ee-ee. *D.*
caryophyllus × *D. plumarius*. After M. C.
W. Allwood (c. 1879–1958),
nurseryman specialising in *Dianthus*.

Dianthus (continued)
superbus soo-*perb*-us. Superb. Fringed
Pink. Europe to Japan.

Diascia dee-*ă*-skee-a *Scrophulariaceae*. From
Gk. *di* (two) and *askos* (a sac), the flowers
have two spurs. Perennial herbs. S. Africa.
barberiae bar-*be*-ree-ie. After Mrs Barber.
Twinspur.
cordata kor-*dah*-ta. Heart-shaped (the
leaves).
rigescens ri-*ges*-enz. Somewhat rigid.

Dicentra di-*ken*-tra *Fumariaceae*. From Gk. *di*
(two) and *kentron* (a spur), the flowers have
two spurs. Prennial herbs.
cucullaria kuk-ew-*lah*-ree-a. Hood-like
(the flowers). Dutchman's Breeches. E N
America.
eximia eks-*i*-mee-a. Distinguished. E
United States.
formosa for-*mō*-sa. Beautiful. W N
America.
spectabilis spek-*tah*-bi-lis. Spectacular.
Bleeding Heart. Japan.

Dichelostemma di-kel-ō-*stem*-a *Liliaceae*
(*Alliaceae*). From Gk. *dicha* (bifid) and *stemma*
(a garland) referring to the stamen
appendages. Cormous herbs.
ida-maia ee-da-*mah*-ya. (= *Brodiaea ida-*
maia) After Ida May Burke of California.
Floral Firecracker. Oregon, California.
pulchellum pul-*kel*-lum. Pretty. Wild
Hyacinth. W N America.

Dicksonia dik-*son*-ee-a *Dicksoniaceae*. After
James Dickson (1738–1822), Scottish
nurseryman and naturalist. Tender Tree
Ferns.
antarctica ăn-*tark*-ti-ka. Of Antarctic
regions. Woolly Tree Fern. Australia,
Tasmania.
fibrosa fi-*brō*-sa. Fibrous (the trunk).
Golden Tree Fern. New Zealand.
squarrosa skwah-*rō*-sa. With the parts
spreading at right angles. New Zealand.

Dictamnus dik-*tăm*-nus *Rutaceae*. Gk. name
for an origanum. Perennial herb.
albus ăl-bus. (= *D. fraxinella*). White, the
flowers vary from white to purple.
Dittany, Burning Bush. S Europe to N
China.

Didiscus caeruleus see *Trachymene caerulea*

Dieffenbachia dee-fan-*bahk*-ee-a. *Araceae*.

Dieffenbachia (continued)
After J. F. Dieffenbach (1790–1863). Tender perennials. Dumb Cane.
  amoena a-*moyn*-a. Pleasant. S America.
  × *bausei bowz*-ee-ee. *D. maculata* × *D. weirii*. After Christian Frederick Bause (c. 1839–95), German nurseryman with the R.H.S. at Chiswick, later with Veitch.
  bowmannii bow-*măn*-ee-ee. After its introducer, David Bowman (1838–68), who collected for Veitch in S America. Brazil.
  imperialis im-pe-ree-*ah*-lis. Showy. Peru.
  maculata măk-ew-*lah*-ta. (= *D. picta*). Spotted (the leaves). C and S America.
  'Exotica' eks-*o*-ti-ka. Exotic. Costa Rica.
  oerstedii ur-*sted*-ee-ee. After Anders Oersted who collected in S America in the 19th century. Mexico, Costa Rica.

Dierama dee-e-*rah*-ma *Iridaceae*. From Gk. *dierama* (a funnel) referring to the shape of the flowers. Cormous perennial.
  pulcherrimum pul-*ke*-ri-mum. Very pretty. Angel's Fishing Rod, Wand Flower. S Africa.

Diervilla dee-er-*vil*-la *Caprifoliaceae*. After Dr. N. Dierville, a French surgeon who introduced the following to Europe about 1700. Deciduous shrub.
  lonicera lon-i-*se*-ra. From *Lonicera* q.v., a related genus. E N America.

Dietes di-*ee*-tees *Iridaceae*. Derivation not found. Rhizomatous perennials.
  iridiodes ee-ri-dee-*oi*-dees (= *Moraea iridiodes*). Iris-like. S Africa to Kenya.

Digitalis di-gi-*tah*-lis *Scrophulariaceae*. From L. *digitus* (a finger) referring to the finger-like flowers. Biennial and perennial herbs.
  dubia dub-ee-a. Dubious (that it belongs to the genus), not typical. Balearic Islands.
  ferruginea fe-roo-*gin*-ee-a. Rusty (markings on the flowers). SE Europe, W Asia.
  grandiflora grăn-di-*flō*-ra. Large-flowered. E and C Europe, W Asia.
  lanata lah-*nah*-ta. Woolly (the racemes). E Europe.
  lutea loo-tee-a. Yellow. W Europe.
  × *mertonensis* mer-ton-*en*-sis. *D. grandiflora* × *D. purpurea*. Of Merton.
  purpurea pur-*pewr*-ree-a. Purple. Foxglove. Europe.

Dill see *Anethum graveolens*

Dimorphotheca di-mor-fō-*thee*-ka *Compositea*. From Gk. *dis* (twice), *morphe* (shape) and *theka* (a fruit) referring to the different kinds of fruit produced by the ray and disc flowers. Tender annuals. S Africa.
  barberiae see *Osteospermum jucundum*
  ecklonis see *Osteospermum ecklonis*
  pluvialis ploo-vee-*ah*-lis. (= *D. annua*). Of rain, it flowers after the rains.
  sinuata sin-ew-*ah*-ta. (= *D. aurantiaca* hort. *D. calendulacea*). Wavy-edged (the leaves). Star of the Veldt.

Dionaea dee-on-*ie*-a *Droseraceae*. Gk. name for Venus. Carnivorous herb.
  muscipula mus-*kip*-ew-la. Fly-catching. Venus's Fly Trap. N and S Carolina (bogs).

Dioscorea dee-os-*ko*-ree-a *Dioscoreaceae*. After Pedanios Dioscorides, 1st century Gk. herbalist. Tender Climbers. Yam.
  discolor dis-ko-lor. Two-coloured (the leaves). Ornamental Yam. Ecuador.
  elephantipes e-le-făn-ti-pays. Like an elephant's foot (the half-exposed tuber). Elephant's Foot, Tortoise Plant. S Africa.

Diospyros dee-*os*-pi-ros *Ebenaceae*. From Gk. *dios* (divine) and *pyros* (wheat) referring to the edible fruits. Deciduous trees.
  kaki kah-ki. The Japanese name. Chinese Persimmon, Kaki. China.
  lotus lō-tus. Gk. *lotos*, which was applied to many plants. Date Plum. China.
  virginiana vir-jin-ee-*ah*-na. Of Virginia. Persimmon. SE United States.

Dipelta di-*pel*-ta *Caprifoliaceae*. From Gk. *di* (two) and *pelta* (a shield) referring to the conspicuous bracts which enclose the fruit. Deciduous shrubs.
  floribunda flō-ri-*bun*-da. Profusely flowering. C and W China.
  ventricosa ven-tri-*kō*-sa. Swollen on one side (the base of the corolla). W China.
  yunnanensis yoo-nan-*en*-sis. Of Yunnan, China.

Diplacus aurantiacus see *Mimulus aurantiacus*
  puniceus see *M. puniceus*
Dipladenia see *Mandevilla*

Dipsacus *dip*-sa-kus *Dipsacaceae*. The Gk. name, from *dipsa* (thirst), water collects in the cup formed by the leaf bases. Biennial herb.
  fullonum fu-*lō*-num. Of fullers, the prickly fruiting heads of *D. sativus* were used by fullers to tease cloth. Teasel. Europe, Asia.

Disanthus dis–*ănth*–us *Hamamelidaceae*. From Gk. *dis* (twice) and *anthos* (a flower), the flowers are born in pairs. Deciduous shrub.
  cercidifolius ker-ki-di-*fo*-lee-us. *Cercis*-leaved. Japan.

Disporum di-spo-rum *Liliaceae* (*Colchicaceae*). From Gk. *dis* (two) and *spora* (a seed), each chamber of the ovary contains two seeds. Perennial herb. Fairy Bells.
  sessile se-si-lee. Stalkless (the leaves). Japan, China.

Distylium di-*stil*-ee-um *Hamamelidaceae*. From Gk. *dis* (two) and *stylos* (a style), the flowers have two styles. Evergreen Shrub.
  racemosum ră-kay-*mo*-sum. With flowers in racemes. S Japan.

Dittany see *Dictamnus albus*

Dizygotheca elegantissima see *Schefflera elegantissima*

Dodecatheon dō-dek-a-*thee*-on *Primulaceae*. From Gk. *dodeka* (twelve) and *thios* (god). Perennial herbs. Shooting Star.
  frigidum fri-gi-dum. Of cold regions. NW North America, NE Asia.
  hendersonii hen-der-*son*-ee-ee. After Louis Fourniquet Henderson (1853–1942). WN America.
  jeffreyi jef-ree-ee. After Jeffrey, see *Pinus jeffreyi*.
  meadia mee-dee-a. After Richard Mead (1673–1754), English physician and botanical patron. E United States.

Dog's-tooth Violet see *Erythronium dens-canis*
Dogwood see *Cornus*
  Common see *C. sanguinea*
  Flowering see *C. florida*
  Pacific see *C. nuttallii*

Dombeya dom-bee-a *Sterculiaceae*. After Joseph Dombey (1742–94), French botanist who collected in S America. Tender, evergreen shrubs.
  burgessiae bur-*jes*-ee-ie. (= *D. mastersii*). After Miss Burgess of Birkenhead. E and S Africa.
  × *cayeuxii* kie-*yuz*-ee-ee. *D. burgessiae* × *D. wallichii*. After M. Henry Cayeux of the Lisbon Botanic Garden who raised it in 1895.

Doritis do-*ree*-tis *Orchidaceae*. From Gk. *dory*

Doritis (continued)
(a spear) referring to the spear-shaped lip. Greenhouse orchid.
  pulcherrima pul-*ke*-ri-ma. (= *Phalaenopsis esmeralda*). Very pretty. SE Asia.

Doronicum do-*rom*-i-kum *Compositae*. Derivation obscure. Perennial herbs. Leopard's Bane.
  austriacum ow-stree-*ah*-cum. Austrian. Europe.
  columnae ko-*lum*-nie. (= *D. caucasicum*. *D. cordatum*). After Fabius Columna (1567–1640). SE Europe, W Asia.
  pardalianches par-da-lee-*ahn*-kees. Gk. name for a poisonous plant, originally thought to be this one, meaning literally, strangling leopards. Europe.
  plantagineum plahn-ta-*gin*-ee-um. *Plantago*-like. Europe.

Dorotheanthus do-ro-thee-*ănth*-us *Aizoaceae*. Dr Martin Heinrich Schwantes named this genus after his mother Dorothea. Succulent annuals. S Africa.
  bellidiformis bel-id-ee-*form*-is. (= *Mesembryanthemum criniflorum*). *Bellis*-like. Livingstone Daisy.
  tricolor tri-ko-lor. (= *Mesembryanthemum tricolor*). Three-coloured (the flowers).

Dorycnium do-*rik*-nee-um *Leguminosae* (*Papilionoidae*). From *doryknion*, Gk. name for a *Convolvulus*. Sub-shrub.
  hirsutum hir-*soo*-tum. Hairy. S Europe.

Douglas Fir see *Pseudotsuga menziesii*
Douglasia vitaliana see *Vitaliana primuliflora*
Dove Tree see *Davidia involucrata*

Draba drah-ba *Cruciferae*. From *drabe*, Gk. name for a related plant. Perennial herbs.
  aizoides ie-zō-*ee*-deez. Like *Aizoon*. Europe.
  aizoon see *D. lasiocarpa*
  alpina ăl-peen-a. Alpine. Europe.
  bruniifolia brun-ee-i-*fo*-lee-a. With leaves like *Brunia*. Caucasus.
  bryoides see *D. rigida* var. *bryoides*
  dedeana dee-dee-*ah*-na. After Dede. N and E Spain.
  imbricata see *D. rigida imbricata*
  lasiocarpa lă-see-ō-*kar*-pa. Woolly-fruited. E Europe.
  mollisima mol-*lis*-i-ma. Very soft. Caucasus.
  polytricha po-*li*-tri-ka. With many hairs. Armenia.
  repens see *D. sibirica*

Draba (continued)
rigida ri-gi-da. Rigid (the leaves).
Caucasus.
  bryoides bry-oy-deez. Moss-like.
  imbricata im-bri-*kah*-ta. (= *D. imbricata*).
  Overlapping (the leaves).
  x *salomonii* sah-lo-*mon*-ee-ee. *D.
  bruniifolia* x *D. dedeana*. After Salomon.
  sibirica si-*bi*-ri-ka. (= *D. repens*). Of
  Siberia. Siberia, Caucasus, Greenland.

Dracaena dra-*kie*-na *Agavaceae* (*Dracaenaceae*).
From Gk. *drakcina* (a dragon). It has also
been suggested that it was named after Sir
Francis Drake. Tender, evergreen shrubs and
trees.
  deremensis de-rem-*en*-sis. Of Derema,
  Tanzania. Tropical Africa.
    'Bausei' *bowz*-ee-ee. After Bause, see
    Dieffenbachia x *bausei*.
  draco drǎ-ko. A dragon. Dragon tree.
  Canary Islands.
  fragrans frah-granz. Fragrant (the flowers).
  Upper Guinea.
    'Lindenii' lin-*den*-ee-ee. After Linden,
    Belgian nurseryman.
    'Massangeana' ma-*son*-zhee-ah-na.
    After M. de Massange.
  goldieana gōl-dee-*ah*-na. After the Rev.
  Hugh Goldie, American missionary in W
  Africa in the late 19th century. Upper
  Guinea.
  hookeriana huk-a-ree-*ah*-na. After Hooker.
  S Africa.
  marginata mar-gi-*nah*-ta. Margined (the
  leaves). Madagascar.
  reflexa re-*fleks*-a. (= *Pleomele reflexa*).
  Reflexed (the inflorescence). Madagascar,
  Mauritius.
  sanderiana sahn-da-ree-*ah*-na. After Henry
  Sander, founder of the famous Sander's
  Nursery of St Albans and Bruges.
  Cameroons.
  surculosa sur-kew-lō-sa. (= *D. godseffiana*).
  Suckering. Tropical W Africa.

Dracocephalum drǎ-kō-*kef*-a-lum *Labiatae*
From Gk. *draco* (a dragon) and *cephale* (a
head) referring to the shape of the flowers.
Herbaceous perennials.
  forrestii fo-rest-ee-ee. After George Forrest,
  see *Abies delavayi forrestii*. W China.
  grandiflorum grǎn-di-*flō*-rum. Large-
  flowered. W Siberia.
  hemsleyanum hemz-lee-*ah*-num. After
  William Botting Hemsley (1843–1924).
  Tibet.
  ruyschiana riesh-ee-*ah*-na. After Frederick

Dracocephalum (continued)
  Ruysch (1638–1731), Prof. of botany at
  Amsterdam. Siberia, N China.

Dracula dra-kew-la *Orchidaceae* From Lat.,
dracula, meaning little dragon, possibly a
reference to the sinister flowers. Epiphytic or
lithophytic orchids. Central America.
  bella bel-la (= *Masdevallia bella*). Pretty.
  Colombia.
  chimaera ki-*mie*-ra (= *Masdevallia chimaera*).
  A monster, perhaps referring to the very
  large flowers. Colombia.

Dracunculus dra-*kun*-kew-lus *Araceae*. L.
name for another plant, meaning a small
dragon. Tuberous perennials. Mediterranean
region.
  muscivorus musk-*i*-vo-rus. (= *Helicodiceros
  muscivorum*). Fly-eating.
  vulgaris vul-*gah*-ris. Common. Dragon
  Arum.

Dragon Arum see *Dracunculus vulgaris*
Dragon Tree see *Dracaena draco*

Drimys drim-is *Winteraceae*. The Gk. word
for acrid, from the taste of the bark.
Evergreen shrubs and trees.
  lanceolata lǎn-kee-ō-*lah*-ta. (= *D.
  aromatica*). Lanceolate (the leaves).
  Mountain Pepper. SE Australia, Tasmania.
  winteri win-ta-ree. After Captain William
  Winter who sailed with Frances Drake and
  collected bark for medicinal and culinary
  purposes. Winter's Bark. C and S Chile.

Dropwort see *Filipendula vulgaris*.

Drosera dro-se-ra *Droseraceae*. From Gk.
droseros (dewy), referring to the appearance
of the leaves. Carnivorous, perennial herbs.
Sundew.
  binata bi-*nah*-ta. Paired, the leaf is forked
  into two segments. SE Australia, New
  Zealand.
  capensis ka-*pen*-sis. Of the Cape of Good
  Hope.
  filiformis fee-lee-*form*-is. Thread-like (the
  leaves). SE United States.

Drunkard's Dream see *Hatiora salicornioides*

Dryadella dri-a-del-a *Orchidaceae* Diminutive
of dryad, a tree nymph. Epiphytic and
lithophytic orchids, formerly included in
Masdevallia. C & S America.
  simula sim-ew-la (= *Masdevallia simula*).
  Simulating, imitating. Colombia.

Dryas *dree*-as *Rosaceae*. Gk. *dryas*, from Dryades, daughters of Zeus and originally nymphs of the oak, the leaves somewhat resemble oak leaves. Evergreen shrubs.

drummondii drū-*mond*-ee-ee. After its discoverer Thomas Drummond (1780–1835). Artic N America.

octopetala ok-tō-*pe*-ta-la. Eight-petalled. Mountain Avens. Arctic and alpine N hemisphere.

x *suendermannii* soon-der-*mahn*-ee-ee. *D. drummondii* x *D. octopetala*. After Franz Sündermann (1864–1946). Arctic N America.

Dryopteris *dree-op-te-ris Aspidiaceae*. The Gk. name from *drys* (oak) and *pteris* (a fern). Buckler Fern.

aemula ie-mew-la. Imitating. Hay-scented Buckler Fern. W Europe.

cristata kris-*tah*-ta. Crested. Crested Buckler Fern. N America, Europe, Asia.

dilatata dil-a-*tah*-ta. (= *D. austriaca*). Expanded (the fronds). Broad Buckler Fern. N America, Europe, Asia.

erythrosora e-rith-rō-*so*-ra. With red sori. China, Japan.

filix-mas fil-iks-mahs. Literally male fern, to distinguish it from the more delicate Lady Fern. Male Fern. Europe, N America.

goldieana gol-dee-*ah*-na. After its discoverer, John Goldie (1793–1886). Giant Wood Fern. E N America.

Duchesnea dew-*shez*-nee-a *Rosaceae*. After Antoine Nicolas Duchesne (1747–1827), French horticulturist. Perennial herb.

indica in-di-ka. (= *Fragaria indica*). Indian. Mock Strawberry. Afghanistan, Himalaya, E Asia.

Dumb Cane see *Dieffenbachia*
Dusty Miller see *Artemisia stelleriana*
Dutchman's Breeches see *Dicentra cucullaria*
Dutchman's Pipe see *Aristolochia macrophylla*
Dyer's Greenweed see *Genista tinctoria*

E

Earth Star see *Cryptanthus*
Easter Lily see *Lilum longiflorum*
Eastern Red Cedar see *Juniperus virginiana*

Eccremocarpus e-krem-ō-*kar*-pus

Eccremocarpus (continued) *Bignoniaceae*. From Gk. *ekkremus* (hanging) and *karpos* (a fruit) referring to the hanging pods. Semi-hardy woody climber or annual.

scaber skǎ-ber. Rough. Chile.

Echeveria e-kee-*ve*-ree-a *Crassulaceae*. After Athanasio Echeverriay Godoy, 18th-century botanical artist. Tender succulents. Mexico.

affinis a-feen-is. Related to.

agavoides a-gahv-*oi*-deez. Like *Agave*.

carnicolor kar-*ni*-ko-lor. Flesh coloured (the flowers).

derenbergii de-ran-*berg*-ee-ee. After J. Derenberg (1873–1928).

elegans ay-le-gahnz. Elegant.

gibbiflora gib-bi-*flō*-ra. With the flowers swollen on one side.

glauca see E. *secunda glauca*

harmsii harmz-ee-ee. After Dr Hermann Harms (1870–1942), German botanist.

leucotricha loo-*ko*-tri-ka. White-haired.

multicaulis mul-tee-*kaw*-lis. Many-stemmed.

nodulosa nō-dew-*lō*-sa. With nodules (papillae on the leaves).

pulvinata pul-vee-*nah*-ta. Cushion-like.

runyonii run-*yon*-ee-ee. After Robert Runyon.

secunda se-*kun*-da. With flowers on one side of the stalk.

glauca glow-ka. (= E. *glauca*). Glaucous (the leaves).

setosa say-tō-sa. Bristly-hairy (the stems).

shaviana shah-vee-ah-na. After the Missouri Botanic Garden (Shaw's Garden, after the founder, Henry Shaw 1800–99) whose staff collected it.

Echinacea e-kee-*nah*-kee-a*Compositae*. From Gk. *echinos* (a hedgehog) referring to the prickly receptacle scales. Perennial herb.

purpurea pur-*pewr*-ree-a. (= *Rudbeckia purpurea*). Purple (the flowers). Cone flower. E United States.

Echinocactus e-keen-ō-*kǎk*-tus *Cactaceae*. From Gk. *echinos* (a hedgehog) and *Cactus* q.v.

grusonii gru-*son*-ee-ee. After Herman Gruson who built up a large collection of cacti. Golden Barrel Cactus. Mexico.

horizonthalonius ho-ri-zon-tha-*lō*-nee-us. Referring to the horizontally-held areoles. SW United States, Mexico.

ingens see E. *platyacanthos*

platyacanthos plat-i-a-*can*-thos (= E. *ingens*) With flat spines.

Echinocereus e-keen-ō-*kay*-ree-us *Cactaceae*.

Echinocereus (continued)
From *Gk. echinos* (a hedgehog) and *Cereus* q.v.

engelmannii eng-gal-*mahn*-ee-ee. After Georg Engelmann (1809–44), German physician and botanist. SW United States, N Mexico.
enneacanthus en-ee-a-*kănth*-us. 9-spined (the areoles). SW United States, N Mexico.
knippelianus ni-pel-ee-*ah*-nus. After Karl Knippel, German cactus dealer. Mexico.
pectinatus pek-tin-*ah*-tus. Comb-like. SW United States, N Mexico.
pentalophus pen-ta-*lof*-us. With five crests. S Texas.
poselgeri pō-zal-*ge*-ree (= *Wilcoxia poselgeri*). After Heinrich Poselger (d. 1883), a German cactus grower. Texas, N Mexico.
pulchellus pul-*kel*-us. Pretty. Mexico.
reichenbachii ri-ken-*băk*-ee-ee. After Reichenbach.
  fitchii fich-ee-ee. After W. R. Fitch. Texas.
rigidissimus ri-gi-*dis*-i-mus. Very rigid. S Arizona, N Mexico.
schmollii schmol-ee-ee. (= *Wilcoxia schmollii*). After F. Schmoll. Lamb's tail cactus. Mexico
stramineus strah-*min*-ee-us. Straw-coloured (the spines)
viridiflorus vi-ri-di-*flō*-rus. With green flowers. SW United States.

Echinochloa e-keen-*o*-klō-a *Gramineae*. From Gk. *echinos* (a hedgehog) and *chloe* (a grass). Annual grass.

crus-galli kroos-*hă*-lee. A cock's spur. Asia.

Echinodorus e-keen-*o*-do-rus *Alismataceae*. From Gk. *echinos* (a hedgehog) and *doros* (a bag) referring to the clustered, spiny fruits. Aquarium plants. Burhead.

andrieuxii an-drie-*ux*-ee-ee (= *E. longistylis*). After Andrieux. Brazil.
berteroi bert-a-*rō*-ee. After Carlo Giuseppe Bertero (1789–1831), Italian physician. S United States, W Indies.
cordifolius kor-di-*fo*-lee-us. With heart-shaped leaves. S United States.
horizontalis hoi-ri-zon-*tah*-lis. Horizontal (the leaves). Amazon.
longistylis see *E. andrieuxii*
magdalenensis măg-da-layn-*en*-sis. Of the Rio Magdalena, Colombia.
paniculatus pan-ik-ew-*lah*-tus. With flowers in panicles. S America.

Echinodorus (continued)
quadricostatus kwod-ree-kos-*tah*-tus. Four-ribbed. Tropical S America.
tenellus te-*nel*-us. Dainty. S America.

Echinofossulocactus see *Stenocactus*
Echinomastus macdowellii see *Thelocactus macdowellii*

Echinops e-*kee*-nops *Compositae*. From Gk. *echinos* (a hedgehog) and *ops* (appearance). Perennial herbs. Globe Thistle.

bannaticus ba-*nă*-ti-kus. Of Banat, N Romania. SE Europe, W Asia.
humilis hu-mi-lis. Low growing. W Asia.
ritro-rit-rō. A S European name. E Europe, W Asia.
ruthenicus roo-*then*-i-kus. Of Ruthenia, SW Russia. E Europe.
sphaerocephalus sfie-rō-*kef*-a-lus. With a spherical head. Europe, W Asia.

Echinopsis e-kee-nop-sis *Cactaceae*. From Gk. *echinos* (a hedgehog) and *-opsis* (appearance). Sea Urchin Cactus.

aurea ow-ree-a. Golden (the flowers). Argentine.
bruchii bruk-ee-ee (= *Lobivia bruchii, Soehrensia bruchii*). After Bruch. N Argentina.
candicans kăn-di-kanz (= *Trichocereus candicans*). White (the flowers). Argentina.
chilensis kil-*en*-sis (= *Trichocereus chiloensis*). Of Chile, Chile.
coquimbana kō-kim-*bah*-ra (= *Trichocereus coquimbanus*). Of Coquimbana. Chile.
chamaecereus kă-mie-*kay*-ree-us (= *Chamaecereus sylvestrii*). From Gk. *chamai* (on the ground) and *Cereus* q.v. referring to the procumbent habit. Peanut Cactus. Argentina.
  eyriesii ie-*reez*-ee-ee. After Alexander Eyries who introduced it from Uruguay in 1830. S America.
ferox fe-rŏks. (= *Lobivia ferox*). Spiny. Bolivia.
hertrichiana her-trik-ee-*ah*-na (= *Lobivia allegraiana, Lobivia hertrichiana*). After William Hertrich. Peru.
keunrichii koon-*rik*-ee-ee (= *Lobivia densispina*). After Keuhnrich. N Argentina (Jujuy).
leucantha loo-*kănth*-a. White-flowered. W Argentina.
marsoneri mah-*son*-e-ree (= *Lobivia jajoina*). After Marson. N Argentina (Jujuy).
multiplex see *E. oxygona*
oxygona ox-i-*gō*-na (= *E. multiplex*) with sharp angles. (the ribs). Brazil, Argentina.

Echinopsis (continued)
rhodotricha rod-*o*-tri-ka. Red-haired.
Paraguay, N Argentina.
schickendantzii shi-kan-*dănts*-ee-ee (=
Trichocereus schickendantzii). After
Schichendantz. Argentina.
spachiana spăch-ee-*ah*-na (= *Trichocereus
spachiana*). After Edouard Spach (1801–79).
Golden Column, White Torch Cactus.
Argentina.

Echium e-kee-um *Boraginaceae*. From *echion*
the Gk. name. Biennial herbs.
plantagineum plahn-ta-*gin*-ee-um. (= *E.
lycopsis*). Like *Plantago*. Europe, Caucasus.
russicum ru-si-kum. (= *E. rubrum*).
Russian. E Europe, W Asia.
vulgare vul-*gah*-ree. Common. Viper's
Bugloss. Europe, Asia.

Edgeworthia ej-*werth*-ee-a *Thymelaeaceae*.
After Michael Pakenham Edgeworth
(1812–81), amateur botanist and plant
collector with the E India Co. Semi-hardy,
deciduous shrub.
chrysanthana kris-*ănth*-a. (= *E. papyrifera*).
With golden flowers. China.

Edraianthus ed-rie-*ănth*-us *Campanulaceae*.
From Gk. *hedraios* (sitting) and *anthos* (a
flower) referring to the sessile flowers.
Perennial herbs.
dalmaticus dăl-*mă*-ti-kus (= E. *caudatus*). Of
Dalmatia.
graminifolius grah-min-i-*fol*-ee-us. Grass-
leaved. SE Europe.
pumilio pew-*mil*-ee-ō. Dwarf. Yugoslavia.
serpyllifolius ser-pi-li-*fo*-lee-us. Thyme-
leaved. Yugoslavia.

Egeria ay-*ge*-ree-a *Hydrocharitaceae*. After a
Roman goddess of water. Aquatic perennial
herb.
densa den-sa (= *Elodea densa*). Dense. S
America.

Egg Plant see *Solanum melongena*
Eglantine see *Rosa rubiginosa*
Egyptian Star Cluster see *Pentas lanceloata*

Eichhornia iek-*horn*-ee-a *Pontederiaceae*. After
J. A. F. Eichhorn (1779–1856), an eminent
Prussian. Tender aquatic perennial.
crassipes krăs-i-pays. With a thick stalk (the
leaves). Water Hyacinth. Tropical
America.

Elaeagnus e-lee-*ăg*-nus *Elaeagnaceae*. A Gk.
name originally applied to a willow, from

Elaeagnus (continued)
helodes (growing in marshes) and *hagnos*
(pure), referring to the white fruit masses
(of the willow). Deciduous and evergreen
shrubs.
angustifolia ăn-gust-i-*fo*-lee-a. Narrow-
leaved. W Asia.
commutata kom-ew-*tah*-ta. Changeable.
Silver Berry. N America.
× *ebbingei* e-*bing*-gee-ee. E. *macrophylla* ×
E. *pungens*. After J. W. E. Ebbinge of
Boskoop.
glabra glă-bra. Glabrous. China, Japan.
macrophylla măk-tō-*fil*-la. Large-leaved.
Korea, Japan.
pungens pung-genz. Spiny. Japan.
'Maculata' măk-ew-*lah*-ta. Blotched (the
leaves).
umbellata um-be-*lah*-ta. The flowers are in
umbel-like clusters. Himalaya, China,
Japan.

Elder see *Sambucus*
    Common see *S. nigra*
    Red-berried see *S. racemosa*
Elecampane see *Inula helenium*

Eleocharis e-lee-*o*-ka-ris *Cyperaceae*. From
Gk. *helodes* (growing in marshes) and *charis*
(grace). Aquatic perennials.
acicularis a-kik-ew-*lah*-ris. Needle-like (the
stems). Hair Grass. N temperate regions.
dulcis dul-kis. Sweet (the edible tuber).
Chinese Water Chestnut. Tropical Asia.
W Africa.

Elephant's Ears see *Caladium*
Elephant's Foot see *Dioscorea elephantipes*

Elettaria e-la-*tah*-ree-a *Zingiberaceae*. From
elettari the Indian name. Tender herb.
cardamomum kar-da-*mō*-mum. A Gk.
name. India.

Eleutherococcus el-oy-therō-*kok*-us *Araliaceae*
(= *Acanthopanax*, in part). From Gk.
eleuthros, free, and *kokkos*, pip, referring to
the pyrenes. Spiny or bristly shrubs and
trees.
henryi hen-ree-ee. After Augustine Henry,
see *Illicium henryi*. C. China.
sieboldianus see-bōld-ee-*ah*-nus. After
Philipp Franz von Siebold (1796–1866),
German doctor who introduced and
named many Japanese plants. China,
Japan.

Elfin Herb see *Cuphea hyssopifolia*
Elk's Horns see *Rhombophyllum nelii*

Elliottia e-lee-*o*-tee-a *Ericaceae*. After
Stephen Elliott (1771–1830) who
discovered the following. Deciduous shrub.
  racemosa ră-kay-*mŏ*-sa. With flowers in
  racemes. SE United States.

Elm see *Ulmus*.
  Belgian see *U* × *hollandica* 'Belgica'
  Camperdown see *U. glabra* 'Camperdownii'
  Cornish see *U. carpinifolia cornubiensis*
  Dutch see *U.* × *hollandica*
  English see *U. procera*
  Exeter see *U. glabra* 'Exoniensis'
  Goodyer's see *U. angustifolia*
  Jersey see *U.* 'Sarniensis'
  Smooth see *U. carpinifolia*
  Wheatley see *U.* 'Sarniensis'
  Wych see *U. glabra*

Elodea e-lŏ-dee-a *Hydrochariataceae*. From
Gk. *helodes* (growing in marshes). Aquatic
herbs.
  canadensis kăn-a-*den*-sis. Of Canada or NE
  North America. N America.
  crispa see *Lagarosiphon major*
  densa see *Egeria densa*

Elsholtzia el-*sholtz*-ee-a *Labiatae*. After
Johann Sigismund Elsholtz (1623–88).
Deciduous shrub.
  stauntonii stawn-*ton*-ee-ee. After Sir
  George Staunton (1737–1801). N China.

Elymus *e*-li-mus *Gramineae*. From *elymos* the
Gk. name for millet. Perennial Grass.
  arenarius see *Leymus arenarius*

Embothrium em-*both*-ree-um *Proteaceae*.
From Gk. *en* (in) and *bothrion* (a small pit),
the anthers are borne in cup-shaped pits on
the perianth segments. Evergreen shrub or
tree.
  coccineum kok-*kin*-ee-um. Scarlet (the
  flowers). Fire Bush. Chile, Argentina.

Emilia em-*ee*-lee-a *Compositae*. Derivation
obscure, presumably commemorative.
Annual herb.
  coccinea kok-*sin*-ee-a (= *E. javanica*).
  Scarlet (the flowers). Tassel
  Flower. Tropics.

Emmenopterys e-men-*op*-te-ris *Rubiaceae*.
From Gk. *emmenes* (enduring) and *pteryx* (a
wing) one lobe of the calyx enlarges into a
conspicuous, leaf-like wing. It has not
flowered in this country. Deciduous tree.
  henryi hen-ree-ee. After Augustine Henry,
  see *Illicium henryi*. C and SW China.

Empetrum em-pe-trum *Empetraceae empetron*
the Gk. name, from *en* (on) and *petros* (a rock)
referring to its habitat. Evergreen shrub.
Crowberry.
  nigrum nig-rum. Black (the fruit). N
  hemisphere.

Encyclia en-*kie*-klee-a *Orchidaceae* From Gk.
enkyklo, to encircle; the lateral lobes of the
lip encircle the column. Tropical and
subtropical America.
  brassavolae bra-*sah*-vo-lie. (= *Epidendrum
  brassavolae*). After *Brassavola* q.v. C
  America.
  citrina ki-*tree*-na (= *Cattleya citrina*).
  Lemon yellow (the flowers). Mexico.
  cochleata kok-lee-*ah*-ta (= *Epidendrum
  cochleatum*). Shell-like (the lip). Tropical
  America.
  fragrans frah-granz (= *Epidendrum fragrans*).
  Fragrant. Mexico to S America.
  mariae mă-ree-ie (= *Epidendrum mariae*).
  After Mrs. Mary Östlund who grew
  Mexican orchids. Mexico.
  polybulbon po-lee-*bul*-bon (= *Epidendrum
  polybulbon*). With many bulbs
  (pseudobulbs). C America, W Indies.
  prismatocarpa pris-măt-ŏ-*kar*-pum (=
  Epidendrum prismatocarpum). With a
  prism-shaped fruit. C America.
  radiata ră-dee-*ah*-ta (= *Epidendrum
  radiatum*). Radiating, the purple, radial
  lines on the lip. C America, Mexico.
  vitellina vi-te-*leen*-a (= *Epidendrum
  vitellinum*). Egg-yolk yellow. Mexico,
  Guatemala.

Endymion see *Hyacinthoides*
Endive see *Cichorium endivia*

Enkianthus eng-kee-*ănth*-us *Ericaceae*. From
Gk. *enkyos* (pregnant) and *anthos* (a flower),
in *E. quinqueflorus* each flower appears to bear
another inside it. Deciduous shrubs.
  campanulatus kăm-păn-ew-*lah*-tus. Bell-
  shaped (the corolla). Japan.
  cernuus ker-new-us. Nodding (the
  racemes). Japan.
    rubens roo-bens. Red (the flowers).
  chinensis chin-*en*-sis. Of China. W. China,
  Upper Burma.
  perulatus pe-ru-*lah*-tus. With conspicuous
  bud scales. Japan.

Epidendrum e-pi-*den*-drum *Orchidaceae*.
From Gk. *epi* (upon) and *dendron* (a tree)
referring to their epiphytic habit.
Greenhouse orchids.
  brassavolae see *Encyclia brassavolae*

Epidendrum (continued)
ciliare ki-lee-*ah*-ree. Edged with hairs (the fringed lip). Tropical America.
cochleatum see *Encyclia cochleata*
difforme di-*form*-ee. Of unusual shape. Florida, Tropical America.
endresii see *Oerstedella endressii*
fragrans see *Encyclia fragrans*
ibaguense ee-ba-*gen*-see. From Ibagué, Colombia. C and S America.
mariae see *Encyclia mariae*
nocturnum nok-*tur*-num. Night flowering. Tropical America.
parkinsonianum par-kin-son-ee-*ah*-num. After John Parkinson (1772–1847). Consul-General in Mexico who sent plants to Kew. C America.
polybulbon see *Encyclia polybulbon*
prismatocarpum see *Encyclia prismatocarpa*
radiatum see *Encyclia radiata*
vitellinum see *Encyclia vitellina*

Epigaea e-pi-*gie*-a *Ericaceae*. From Gk. *epi* (on) and *gaia* (the earth) referring to their creeping habit. Evergreen shrubs.
asiatica ah-see-*ah*-ti-ka. Asian. Japan.
repens ree-penz. Creeping. Trailing Arbutus. E N America.

Epilobium e-pi-*lo*-bee-um *Onagraceae*. From Gk. *epi* (upon) and *lobos* (a pod), the corolla is borne on the end of the ovary. Perennial herbs. Willow Herb.
canum kah-num (= *Zauschneria cana*). Grey. California.
chlorifolium klō-ri-*fo*-lee-um. With leaves like *Chlora* (now *Blackstonia*). New Zealand.
kaikourese kie-ku-*ren*-see. Of Kaikouira, New Zealand.
dodonaei dō-do-*nie*-ee. (= *E. rosmarinifolium*). After Rembert Dodoens (1517–85), Flemish physician and herbalist. Europe.
fleischeri flie-sha-ree. After M. Fleischer (1861–1930). Alps.
glabellum gla-*bel*-um. Nearly glabrous. New Zealand.
latifolium lah-tee-*fo*-lee-um. Broad-leaved. N America, Europe, Asia.

Epimedium e-pi-*may*-dee-um *Berberidaceae*. From *epimedion*, Gk. name for another plant. Perennial herbs. Barrenwort. Bishop's Mitre.
alpinum ăl-*peen*-um. Alpine. S Europe.
× *cantabrigiense* kăn-ta-brig-ee-en-se. *E. alpinum* × *E. pubigerum*. Of Cambridge where it was found by W. T. Stearn and R. Thoday in 1950.

Epimedium (continued)
grandiflorum gränd-i-*flō*-rum. Large-flowered. E Asia.
perralderianum pe-ral-de-ree-*ah*-num. After Henri René le Tourneux de la Perraudière (1831–61), French naturalist. Algeria.
pubigerum pew-bi-*ge*-rum. Hairy. Balkans, Caucasus, W Asia.
×*rubrum* rub-rum. *E. alpinum* × *E. grandiflorum*. Red (the sepals).
× *versicolor* ver-*si*-ko-lor. *E. grandiflorum* × *E. pinnatum colchicum*. Variously coloured (the flowers).
× *warleyense* wor-lee-*en*-see. *E. alpinum* × *E. pinnatum colchicum*. Of Warley Place, see *Ceratostigma willmottianum*.
× *youngianum* yung-ee-*ah*-num. *E. diphyllum* × *E. grandiflorum*. After Young. 'Niveum' ni-vee-um. Snow-white (the flowers).

× **Epiphronitis** e-pi-*frō*-ni-tis *Orchidaceae*. Intergeneric hybrid, from the names of the parents. *Epidendrum* × *Sophronitis*. Greenhouse orchids.

Epiphyllum e-pi-*fil*-lum *Cactacea*. From Gk. *epi* (upon) and *phyllon* (a leaf), the flattened, green stems which bear the flowers, resemble leaves.
ackermannii see *Noplaxochia ackermannii*
anguliger ang-*gew*-li-ger. Hooked (the stems). S Mexico.
caidatum kaw-*dah*-tum. Prolonged into a slender tail (the shoots). S Mexico.
chrysocardium kris-ō-*kar*-dee-um. With a golden heart (the yellow filaments in the centre of the flower). S Mexico.
crenatum kray-*nah*-tum. With shallow, rounded teeth (on the shoots). C America.

Epipremnum e-pi-*prem*-num *Araceae*. From Gk. *epi* (upon) and *premnum* (a tree stump) referring to the epiphytic habit. Tender, evergreen climber.
aureum ow-ree-um (= *Scindapsus aureus*). Golden (the variegated leaves). Devil's Ivy. Solomon Islands.

Episcia e-*pis*-kee-a *Gesneriaceae*. From Gk. *episkios* (shaded), they grow in shady places. Tender herbs.
cupreata kew-pree-*ah*-ta. Coppery (the leaves). Flame Violet. N S America.
dianthiflora see *Alsobia dianthiflora*
lilacina li-la-*keen*-a Lilac (the flowers). C America.
reptans rep-tănz. Creeping. N S America.

Eragrostis e-ra-*gros*-tis *Gramineae*. From Gk.
eros (love) and *agrostis* (a grass). Annual
grasses.
  amabilis see *E. tenella*
  elegans ay-le-gahnz. Elegant. Love Grass.
  Brazil.
  tef tef. The native name. NE Africa.
  tenella ten-*ell*-a (= *E. amabilis*). Tender,
  delicate. Love Grass. SE Asia

Eranthis e-răn-this *Ranunculaceae*. From Gk.
er (spring) and *anthos* (a flower), referring to
the early flowers. Tuberous herbs.
  cilicia ki-*li*-kee-a. Of Cilicia, S Turkey. W
  Asia.
  hyemalis hee-e-mah-lis. Of winter
  (flowering). Winter Aconite. S Europe. ×
  tubergenii tew-ber-*gen*-ee-ee. *E. cilicia* × *E.
  hyemalis*. After the Dutch bulb nursery,
  van Tubergen.

Eremurus e-ray-*mew*-rus *Liliaceae*
(*Asphodelaceae*) Gk. *eremia* (a desert) and *oura*
(a tail) referring to their habitat and the shape
of the inflorescence. Perennial herbs. Foxtail
Lily.
  aitchisonii ai-chis-ō-nee-ee (= *E. elwesii*).
  After J. E. T. Aitchison (1835–1898),
  botanist. C Asia, Afghanistan.
  elwesii see *E. aitchisonii*
  himalaicus hi-ma-*lah*-i-kus. Of the
  Himalaya.
  olgae ol-gie. After Olga Fedtschenko
  (1845–1921). N Iran, Asia.
  robustus rō-*bust*-us. Robust. C Asia.
  stenophyllus sten-ō-*fil*-lus. (= *E. bungei*).
  Narrow-leaved. C Asia, Iran.

Erica e-*ree*-ka *Ericaceae*. The classical name,
probably for *E. arborea*. Hardy and tender,
evergreen shrubs. Heath.
  arborea ar-*bo*-ree-a. Tree-like. Tree Heath.
  S Europe, N and E Africa, W Asia.
  'Alpina' ăl-*peen*-a. Alpine. Spain.
  australis ow-*strah*-lis. Southern. Spanish
  Heath. Spain, Portugal.
  canaliculata kăn-ah-lik-ew-*lah*-ta.
  Channelled (the leaves). S Africa.
  carnea kar-nee-a (= *E. herbacea*). Flesh-
  coloured. C and E Europe (mountains).
  cinerea ki-*ne*-ree-a. Grey. Bell Heather. W
  Europe.
  × *darleyensis* dar-lee-*en*-sis. *E. erigena* × *E.
  carnea*. Of Darley Dale, Derbyshire, where
  it was raised.
  erigena e-ri-*gen*-a. (= *E. mediterranea*).
  Irish. W Europe.
  gracilis gră-ki-lis. Graceful. S Africa.
  herbacea see *E. carnea*

Erica (continued)
  hiemalis hee-e-*mah*-lis. Of winter
  (flowering). ? S Africa.
  mediterranea see *E. erigena*
  terminalis ter-mi-*nah*-lis. Terminal (the
  flowers). SW Europe.
  tetralix tet-ra-liks. *tetralice*, an Athenian
  name for *Erica*. Cross-leaved Heath. N
  and W Europe.
  vagans vă-gănz. Wandering, the spreading
  habit. SW Europe, Cornwall.
  ventricosa ven-tri-*kō*-sa. Swollen on one
  side (the corolla). S Africa.
  × *watsonii* wot-*son*-ee-ee. *E. ciliaris* × *E.
  tetralix*. After H. C. Watson who
  discovered it near Truro. NW Europe.
  × *williamsii* wil-*yămz*-ee-ee. *E. tetralix* × *E.
  vagans*. After P. D. Williams who
  introduced it to cultivation in 1910.
  Cornwall.

Erigeron e-ri-ge-ron *Compositae*. The
classical name of a plant, probably groundsel,
from Gk. *eri* (early) and *geron* (an old man),
referring to the fluffy, white seed heads.
Perennial herbs. Fleabane.
  aurantiacus ow-răn-tee-*ah*-kus. Orange
  (the flowers). Turkestan.
  aureus ow-ree-us. Golden (the flowers).
  NW United States.
  compositus kom-*po*-si-tus. Compound (the
  leaves).
  glaucus glow-kus. Glaucous (the leaves).
  Beach Aster. W United States.
  karvinskianus kar-vin-skee-*ah*-nus. (= *E.
  mucronatus*). After Wilhelm Friedrich
  Karwinski von Karwin (1780–1855), who
  collected in S America. Mexico to
  Venezula.
  leiomerus lay-o-me-rus. With smooth parts.
  W United States.
  simplex sim-pleks. Simple (the stem, i.e.
  unbranched). W United States.
  speciosus spek-ee-ō-sus. Showy. W N
  America.
    macranthus ma-*krănth*-us. (= *E.
    macranthus*). Large-flowered.

Erinacea e-ri-*nah*-kee-a *Leguminosae*
(*Papilionoidae*). L. for resembling a
hedgehog, referring to its habit. Deciduous,
spiny shrub.
  anthyllis ăn-thil-lis (= *E. pungens*). Gk.
  name for the related kidney vetch.
  Hedgehog Broom. Spain, N Africa.

Erinus e-ri-nus *Scophulariaceae*. Gk. name for
another plant. Perennial herb.

Erinus (continued)
  alpinus ăl-*peen*-us. Alpine. Fairy Foxglove.
  Europe.

Eriobotrya e-ree-ō-*bot*-ree-a *Rosaceae*. From
Gk. *erion* (wool) and *botrys* (a bunch of
grapes), referring to the woolly inflorescence.
Evergreen, semi-hardy shrub or small tree.
  japonica ja-*pon*-i-ka. Of Japan. Loquat.
  China, Japan.

Eritrichium e-ri-trik-ee-um *Boraginaceae*.
From Gk. *erion* (wool) and *thrix* (hair).
Perennial herbs.
  canum kah-num (= *E. rupestre pectinatum*).
  Grey. Afghanistan, Himalaya, Tibet.
  nanum nah-num. Dwarf. Fairy Forget-me-
  not, King of the Alps. Alps.
  rupestre see *E. canum*.

Erodium e-rō-dee-um *Geraniaceae*. From Gk.
erodios (a heron) referring to the shape of
the fruits. Perennial herbs. Heron's Bill.
  absinthoides ăb-sinth-*oi*-deez. Like *Artemisia
  absinthium* W Asia.
  chrysanthum kris-*ănth*-um. With golden
  flowers. Greece.
  corsicum kor-si-kum. Of Corsica. Corsica,
  Sardinia.
  foetidum fee-tid-um (= *E. petraeum*)
  Malodorous. Pyrenees, S France.
  glandulosum glăn-dew-*lō*-sum. (= *E.
  macradenum*). Glandular. Pyrenees, N
  Spain.
  guttatum gu-*tah*-tum. Spotted (the petals).
  SW Mediterranean region.
  manescavi măn-es-*kah*-vee. After Manescau
  (died 1875), Italian merchant and
  naturalist. Pyrennees.
  petraeum see *E. foetidum*
  reichardii rie-*kard*-ee-ee. (= *E.
  chamaedryoides*). After Reichard. Balearic
  Islands.
  × *variable* vă-ree-*ah*-bi-le. *E. corsicum* × *E.
  reichardii*, Variable.
    'Roseum' ro-see-um. (= *E.
    chamaedryoides* 'Rosem'). Rose-
    coloured (the flowers).

Eryngium e-*ring*-gee-um *Umbelliferae*. From
eryggion the Gk. name for *E. campestre*.
Perennial herbs.
  agavifolium a-gah-vi-*fo*-lee-um. *Agave*-
  leaved. Argentina.
  alpinum ăl-*peen*-um. Alpine. Europe.
  amethystinum ă-me-*thist*-i-num. Violet (the
  flowers). Europe.
  bourgatti bour-*găt*-ee-ee. After M. Bourgat

Eryngium (continued)
  who collected in the Pyrenees.
  Mediterranean region.
  giganteum gi-*găn*-tee-um. Very large.
  Caucasus.
  maritimum ma-*ri*-ti-mum. Growing near
  the sea. Sea Holly. Europe (coasts).
  × *oliverianum* o-li-va-ree-*ah*-num. After
  Oliver.
  planum plah-num. Flat (the leaves). Europe,
  Asia.
  × *tripartitum* tri-*par*-tee-tum. Three parted
  (the leaves). Cult. Hybrid of unknown
  origin.

Erysimum e-*ri*-si-mum *Cruciferae*. From
erysimon the Gk. name. Annual, biennial and
perennial herbs.
  allionii ah-lee-*ōn*-ee-ee. After Carlo
  Allioni (1705–1804), Italian botanist.
  Siberian Wallflower. The identity and
  origin of plants grown under this name
  is uncertain.
  asperum a-*spe*-rum. Rough (the leaves). W
  and C N America.
  capitatum kap-i-*tah*-tum. In a dense head
  (the flowers). W N America.
  cheiri kay-ree. See above (*E. allionii*).
  Wallflower. S Europe.
  helveticum hel-*vay*-ti-kum. Of Switzerland.
  Pyrenees to SE Europe.
  linifolium leen-i-*fo*-lee-um. *Linum*-leaved.
  Spain, Portugal.
  murale see *E. cheiri*
  perofskianum pe-rof-skee-*ah*-num. After V.
  A. Perofsky (1794-c. 1857). Afghanistan,
  Pakistan.
  pulchellum pul-*kel*-um. (= *E. rupestre*).
  Pretty. Greece, W Asia.

Erythrina e-rith-*reen*-a *Leguminosae*
(*Papilionoideae*). From Gk. *erythros* (red)
referring to the colour of the flowers. Tender
shrub or tree.
  crista-galli kris-ta-gă-lee. Cock's comb. Coral
  Tree. S America.

Erythronium e-rith-*ron*-ee-um *Liliaceae*. From
erythronion the Gk. name for another plant.
Perennial herbs.
  albidum ăl-bi-dum. White (the flowers). N
  America.
  americanum a-me-ri-*kah*-num. American.
  Yellow Adder's Tongue. E N America.
  californicum kăl-i-*forn*-i-kum. Of
  California.
  citrinum ki-*tree*-num. Lemon-yellow (the
  flowers). W N America.

Erythronium (continued)
dens-canis dens-*kǎ*-nis. A dog's tooth.
Dog's-tooth Violet. S Europe, Turkey.
grandiflorum grǎnd-i-*flō*-rum. Large-
flowered. W N America.
hendersonii hen-der-*son*-ee-ee. After Louis
Fourniquet Henderson (1853–1942). W
N America.
howellii how-*el*-ee-ee. After Thomas
Howell (1842–1912). W N America.
multiscapoideum mul-tee-ska-*poi*-dee-um.
With many scapes. N California.
oregonum o-ree-*gō*-num. Of Oregon. W N
America.
revolutum re-vo-*loo*-tum. Turned back (the
perianth lobes). Trout Lily. W N
America.
tuolumnense too-o-lum-*nen*-see. Of
Tuolumne County, California. C
California.

Erythrorhipsalis see *Rhipsalis*

Escallonia es-ka-*lon*-ee-a *Grossulariaceae* After
Senor Escallon, a Spanish traveller in S
America. Evergreen shrubs.
bifida bi-fi-da. (= *E. montevideinsis*). Split
into two (the leaf apex). E S America.
'Edinensis' e-din-*en*-sis. Of Edinburgh,
where it was raised.
× *exoniensis* eks-ō-nee-*en*-sis. Of Exeter,
where it was raised by Messrs. Veitch.
'Iveyi' ie-vee-ee. *E. bifida* × *E.* × *exoniensis*.
After Mr Ivey, gardener at Caerhays who
found it.
laevis lie-vis. Smooth (the leaves). Brazil.
'Langleyensis' lang-lee-*en*-sis. Of Langley,
where it was raised in 1893.
rubra ru-bra. (= *E. punctata*). Red (the
flowers). Chile, Argentina.
macrantha ma-*krǎnth*-a. (= *E. macrantha*).
Large-flowered. Chiloe.
virgata vir-*gah*-ta. Twiggy. Chile,
Argentina.

Eschscholzia esh-*sholts*-ee-a *Papaveraceae*.
After Johann Friedrich Eschscholz
(1793–1831), Russian botanist. Annual
herbs.
caespitosa kie-spi-*tō*-sa. Tufted. California.
californica kǎl-i-*forn*-i-ka. Of California.
Californian Poppy.

Espostoa es-*po*-stō-a *Cactaceae*. After Nicolas
Esposto, a botanist in Lima, Peru.
lanata lah-*nah*-ta. Woolly. Peru, Ecuador.

Euanthe ew-*anth*-ee *Orchidaceae*. From Gk.
euanthes, blooming; the inflorescence is

Euanthe (continued)
spectacular. Epiphytic orchid formerly
included in *Vanda*.
sanderiana sahn-da-ree-*ah*-na (= *Vanda
sanderiana*). After the Sander orchid
nursery. Philippines.

Eucalyptus ew-ka-*lip*-tus *Myrtaceae*. From Gk.
eu (well) and *kalypto* (to cover) referring to
the calyx which forms a lid over the flowers
in bud. Hardy to tender evergreen trees.
citriodora kit-ree-o-*dō*-ra. Lemon-scented.
Lemon-scented Gum. Queensland.
coccifera kok-*ki*-fe-ra. Berry-bearing.
Mount Wellington Peppermint,
Tasmanian Snow Gum. Tasmania.
cordata kor-*dah*-ta. With heart-shaped
leaves. Silver Gum. Tasmania.
dalrympleana dǎl-rim-plee-*ah*-na. After
Dalrymple. Broad-leaved Kindling Bark.
Tasmania, SE Australia.
glaucescens glow-*kes*-enz. Somewhat
glaucous (the foliage). Tingiringi Gum.
SE Australia.
globulus glob-ew-lus. Like a small globe (the
buds). Tasmanian Blue Gum. Tasmania.
gunni gǔn-ee-ee. After Ronald Campbell
Gunn (1808–81), a magistrate and
botanist in Tasmania. Tasmania.
parvifolia par-vi-*fo*-lee-a. Small-leaved.
Small-leaved Gum. New South Wales.
pauciflora paw-si-*flō*-ra. Few-flowered.
Cabbage Gum. Tasmania, SE Australia.
niphophila ni-*fo*-fi-la. Snow-loving.
Snow Gum. SE Australia.
perriniana pe-rin-ee-*ah*-na. After George
Samuel Perrin (1849–1900), a forester in
SE Australia. Spinning Gum. SE Australia.
pulverulenta pul-ve-ru-*len*-ta. Dusty,
referring to the glaucous bloom on the
leaves. New South Wales.
urnigera ur-*ni*-ge-ra. Urn-bearing (the fruit
is urn-shaped). Urn Gum. Tasmania.
viminalis vee-mi-*nah*-lis. With osier-like
shoots. Ribbon Gum. Tasmania, SE
Australia.

Eucomis ew-*kom*-is *Liliaceae (Hyacinthaceae)*.
From Gk. *eu* (good) and *kome* (hair)
referring to the attractive flower heads.
Bulbous herbs. S Africa.
bicolor bi-ko-lor. Two-coloured (the
flowers).
comosa ko-*mō*-sa. With a tuft (of leafy
bracts at the apex of the raceme).

Eucommia ew-*kom*-ee-a *Eucommiaceae*. From
Gk. *eu* (good) and *kommi* (gum). It is the

Eucommia (continued)
only hardy tree that can produce rubber.
Deciduous tree.
ulmoides ul-*moi*-deez. Like *Ulmus*. China.

Eucryphia ew-*krif*-ee-a *Eucyphiaceae*. From
Gk. *eu* (well) and *kryphios* (covered), the
sepals form a cap over the flower bud. Hardy
and semi-hardy trees and shrubs.
cordifolia kor-di-*fo*-lee-a. With heart-
shaped leaves. Chile.
glutinosa gloo-ti-*no*-sa. Sticky. Chile
× *intermedia* in-ter-*me*-dee-a. *E. glutinosa* ×
E. lucida. Intermediate (between the
parents).
lucida loo-ki-da. Glossy (the leaves).
Tasmania.
milliganii mil-li-*găn*-ee-ee. After Joseph
Milligan (1807-c. 1883), a Scottish
surgeon who collected in Tasmania.
Tasmania.
× *nymansensis* nie-manz-*en*-sis. *E. cordifolia*
× *E. glutinosa*. Of Nymans, Sussex where
it originated.
    'Nymansay' nie-manz-ay. Derived from
    one of two clones of the above first
    exhibited, Nymans A and Nymans B.

Euodia see *Tetradium* (= *Evodia*)

Euonymus ew-*on*-i-mus *Celastraceae*. The L.
name. Deciduous and evergreen trees and
shrubs.
alatus ah-*lah*-tus. Winged (the shoots).
China, Japan.
europaeus oy-rō-*pie*-us. European. Spindle
Tree. Europe, W Asia.
fortunei for-*tewn*-ee-ee. After Robert
Fortune, see *Fortunella*, E Asia.
    'Coloratus' ko-lo-*rah*-tus. Coloured,
    the leaves colour in winter.
    'Kewensis' kew-*en*-sis. Of Kew, to
    where it was first introduced.
    radicans rah-di-kănz. With rooting stems.
grandiflorus grănd-i-*flō*-rus. Large-
flowered. Himalaya, W China.
salicifolius să-li-ki-*fo*-lee-us. *Salix*-
leaved.
hamiltonianus hă-mil-ton-ee-*ah*-nus. After
Francis Buchanan-Hamilton, Scottish
surgeon and botanist. Himalaya, E Asia.
    sieboldianus see-bōld-ee-*ah*-nus. After
    Siebold, see *Acanthopanax sieboldianus*.
    E Asia.
japonicus ja-*pon*-i-kus. Of Japan.
nanus nah-nus. Dwarf. Caucasus to China.
    turkestanicus tur-kes-*tahn*-i-kus. Of
    Turkestan. C Asia.

Euonymus (continued)
oxyphyllus oks-ee-*fil*-lus. With sharp-
pointed leaves. Japan, Korea.
phellomanus fel-ō-*mahn*-us. With corky
shoots. China.
planipes plahn-i-pays. (= *E. sachalinensis*
hort.). With a flat stalk. NE Asia.
wilsonii wil-*son*-ee-ee. After Ernest
Wilson, see *Magnolia wilsonii*.

Eupatorium ew-pa-*to*-ree-um *Compositae*.
The Gk. name, from Eupator, King of
Pontus. Perennial herbs and tender shrubs.
cannabinum kăn-a-*been*-um. Like *Cannabis*
(the leaves). Hemp Agrimony. Europe, N
Africa to C Asia.
coelestinum see *Conoclinum coelestinum*
ligustrinum see *Ageratina ligustrina*
maculatum măk-ew-*lah*-tum. Spotted (the
leaves). Joe-pye Weed. E N America.
purpureum pur-*pewr*-ree-um. Purple (the
flowers). Joe-pye Weed. E N America.
rugosum see *Ageratina altissima*

Euphorbia ew-*for*-bee-a *Euphorbiaceae*. The
classical name after Euphorbus, physician to
Juba, king of Mauritania. Annuals,
perennials, sub-shrubs, tender shrubs and
succulents.
bubalina bew-ba-*leen*-a. Of the African
gazelle. S Africa.
caput-medusae kă-put-may-*dew*-sie.
Medusa's head, to which this plant has
been likened. Medusa's Head. S Africa,
characias ka-*ră*-kee-ahs. L. name of a plant.
Mediterranean region.
    wulfenii wul-*fen*-ee-ee. (= *E. wulfenii*).
    After Wulfen, see *Wulfenia*. SE Europe.
cyathophora see-ăth-ō-*for*-ra. (= *E.
heterophylla* hort.). Cup-bearing. Fire on
the Mountain, Annual Poinsettia. E
United States, Mexico.
cyparissias kew-pa-*ris*-ee-as. Cypress-like.
Cypress Spurge. Europe.
echinus e-*keen*-us. Spiny. Morocco.
epithymoides see *E. polychroma*.
fulgens *ful*-genz. Shining (the bracts).
Mexico.
grandicornis grănd-i-*kor*-nis. With large
horns (on the shoots). S Africa.
griffithii gri-*fith*-ee-ee. After Griffith, see
Ceratostigma griffithii. Himalaya, Tibet.
hermentiana see *E. trigona*.
heterophylla hort. see *E. cyathophora*
lathyris lă-thi-ris. The classical name.
Caper Spurge, Mole Plant. Europe.
mammillaris mă-mi-*lah*-ris. Bearing
nipples. Corncob Cactus. S Africa.
marginata mar-gi-*nah*-ta. Margined (the

Euphorbia (continued)
upper leaves). Snow on the Mountain.
United States.
meloformis may-lō-*form*-is. Melon-shaped.
S Africa.
milii mil-ee-ee. Said to be after M.
Millius, Governor of the Isle of Bourbon where
it was grown. Crown of Thorns.
Madagascar.
'Splendens' *splen*-dens. (= *E. splendens*).
Splendid.
myrsinites mur-sin-*ee*-teez. Like *Myrsine*.
Europe.
obesa o-*bay*-sa. Fat. Baseball Cactus. S
Africa.
palustris pa-*lus*-tris. Growing in marshes.
Europe.
polychroma po-li-*krō*-ma (= *E.
epithymoides*). Of many colours; the
yellow-green flowers are often tinted with
red and violet.
pseudocactus soo-dō-*kăk*-tus. False cactus. S
Africa.
pulcherrima pul-*ke*-ri-ma. Very pretty.
Poinsettia. C America.
resinifera ray-see-*ni*-fe-ra. Resin-bearing.
Morocco.
robbiae rob-ee-ie. After Mary Ann Robb
(1829–1912) who introduced it. Mrs
Robb's Bonnet. W Asia.
seguieriana seg-wee-e-ree-*ah*-na. After Jean
Francis Seguier (1703–84), French
botanist. Europe.
   niciciana nee-cheech-ee-*ah*-na. After
   Nicic. SE Europe.
sikkimensis sik-im-*en*-sis. Of Sikkim.
Himalaya, Tibet.
splendens see *E. milii* 'Splendens'
submammillaris sub-mă-mi-*lah*-ris. With
small nipples. S Africa.
trigona tri-*gō*-na. (= *E hermentiana*). Three-
angled (the stem). SW Africa.
valida vă-li-da. Robust. S Africa.
wulfenii see *E. characias wulfenii*.

Eurya *ew*-ree-a *Theaceae*. From Gk. *euru*
(broad). Evergreen shrub.
   japonica ja-*pon*-i-ka. Of Japan. Himalaya,
   Japan, SE Asia.

Euryops *ew*-ree-ops *Compositae*. From Gk. *eu*
(well) and *ops* (appearance). Evergreen
shrubs.
   acraeus a-*krie*-us. (= *E. evansii* hort.).
   Growing in high places. Drakensberg
   Mountains, S Africa.
   pectinatus pek-ti-*nah*-tus. Comb-like (the
   pinnate leaves). S Africa.

Evening Primrose see *Oenothera biennis*
Everlasting see *Helichrysum bracteatum*

Exacum *eks*-a-kum *Gentianaceae*. From *exacon*
the Gallic name for *Centaurium*. Tender
annual or biennial.
   affine a-*fee*-nee. Related to. Persian Violet.
   Socotra.

Exochorda eks-ō-*kor*-da *Rosaceae*. From Gk.
exo (outside) and *chorda* (a chord), referring
to fibres outside the placenta in the ovary.
Deciduous shrubs.
   giraldii ji-*răl*-dee-ee. After Giraldi who
   introduced it, see *Callicarpa bodinieri
   giraldii*. NW China.
      wilsonii wil-*son*-ee-ee. After Ernest
      Wilson who introduced it in 1907, see
      Magnolia wilsonii.
   korolkowii ko-rol-*kov*-ee-ee. After
   Korolkow, see *Crocus korolkowii*. Turkestan.
   × *macrantha* ma-*krănth*-a. *E. korolkowii* ×
   E. racemosa. Large-flowered.
   racemosa ră-kay-*mō*-sa. With flowers in
   racemes. N China.

F

Fabiana fah-bee-*ah*-na *Solanaceae*. After
Archbishop Francisco Fabian y Fuero
(1719–1801). Evergreen, semi-hardy shrub.
   imbricata im-bri-*kah*-ta. Closely
   overlapping (the leaves).

Fagus *fah*-gus *Fagaceae*. The L. name.
Deciduous trees. Beech.
   orientalis o-ree-en-*tah*-lis. Eastern
   Caucasus, W Asia.
   sylvatica sil-*vă*-ti-ka. Common Beech.
   Europe.
      'Asplenifolia' a-splay-ni-*fo*-lee-a. With
      leaves like *Asplenium*.
      'Pendula' *pen*-dew-la. Pendulous.
      purpurea pur-*pewr*-ree-a. Purple (the
      leaves). Copper Beech.
      'Riversii' ri-*verz*-ee-ee. After Messrs
      Rivers who raised it.
      'Rohanii' rō-*hahn*-ee-ee. After Prince
      Camille de Rohan on whose estate it
      was raised.
      'Zlatia' *zlah*-tee-a. Golden (the leaves),
      from Serbian *zlatos*.

Maids of France see *Ranunculus aconitifolius*
'Flore Pleno'
Fairy Bells see *Disporum*.

Fairy Forget-me-not see *Eritrichium nanum*
Fairy Foxglove see *Erinus alpinus*
False Acacia see *Robinia pseudacacia*
False African Violet see *Streptocarpus saxorum*
False Aralia see *Schefflera elegantissima*
False Hellebore see *Veratrum*
False Spikenard see *Smilacina racemosa*
Fameflower see *Talinum*
Fanwort see *Cabomba*

Farfugium far-*few*-gi-um Compositae. The name used by Pliny. Evergreen herb. E Asia.
  japonica ja-*pon*-ik-a (= *Ligularia tussilago*). Of Japan.

× *Fatshedera* fåts-*he*-de-ra Araliaceae. Intergeneric hybrid, from the names of the parents. *Fatsia* × *Hedera*. Evergreen, semi-scandent shrub.
  lizei lee-*zay*-ee. *Fatsia japonica* 'Moseri' × *Hedera hibernica*. After Messrs. Lizé Frères of Nantes who raised it in 1910.

Fatsia fåts-ee-a Araliaceae. From a Japanese name. Evergreen shrub.
  japonica ja-*pon*-i-ka. (= *Aralia sieboldii*). Of Japan. False Castor-oil Plant.

Faucaria fow-*kah*-ree-a Aizoaceae. From L. *faux* (a gullet), the paired, toothed leaves resemble open jaws. Tender succulents. S Africa.
  tigrina tig-*reen*-a. Tiger-like. Tiger's Jaws.
  tuberculosa tew-ber-kew-*lō*-sa. With tubercles, the white spots on the leaves.

Feather Grass see *Stipa pennata*.

Feijoa fie-*hō*-a Myrtaceae. After Don de Silva Feijo, 19th century Brazilian botanist. Evergreen semi-hardy shrub.
  sellowiana se-lō-ee-*ah*-na. After its discoverer, Friedrich Sellow (Sello) (1789–1831), a German botanist who collected in S America. S America.

Felicia fe-*lik*-ee-a Compositae. After Felix, a German official. Annuals and sub-shrubs. S Africa.
  amelloides ā-mel-*oi*-deez. Like *Aster amellus*.
  amoena am-*een*-a (= *F. pappei*). Delightful. S Africa.
  bergeriana ber-ga-ree-*ah*-na. After Berger. Kingfisher Daisy.
  pappei see *F. amoena*.
  rosulata ros-ew-*lah*-ta. (= *Aster natalensis*). With leaves in a rosette.

Felt Bush see *Kalanchoe beharensis*
Fennel see *Foeniculum vulgare*
  Florence see *F. vulgare azoricum*
  Giant see *Ferula communis*
Fern, Alpine Lady see *Athyrium distentifolium*
  American Sword see *Polystichum munitum*
  Arctic Bladder see *Cystopteris dickieana*
  Asparagus see *Asparagus setaceus*
  Berry Bladder see *Cystopteris bulbifera*
  Bird's Nest see *Asplenium nidus*
  Bladder see *Cystopteris*
  Boston see *Nephrolepis exaltata* 'Bostoniensis'
  Brazil Tree see *Blechnum brasiliense*
  Brittle Bladder see *Cystopteris fragilis*
  Broad Buckler see *Dryopteris dilatata*
  Buckler see *Dryopteris*
  Button see *Pellaea rotundifolia*
  Christmas see *Polystichum acrostichoides*
  Cinnamon see *Osmunda cinnamonea*
  Crested Buckler see *Dryopteris cristata*
  Crown see *Blechnum discolor*
  Deer's Foot see *Davallia canariensis*
  Elk's Horn see *Platycerium bifurcatum*
  Erect Sword see *Nephrolepis cordifolia*
  Filmy see *Hymenophyllum*
  Floating see *Salvinia auriculata*
  Giant Wood see *Dryopteris goldieana*
  Golden Tree see *Dicksonia fibrosa*
  Hairy Lip see *Cheilanthes lanosa*
  Hammock see *Blechnum occidentale*
  Hard see *Blechnum spicant*
  Hard Shield see *Polystichum aculeatum*
  Hare's Foot see *Polypodium vulgare*
  Hart's Tongue see *Asplenium scolopendrium*
  Hay-scented Buckler see *Dennstaedtia punctilobula*
  Hay-scented see *Dryopteris aemula*
  Hen and Chicken see *Asplenium bulbiferum*
  Holly see *Polystichum acrostichoides, Cyrtomium falcatum*
  Interrupted see *Osmunda claytoniana*
  Japanese Painted see *Athyrium nipponicum* 'Pictum'
  Lady see *Athyrium filix-femina*
  Lip see *Cheilanthes*
  Maidenhair see *Adiantum*
  Male see *Dryopteris filix-mas*
  Mountain Bladder see *Cystopteris montana*
  Necklace see *Asplenium bulbiferum*
  Oak see *Gymnocarpium dryopteris*
  Ostrich-feather see *Matteuccia struthiopteris*
  Palm Leaf see *Blechnum capense*

Fern (continued)
  Parsley see *Cryptogramma crispa*
  Rabbit's Foot see *Davallia fejeensis*
  Rib see *Blechnum brasiliense*
  Royal see *Osmunda regalis*
  Rusty Back see *Asplenium ceterach*
  Sensitive see *Onoclea sensibilis*
  Soft Shield see *Polystichum setiferum*
  Squirrel's Foot see *Davallia mariesii, D. trichomanoides*
  Stag's Horn see *Platycerium bifurcatum*
  Sword see *Nephrolepis*
  Tunbridge Filmy see *Hymenophyllum tunbrigense*
  Walking see *Asplenium rhizophyllus*
  Wilson's Filmy see *Hymenophyllum wilsoni*
  Woolly Rock see *Cheilanthes distans*
  Woolly Tree see *Dicksonia antarctica*

Ferocactus fe-rō-kăk-tus *Cactaceae*. From L. *ferox* (savage) referring to the spines, and *Cactus* q.v.
  acanthodes see *F. cylindraceus*
  cylindraceus sil-in-dray-see-us (= *F. acanthodes*). Cylindrical. SW United States, N Mexico.
  hamatacanthus hah-mah-ta-kănth-us. (= *Hamatocactus hamatacanthus*). With hooked spines. SW United States, N Mexico.
  latispinus lah-tee-speen-us. With broad spines. Mexico.
  setispinus see *Thelocactus setispinus*
  viridescens vi-ri-des-enz. Greenish (the flowers). California.
  wislizenii wiz-li-zen-ee-ee. After Wislizenius. SW United States, N Mexico.

Ferula fe-ru-la *Umbelliferae*. The L. name. Perennial herbs.
  communis kom-ew-nis. Common. Giant Fennel. S Europe, W Asia.
  tingitana ting-gi-tah-na. Of Tangier. N Africa, W Asia.

Festuca fes-too-ka *Gramineae*. L. name for a grass stalk. Perennial grasses. Fescue.
  alpina ăl-peen-a. Alpine. Alps.
  amethystina ă-me-this-ti-na. Violet. C Europe.
  glacialis glă-kee-ah-lis. Growing in icy places. Pyrenees.
  glauca glow-ka. Glaucous (the leaves). S France.

Ficus fee-kus *Moraceae*. The L. name for *F. carica*. Deciduous and tender, evergreen trees and shrubs.

Ficus (continued)
  benghalensis beng-ga-len-sis. Of Benghal. Banyan Tree. India, Pakistan.
  benjamina ben-ja-meen-a. From *benjan*, the Indian name. Weeping Fig. Himalaya, SE Asia to N Australia.
  carica kah-rik-ka. Of Caria, W Asia. Common Fig.
  diversifolia see *F. deltoidea*.
  deltoidea del-toi-dee-a. (= *F. diversifolia*). Deltoid i.e. shaped like the Gk. letter delta (△) (the leaves). Mistletoe Fig. Malaysia.
  elastica e-lăs-ti-ka. Producing elastic. Rubber Plant. Himalaya. SE Asia.
  lyrata li-rah-ta. Fiddle-shaped (the leaves). Tropical Africa.
  microcarpa my-krō-kah-pa (= *F. retusa*). With small fruits. SE Asia.
  pumila pew-mi-la. (= *F. repens*). Dwarf. Creeping Fig. E Asia.
  radicans see *F. sagittata*
  religiosa ray-lig-ee-ō-sa. Sacred, a sacred tree in India. India, SE Asia.
  repens see *F. pumila*
  retusa see *F. microcarpa*.
  rubiginosa roo-bi-gi-nōsa. Rusty (the hairs on the underside of the leaf). Rusty Fig. New South Wales.
  sagittata săg-i-tah-ta. (= *F. radicans*). Shaped like an arrow-head (the leaves). E Asia.

Fiddler's Trumpets see *Sarracenia leucophylla*
Fig, Common see *Ficus carica*
  Creeping see *F. pumila*
  Mistletoe see *F. deltoidea*
  Rusty see *F. rubiginosa*
  Weeping see *F. benjamina*
Filbert see *Corylus maxima*

Filipendula fi-li-pen-dew-la *Rosaceae*. From L. *filum* (a thread) and *pendulus* (hanging) referring to the threads connecting the root tubers. Perennial herbs.
  kamtschatica kămt-shă-ti-ka. Of Kamtchatka. NE Asia.
  palmata pahl-mah-ta. Lobed like a hand (the leaves). NE Asia.
  purpurea pur-pewr-ree-a. Purple (the flowers). Japan.
    'Nana' nah-na. (= *F. digitata* 'Nana'). Dwarf.
  rubra rub-ra. Red (the flowers). Queen of the Prairie. E United States.
  ulmaria ul-mah-ree-a. *Ulmus*-like (the leaflets). Meadowsweet. Europe, Asia.
  vulgaris vul-gah-ris. Common. Dropwort. Europe, N Africa, W Asia, Siberia.

Finger Aralia see *Schefflera elegantissima*
Finger-nail Plant see *Neoregelia spectabilis*
Finocchio see *Foeniculum vulgare azoricum*
Fir see *Abies*
  Alpine see *A. lasiocarpa*
  Balsam see *A. balsamea*
  Caucasian see *A. nordmanniana*
  European Silver see *A. alba*
  Flaky see *A. squamata*
  Giant see *A. grandis*
  Greek see *A. cephalonica*
  Himalayan see *A. spectabilis*
  Korean see *A koreana*
  Nikko see *A. homolepis*
  Noble see *A. procera*
  Pacific Silver see *A. amabilis*
  Red see *A. magnifica*
  Red Silver see *A. amabilis*
  Santa Lucia see *A. bracteata*
  White see *A. concolor*
Fire Bush see *Embothrium coccineum*
Fire-on-the-Mountain see *Euphorbia cyathophora*
Firecracker Flower see *Crossandra infundibuliformis*
Firecracker Vine see *Manettia luteo-rubra*
Firethorn see *Pyracantha*
Firewheel Tree see *Stenocarpus sinuatus*

Fittonia fi-*ton*-ee-a *Acanthaceae*. After Elizabeth and Sarah Mary Fitton. Tender perennial herbs. NE South America.
  gigantea gi-*gán*-tee-a. Very large.
  verschaffeltii vair-sha-*felt*-ee-ee. After M. Verschaffelt, a 19th century Belgian nurseryman.
    argyroneura ar-gi-ro-*newr*-ra. Silver-veined. Mosaic Plant.

Fitzroya fitz-*roy*-a *Cupressaceae*. After Captain Robert Fitzroy (1805–65), commander of The Beagle during Darwin's voyage. Evergreen conifer.
  cupressoides kew-pres-*oi*-deez *Cupressus*-like. Chile, Argentina.

Five Fingers see *Syngonium auritum*
Flame Creeper see *Tropaeolum speciosum*
Flame Nettle see *Solenostemon*
Flame of the Woods see *Ixora coccinea*
Flame Plant see *Anthurium scherzerianum*
Flame Violet see *Episcia cupreata*
Flaming Sword see *Vriesia splendens*
Flamingo Flower see *Anthurium scherzerianum*
Flax see *Linum usitatissimum*
  Golden see *L. flavum*
  Tree see *L. arboreum*
  Yellow see *Reinwardtia indica*

Fleabane see *Erigeron*
Floral Firecracker see *Dichelostemma idamaia*
Floss Flower see *Ageratum conyzoides*
Flower of an Hour see *Hibiscus trionum*
Flower of the Western Wind see *Zephyranthes candida*
Flowering Rush see *Butomus umbellatus*
Foam Flower see *Tiarella cordifolia*

Foeniculum fee-*nik*-ew-lum *Umbelliferae*. The L. name. Perennial herbs.
  vulgare vul-*gah*-ree. Common. Fennel. S Europe.
    azoricum a-*zo*-ri-kum. Of the Azores. Florence Fennel, Finocchio.

Fokienia fo-kee-*en*-ee-a *Cupressaceae*. From Fukien (now Fujien), China where it grows. Semi-hardy, evergreen conifer.
  hodginsii hoj-*inz*-ee-ee. After Captain A. Hodgins who discovered it.

Fontinalis fon-ti-*nah*-lis *Fontinalaceae*. From L. *fontinalis* (of springs or fountains) referring to its habitat. Aquatic moss.
  antipyretica ăn-ti-pi-*ret*-i-ka. Against fire, it was packed around chimneys in wooden houses to prevent fire. N hemisphere.

Forest Lily see *Veltheimia bracteata*
Forget-me-not see *Myosotis*

Forsythia for-*sieth*-ee-a *Oleaceae*. After William Forsyth (1737–1804) a Scottish gardener who became superintendent of the Royal Garden of Kensington Palace. Deciduous shrubs.
  × *intermedia* in-ter-*med*-ee-a. *F. suspensa* × *F. viridissima*. Intermediate (between the parents).
  ovata ŏ-*vah*-ta. Ovate (the leaves). Korean Forsythia. Korea.
  suspensa sus-*pens*-a. Hanging (the flowers). Golden Bell. China.
  viridissima vi-ri-*di*-si-ma. Most green (the shoots). China.
    'Bronxensis' broks-*en*-sis. Of the Bronx. It was grown at the New York Botanic Garden.

Fortunella for-tew-*nel*-a *Rutaceae*. After Robert Fortune (1812–80) who collected in China. He introduced the tea plant from China into India.
  japonica ja-*pon*-i-ka. Of Japan. Kumquat. S China.

Fothergilla fo-tha-*gil*-a *Hamamelidaceae*. After

Fothergilla (continued)
Dr John Fothergill (1712–80), English
physician who grew American plants.
Deciduous shrubs. SE United States.
   gardenii gar-*den*-ee-ee. After Dr Garden
   who discovered it, see *Gardenia*.
   major mah-yor. (= *F. monticola*). Larger.

Fountain Grass see *Pennisetum setaceum*
Fountain Plant see *Amaranthus tricolor*
'Salicifolius'
Four o'clock Plant see *Mirabilis jalapa*
Foxglove see *Digitalis purpurea*
Foxtail Grass see *Alopecurus pratensis*
Foxtail Lily see *Eremurus*

Fragaria fra-*gah*-ree-a *Rosaceae*. From *fraga*
the L. name, referring to the scent of the
fruit. Perennial herbs. Strawberry.
   × *ananassa* ǎn-a-*nǎs*-a. *F. chiloensis* × *F.*
   virginiana. From *Ananas* q.v. it was
   originally referred to as the pine or
   pineapple strawberry. Garden Strawberry.
   indica see *Duchesnea indica*
   moschata mos-*kah*-ta. Musk-scented.
   Hautbois Strawberry. Europe, W Asia.
   vesca ves-ka. Little. Alpine Strawberry.
   Europe, Asia, EN America.

Francoa frang-*kō*-a *Saxifragaceae*. After
Francisco Franco a 16th-century Spanish
physician. Semi-hardy perennial herb.
   sonchifolia son-ki-*fo*-lee-a. *Sonchus*-leaved.
   Bridal Wreath. Chile.

Frangipani see *Plumeria rubra*

Frankenia frang-*ken*-ee-a *Frankeniaceae*. After
Johan Frankenius (1590–1661). Evergreen
sub-shrub.
   thymifolia tiem-i-*fo*-lee-a. *Thymus*-leaved.
   N Africa.

Franklinia frank-*lin*-ee-a *Theaceae*. After
Benjamin Franklin (1706–90), the
American statesman. Deciduous tree.
   alatamaha a-lah-ta-*mah*-ha. (*Gordonia*
   alatamaha). Of the Altamaha River,
   Georgia, United States, near which it
   grew before extinction.

Fraxinus fraks-i-nus *Oleaceae*. The L. name.
Deciduous trees. Ash.
   americana a-me-ri-*kah*-na. American.
   White Ash. EN America.
   angustifolia ang-gus-ti-*fo*-lee-a (= *F.*
   oxycarpa). Narrow-leaved. Narrow-leaved
   Ash. W Mediterranean region, N Africa.

Fraxinus (continued)
   excelsior eks-*kel*-see-or. Taller. Common
   Ash. Europe, Caucasus.
      'Jaspidea' yas-*pid*-ee-a. Jasper-like (the
      yellow shoots).
   ornus or-nus. L. name for the mountain
   ash. Manna Ash. S Europe, W Asia.
   velutina vel-cw-*teen*-a. Velvety (the leaves
   and shoots of some forms). Arizona Ash.
   SW United States, Mexico.

Freckle Face see *Hypoestes phyllostachya*

Freesia freez-ee-a *Iridaceae*. After Friedrich
Heinrich Theodor Freese (died 1876), a
German physician. Tender or semi-hardy
cormous perennials. S Africa.
   alba ǎl-ba. (= *F. refracta alba*). White (the
   flowers).
   × *hybrida* hib-ri-da. Hybrid. Common
   Freesia.
   refracta re-*frǎk*-ta. Broken.

Fremontia see *Fremontodendron*

Fremontodendron free-mont-ō-*den*-dron
Sterculiaceae. (*Fremontia*). After Major-
General John Charles Fremont (1813–90),
explorer and plant collector in the W
United States who discovered *F. californicum*.
   californicum kǎl-i-*forn*-i-kum. Of
   California.
   mexicanum meks-i-*kah*-num. Mexican. S
   California, Mexico.

French Bean see *Phaseolus vulgaris*
French Honeysuckle see *Hedysarum*
coronarium
Friendship Plant see *Billbergia nutans, Pilea*
involucrata
Fringe Tree see *Chionanthus virginicus*
   Chinese see *C. retusus*

Fritillaria fri-ti-*lah*-ree-a *Liliaceae*. From L.
fritillus (a dicebox) referring to the
chequered flowers. Bulbous perennials.
Fritillary.
   acmopetala ǎk-mō-*pe*-ta-la. With anvil-
   shaped petals. W Asia.
   assyriaca hort. see *F. uva-vulpis*
   bithynica bi-*thin*-i-ka. (= *F. schliemanii*). Of
   Bithynia (NW Turkey).
   camschatkensis kǎm-shǎt-*ken*-sis. Of
   Kamtchatka. WN American, N Japan.
   caucasica kaw-*kǎs*-i-ka. Of the Caucasus.
   Turkey, Iran, Caucasus.
   crassifolia krǎs-i-*fō*-lee-a. Fleshy-leaved.
   Anatolia.

Fritillaria (continued)
  kurdica kŭr-dik-a (= *F. lanceolata*). Of
  Kurdistan, (SE Turkey and Iraq).
  davisii day-*vis*-ee-ee. After Peter Hadland
  Davis (born 1918). Greece.
  gracilis see *F. messanensis gracilis*
  graeca grie-ka. Of Greece. S Greece.
    thessala thes-a-la. Of Thessalia, N
    Greece. S Europe.
  hispanica his-păn-i-ka (= *F. lusitanica*).
  From Spain. Spain, Portugal.
  imperialis im-pe-ree-*ah*-lis. Showy. Crown
  Imperial. Turkey to Kashmir.
  involucrata in-vo-loo-*krah*-ta. With an
  involucre, the upper three leaves are
  whorled. S France, NW Italy.
  lanceolata see *F. crassifolia* var *kurdica*
  latifolia lah-tee-*fo*-lee-a. Broad-leaved.
  Caucasus, NE Turkey.
  lusitanica see *F. hispanica*
  meleagris mel-ee-*ah*-gris. Spotted (the
  flowers). Europe.
  messanensis mes-an-*en*-sis. Of Messina,
  Sicily. S Europe, N Africa.
    gracilis gră-ki-lis. (= *F. gracilis*). Graceful
    Yugoslavia, Albania.
  pallidiflora pa-li-di-*flo*-ra. Pale-flowered. C.
  Asia.
  persica per-si-ka. Or Iran (Persia). W Asia.
  pluriflora ploo-ri-*flo*-ra. Many-flowered.
  California.
  pontica pon-ti-ka. Of Pontus (N Turkey).
  SE Europe, NW Turkey.
  pudica pud-*ee*-ka. Modest. W N America.
  pyrenaica pi-ray-*nah*-i-ka. Of the Pyrenees.
  Pyrenees, NW Spain.
  raddeana rah-dee-*ah*-na. After Gustav
  Ferdinand Richard Radde (1831–1903) of
  Gdansk. Iran, C Asia.
  recurva re-*kur*-va. Curved back (the
  perianth lobes). W United States.
  roylei royl-ee-ee. After John Forbes Royle
  (1798–1858), a surgeon with the E India
  Co. who collected in India and the
  Himalaya. Himalaya.
  ruthenica roo-*then*-i-ka. Of Ruthenia, SW
  Russia. SE Europe, W Asia.
  schiemanii see *F. bithynica*
  uva-vulpis oo-va-*vul*-pis. Fox's grape, a
  translation of the Kurdish name *tarai
  raiwa*. W Asia.
  verticillata ver-ti-ki-*lah*-ta. Whorled (the
  leaves). C Asia, China.

Frog's Bit see *Hydrocharis morsus-ranae*

Fuchsia *fuks*-ee-a *Onagraceae*. After Leonhart
Fuchs (1501–66), a German physician and

Fuchsia (continued)
herbalist. Deciduous and evergreen, semi-
hardy and tender shrubs.
  'Corallina' ko-ra-*leen*-a. (= 'Exoniensis')
  Coral red (the flowers).
  excorticata eks-kor-ti-*kah*-ta. With peeling
  bark. New Zealand.
  magellanica mă-ge-*lăn*-i-ka. From the
  regionof the Magellan Straits. S Chile, S
  Argentina.
    gracilis gră-ki-lis. Graceful.
  molinae mo-*leen*-ie.
  After Juan Ignacio Molina (1740–1829).
  'Versicolor' ver-*si*-ko-lor. Variously
  coloured (the flowers).
  procumbens prō-*kum*-benz. Prostrate.
  Trailing Fuchsia. New Zealand.
  'Riccartonii' ri-kar-*ton*-ee-ee. Of
  Riccarton, Scotland where it was raised.

Furze see *Ulex.*

G

Gagea *gayj*-ee-a *Liliaceae*. After Sir Thomas
Gage (1761–1820), Suffolk botanist whose
grandfather is commemorated in the
greengage. Bulbous perennials.
  fistulosa fist-u-*lo*-sa. With hollow leaves.
  Europe, Caucasus.
  lutea loo-tee-a. Yellow (the flowers).
  Europe.

Gaillardia gay-*lard*-ee-a *Compositae*. After
Gaillard de Charentonneau, 18th-century
French magistrate and botanical patron.
Blanket Flower.
  × *grandiflora* grănd-i-*flo*-ra. G. aristata × G.
  pulchella. (= G. aristata hort.). Large-
  flowered.
  pulchella pul-*kel*-a. Pretty. United States,
  Mexico.

Galanthus ga-*lănth*-us *Amaryllidaceae*. From
Gk. *gala* (milk) and *anthos* (a flower)
referring to the colour of the flowers.
Bulbous herbs. Snowdrop.
  allenii a-*len*-ee-ee. After James Allen, a
  snowdrop grower who found it. Possibly
  a hybrid.
  caucasius kaw-*kă*-si-cus. Of the Caucasus.
  elwesii el-*wez*-ee-ee. After H. J. Elwes
  (1846–1922), English sportsman and
  naturalist who introduced it. SE Europe,
  W Turkey.

Galanthus (continued)
fosteri fos-ta-ree. After Sir Michael Foster
(1836–1907). Turkey, Lebanon.
ikariae i-*kah*-ree-ie. Of Ikaria, an Aegean
island. Caucasus, N Turkey.
latifolius lăt-i-*fō*-lee-us (= *G. platyphyllus*)
Broad-leaved. Turkey, Caucasus.
nivalis ni-*vah*-lis. Of the snow. Common
Snowdrop. Europe.
plicatus pli-*kah*-tus. Pleated (the leaves). E
Europe.
    byzantinus bi-zan-*teen*-us. Of Istanbul
    (Byzantium). NW Turkey.

Galax *gă*-lăx *Diapensiaceae*. From Gk. *gala*
(milk) referring to the white flowers.
Perennial herb.
urceolata ur-kee-ō-*lah*-ta. (= *G. aphylla*).
Urn-shaped. E United States.

Galega ga-*lee*-ga *Leguminosae (Papilionoidae)*
From Gk. *gala* (milk) it was fed to goats to
improve milk flow. Perennial herbs.
officinalis o-fi-ki-*nah*-lis. Sold as a herb.
Goat's Rue. Europe, W Asia.
orientalis o-ree-en-*tah*-lis. Eastern.
Caucasus.

Galeobdolon see *Lamium galeobdolon*.

Galingale see *Cyperus longus*

Galium *gă*-lee-um *Rubiaceae*. From Gk. *gala*
(milk). *G. verum* (lady's bedstraw) has been
used to curdle milk. Perennial herb.
odoratum o-dō-*rah*-tum. (= *Asperula
odorata*). Scented. Sweet Woodruff.
Europe, N Africa, W Asia.

Galtonia gawl-*ton*-ee-a *Liliaceae*. After Sir
Frances Galton (1812–1911). Bulbous
herbs. S Africa.
candicans *kăn*-di-kănz. (= *Hyacinthus
candicans*). White (the flowers). Summer
Hyacinth.
princeps *pring*-keps. Most distinguished.

Gardener's Garters see *Phalaris arundinacea*
'Picta'

Gardenia gar-*den*-ee-a *Rubiaceae*. After Dr
Alexander Garden (1730–91), a Scottish
physician and botanist who lived in S
Carolina. Tender, evergreen shrub.
jasminoides see *G. augusta*.
augusta ow-*gŭs*-ta (= *G. jasminoides*).
Stately, noble, E Asia.
Cape Jessamine.

Garland Flower see *Daphne cneorum,
Hedychium coronarium*
Garlic see *Allium sativum*.

Garrya *gă*-ree-a *Garryaceae*. After Nicholas
Garry of the Hudson's Bay Co. Evergreen
shrubs or small trees.
elliptica e-*lip*-ti-ka. Elliptic (the leaves).
W United States
× *thuretti* thu-*ray*-tee-ee. *G. elliptica* × *G.
fadyenii*. After Gustave Thuret who raised
it in about 1862.

Gasteria găs-*te*-ree-a *Liliaceae* (*Aloeaceae*).
From Gk. *gaster* (a belly) referring to the
swollen base of the corolla tube. Tender
succulents S Africa.
bicolor bi-ko-lor (= *G. maculata, G.
marmorata*). Two-coloured (the flowers).
liliputiana li-lee-put-ee-*ah*-na. Of
Lilliput i.e. very small.
brevifolia brev-i-*fo*-lee-a. Short-leaved.
carinata kar-in-*ah*-ta. Keeled (the leaves). S
Cape.
verrucosa ve-roo-*kō*-sa. Warty (the leaves).
Wart Gasteria.
disticha dis-ti-ka. (= *G. lingua*). Two-
ranked (the leaves).
maculata see *G. bicolor*
marmorata see *G. bicolor*
trigona tri-*gō*-na. Three-angled.

× **Gaulnettya** gawl-*netee*-a *Ericaceae*.
Intergeneric hybrid, from the names of the
parents. *Gaultheria* × *Pernettya*. These genera
have been merged. See *Gautheria* ×
wisleyensis.

Gaultheria gawl-*the*-ree-a *Ericaceae*. After Dr
Gaulthier (c. 1708–58), a Canadian botanist
and physician.
leucocarpa loo-kō-*kar*-pa (= *Pernettya
leucocarpa*). White-fruited.
Chile, Argentina.
mucronata mew-kron-*ah*-ta (= *Pernettya
mucronata*). Mucronate (the leaves). Chile,
Argentina.
prostrata pros-*trah*-ta (= *Pernettya prostrata*).
Prostrate. Andes.
procumbens prō-*kum*-benz. Prostrate.
Creeping Wintergreen. E N America.
shallon *shă*-lon. The native name. Sabal,
Shallon. W N America.
× *wisleyensis* wiz-lee-en-sis (= × *Gaulnettya
wisleyensis*).
G. shallon × *G. mucronata*. Of Wisley
where it arose.

Gay Feather see *Liatris*.

Gazania ga-*zah*-nee-a *Compositae*. After Theodore of Gaza (1398–1478). Semi-hardy perennials.
rigens ri-gayns. (= *G. splendens*). Rigid. Treasure Flower. Cult.
leucolaena loo-ko-*lie*-na. (= *G. uniflora*). White-cloaked, the white-hairy leaves. S Africa.

Gean see *Prunus avium*

Gelsemium gel-*sem*-ee-um *Loganiaceae*. From *gelsomino* the Italian for jasmine. Semi-hardy, evergreen, twining shrub.
sempervirens sem-per-*vi*-rens. Evergreen. Yellow Jessamine. S United States, C America.

Genista ge-*nis*-ta *Leguminosae*. The L name. Deciduous or nearly leafless shrubs. Broom.
aetnensis iet-*nen*-sis. Of Mt Etna, Sicily. Mt Etna Broom. Sicily, Sardinia.
delphinensis see *G. sagittalis*
fragrans hort. see *G. × spachianus*
hispanica hi-*spah*-ni-ka. Spanish. Spanish Gorse. SW Europe.
januensis yăn-ew-*en*-sis. Of Genoa. Genoa Broom. Italy, SW Europe.
lydia li-dee-a. Of Lydia (W Turkey). SE Europe, W Asia.
monspessulanus mon-spes-ew-*lah*-nus. (= *Cytisus monspessulanus*) Of Montpelier. Montpelier Broom. S Europe, N Africa, Syria.
pilosa pi-*lō*-sa. With long, soft hairs. Europe.
sagittalis să-gi-tah-lis (= *G. delphinensis*). Arrow-shaped. S & C Europe, Balkans.
× *spachianus* spah-chee-*ah*-nus (= *Cytisus* × *spachianus*). *G. stenopetala* × *G. canariensis*. After Edouard Spach (1801–79), French botanist.
sylvestris sil-*ves*-tris. Of woods. Dalmatian Broom. SE Europe.
tenera te-ne-ra. Tender, delicate. Madeira, Tenerife.
tinctoria tink-*to*-ree-a. Used in dyeing. Dyer's Greenweed. Europe to Siberia.

Gentiana gen-tee-*ah*-na *Gentianaceae*. After Gentius, a king of Illyria in the 2nd century B.C., who is said to have discovered the medicinal properties of *G. lutea*. Perennial herbs. Gentian.
acaulis a-*kaw*-lis. (= *G. excisa*). Without a stem. Trumpet Gentian. S Europe.
andrewsii ăn-*drooz*-ee-ee. After H. C. Andrews. NE North America.

Gentiana (continued)
angustifolia ang-gus-ti-*fo*-lee-a. Narrow-leaved. Alps.
asclepiadea a-sklay-pee-*ah*-dee-a. Like *Asclepias*. Willow Gentian. Europe, W Asia.
bellidifolia bel-i-di-*fo*-lee-a. *Bellis*-leaved. New Zealand.
cachemirica cash-*me*-ri-ka. Of Kashmir.
clusii klooz-ee-ee. After Charles de l'Ecluse (Carolus Clusius) (1526–1609), Flemish botanist, Alps.
dinarica di-*nah*-ri-ka. Of the Dinaric Alps. SW Yugoslavia, Albania.
excisa see *G. acaulis*
farreri fă-ra-ree. After Farrer, see *Viburnum farreri*. NW China.
gracilipes gra-*kil*-i-pays. Slender-stalked (the flowers). NW China.
hexaphylla heks-a-*fil*-la. Six-leaved, the leaves are in whorls of six. Tibet.
lutea loo-tee-a. Yellow (the flowers). Yellow Gentian. Europe.
ornata or-*nah*-ta. Decorative. Himalaya, Tibet.
pneumonanthe new-mon-*anth*-ee. Lung flower, from supposed medicinal properties. Marsh Gentian. Europe, Caucasus.
punctata punk-*tah*-ta. Spotted (the flowers). Europe.
pyrenaica pi-ray-*nah*-i-ka. Of the Pyrenees. Europe, Asia.
saxosa saks-*ō*-sa. Growing in rocky places. New Zealand.
septemfida sep-*tem*-fi-da. Divided into seven (the corolla). W Asia.
   lagodechiana lă-gōdek-ee-*ah*-na. Of Lagodechi. Caucasus.
sino-ornata see-nō-or-*nah*-ta. The Chinese *G. ornata*. W. China, Tibet.
veitchiorum veech-ee-o-rum. After the Veitch nursery. W China.
verna ver-na. Of spring (flowering). Spring Gentian, Star Gentian. Europe, Asia.

Geogenanthus gay-o-gen-*anth*-us *Commelinaceae*. From Gk. *ge* (earth) *genea* (birthplace) and *anthos* (a flower), the flowers are borne almost at ground level. Tender perennial herbs.
undatus un-*dah*-tus. Wavy-edged (the leaves). Seersucker Plant. Brazil, Peru.

Geranium ge-*ră*-nee-um *Geraniaceae*. From *geranion* the Gk. name, from *geranos* (a crane) referring to the beak-like fruits. Perennial herbs. Cranesbill. For the common bedding geranium, see *Pelargonium*.
armenum see *G. psilostemon*

Geranium (continued)
candicans hort. See *G. lambertii*
cinereum ki-*ne*-ree-um. Grey (the leaves).
Pyrenees.
  subcaulescens sub-kawl=*es*-enz. With a
short stem. E Europe.
dalmaticum dăl-*mat*-i-kum. Of Dalmatia.
endressii en-*dres*-ee-ee. After P. A. C.
Endress (1806–31) who collected in the
Pyrenees. SW France, N Spain.
farreri fă-re-ree. After Farrer, see
Viburnum farreri, China.
grandiflorum hort. see *G. himalayense*
himalayense hi-mah-lay-*en*-see. (= *G.
grandiflorum* hort). Of the Himalaya. C Asia,
Himalaya.
lambertii lăm-*bert*-ee-ee. (= *G. candicans*
hort.). After Lambert. Himalaya, Tibet.
macrorrhizum măk-rō-*ree*-zum. With a
large root. E and SE Europe.
× *magnificum* mahg-*ni*-fi-kum. *G. ibericum*
× *G. platypetalum*. Splendid.
malviflorum măl-vi-*flō*-rum. *Malva*-
flowered. S Spain.
nodosum nō-*dō*-sum. With conspicuous
nodes. S Europe.
phaeum *fie*-um. Dusky. Mourning Widow,
Dusky Cranesbill. Europe.
pratense prah-*tayn*-see. Of meadows.
Europe, C Asia, Himalaya.
procurrens prō-*ku*-renz. Spreading.
Himalaya, Assam.
psilostemon see-lo-ste-mon. (= *G.
armenum*). With glabrous stamens.
Armenia.
renardii re-*nar*-dee-ee. After Charles
Claude Renard (1809–86). Caucasus.
sanguineum sang-*gwin*-ee-um. Bloody (the
flowers). Bloody Cranesbill. Europe,
Caucasus.
striatum stri-*ay*-tum (= *G. sanguineum* var.
lancastriense). Striped (the flowers are
conspicuously veined).
sylvaticum sil-*vă*ti-kum. Of woods. Wood
Cranesbill. Europe, N Asia.
versicolor ver-*si*-ko-lor. Variously coloured.
S and SE Europe.
wallichianum woo-lik-ee-*ah*-num. After
Wallich who introduced it in 1819, see
Pinus wallichiana. Afghanistan, Himalaya.

Gerbera *ger*-ba-ra *Compositae*. After Traugott
Gerber (died 1743), a German naturalist.
Tender perennial.
jamesonii jaym-*son*-ee-ee. After Jameson.
Barberton Daisy, Transvaal Daisy.
Transvaal.

German Ivy see *Delairea odorata*

Germander see *Teucrium*.
  Shrubby see *T. fruticans*.
  Wall see *T. chamaedrys*.

Gesneria ges-*ne*-ree-a *Gesneriaceae*. After
Conrad von Gessner (1516–65), Swiss
naturalist. Tender perennial herbs.
cuneifolia kew-nee-i-*fo*-lee-a. With leaves
tapered to the base. Cuba, Puerto Rica.

Geum *gay*-um *Rosaceae*. The L. name.
Perennial herbs.
× *borisii* bo-*ris*-ee-ee. *G. bulgaricum* × *G.
reptans*. After Boris.
bulgaricum bul-*gah*-ri-kum. Of Bulgaria.
SE Europe.
chiloense ki-lō-*en*-se. Of Chiloe.
montanum mon-*tah*-num. Of mountains.
S Europe.
rivale ree-*vah*-lee. Growing by streams.
Water Avens. Europe, Asia, N America.
reptans rep-tănz. Creeping Europe.

Gevuina ga-*veen*-a *Proteaceae*. The native
name. Semi-hardy, evergreen shrub or tree.
avellana ă-ve-*lah*-na. Like *Corylus avellana*
(the nuts). Chilean Hazel. Chile.

Gherkin see *Cucumis sativus*
Ghost Plant see *Graptopetalum paraguayense*
Giant Caladium see *Alocasia cuprea*
Giant Elephant's Ear see *Alocasia
macrorrhiza*
Giant Hogweed see *Heracleum
mantegazzianum*
Giant Reed see *Arundo donax*

Gilia *gi*-lee-a *Polemoniaceae*. After Filippo
Luigi Gilii (1756–1821), Italian astronomer.
Annual herbs.
achilleifolia a-ki-lee-i-*fo*-lee-a. *Achillea*-
leaved. S California, N Mexico
capitata kă-pi-*tah*-ta. In a dense head (the
flowers). Blue Thimble Flower. W N
America.
lutea see *Linanthus androsaceus luteus*
rubra see *Ipomopsis rubra*
tricolor tri-ko-lor. Three-coloured (the
flowers). Bird's Eyes. California.

Gillenia gi-*len*-ee-a *Rosaceae*. After Arnold
Gille (Gillenius), 17th-century German
botanist. Perennial herb.
trifoliata tri-fo-lee-*ah*-ta. With three leaves
(leaflets). Indian Physic. E N America.

Ginger see *Zingiber officinale*
  Wild see *Asarum*
Ginger Lily see *Hedychium*

Ginger Lily (continued)
Scarlet see *H. coccineum.*

Ginkgo *gink*-gō *Ginkgoaceae.* From Japanese *ginkyo* (silver apricot), originally from old Chinese *ngin-ghang.* Deciduous tree.
biloba bi-*lō*-ba. Two-lobed (the leaves). Maidenhair Tree. China.

Gladdon see *Iris foetidissima*

Gladiolus gla-*dee*-o-lus *Iridaceae.* L. name for a small sword, referring to the leaves. Cormous perennials.
blandus see *G. carneus*
cardinalis kar-di-*nah*-lis. Scarlet (the flowers). S Africa.
carneus kar-nee-us. (= *G. blandus*). Flesh-coloured (the flowers). S Africa.
× *colvillii* kol-*vil*-ee-ee. *G. cardinalis* × *G. tristis.* (= *G. nanus* hort.) After James Colvill (1777–1832), who grew S African bulbs at Chelsea.
callianthus kal-i-*anth*-us (= *Acidanthera bicolor, A. murielae*). With beautiful flowers. Tropical E Africa.
communis ko-*mew*-nis. Common, growing in community. S Europe.
   byzantinus bi-zan-*teen*-us. Of Istanbul (Byzantium). S Europe, N Africa.
dalenii dah-*len*-ee-ee (= *G. natalensis, G. primulinus, G. psittacinus*). After Dalen. A widespread and variable species. Tropical and South Africa.
cuspidatus see *G. undulatus*
× *gandavensis* găn-da-*ven*-sis. Of Ghent.
grandis see *G. liliaceus*
× *hortulanus* hort-ew-*lah*-nus. Of gardens.
illyricus i-*li*-ri-kus. Of Illyria. Europe, Caucasus.
imbricatus im-bri-*kah*-tus. Overlapping. E Europe, W Asia.
liliaceus lee-lee-*ah*-kee-us (= *G. grandis*). Like *Lilium.* S Africa.
nanus hort. see *G.* × *colvillii*
natalensis see *G. dalenii*
primulinus see *G. dalenii*
psittacinus see *G. dalenii*
recurvus re-*kur*-vus. Curved back (the perianth lobes). S Africa.
tristis tris-tis. Sad. Natal.
undulatus un-dew-*lah*-tus. (= *G. cuspidatus*). Wavy margined (the perianth lobes). S Africa.

Glaucidium glow-*kid*-ee-um *Ranunculaceae.* From *Glaucium* q.v. referring to the similar flowers. Perennial herb.

Glaucidium (continued)
palmatum pahl-*mah*-tum. Lobed like a hand (the leaves). Japan.

Glaucium glow-kee-um *Papaveraceae.* From Gk. *glaukos* (grey-green) referring to the colour of the leaves. Annual, biennial or perennial herbs.
corniculatum kor-nik-ew-*lah*-tum. Horned. Red Horned Poppy. Europe, W. Asia.
flavum flah-vum. Yellow. Yellow Horned Poppy. Europe, N Africa, W Asia.

Glechoma glay-*kō*-ma *Labiatae.* From Gk. *glechon,* a kind of mint, referring to the scent of the leaves. Perennial herb.
hederacea he-de-*rah*-kee-a. Like *Hedera.* Ground Ivy. Europe, Asia.

Gleditsia gle-*dits*-ee-a *Leguminosae* (*Caesalpinoideae*). After Gottlieb Gleditsch (died 1786), German botanist. Deciduous trees.
caspica kăs-pi-ka. Of the region of the Caspian Sea. Caspian Locust.
triacanthos tree-a-*kănth*-os. Three-spined. Honey Locust. N America.

Globe Amaranth see *Gomphrena globosa*
Globe Daisy see *Globularia*
Globe Thistle see *Echinops*
Globeflower see *Trollius*

Globularia glob-ew-*lah*-ree-a *Globulariacae.* From L. *globulus* (a small ball) referring to the flower heads. Perennial herbs and sub-shrubs. Globe Daisy.
cordifolia kor-di-*fo*-lee-a. With heart-shaped leaves. Alps.
incanescens in-kah-*nes*-enz. Greyish (the leaves). Italy.
meridionalis me-ree-dee-o-*nah*-lis. (= *G. bellidifolia*). Flowering at mid-day. Alps to SE Europoe.
nudicaulis new-di-*kaw*-lis. Bare-stemmed. Alps, Pyrenees.
punctata punk-*tah*-ta. (= *G. elongata*). Spotted. Europe.

Gloriosa glō-ree-*ō*-sa *Liliaceae.* From L. *gloriosus* (glorious). Tender, climbing lilies. Glory Lily.
superba soo-*per*-ba. Superb. Tropical Africa, Asia.
'Rothschildiana' roths-chield-ee-*ah*-na. After Lionel Walter, 2nd Baron Rothschild (1868–1937). Tropical Africa.

Glory Bower see *Clerodendrum speciossissimum*

Glory Bush see *Tibouchina urvilleana*
Glory of Texas see *Thelocactus bicolor*
Glory Lily see *Gloriosa*
Glory Pea see *Clianthus puniceus*

Gloxinia gloks-*in*-ee-a *Gesneriaceae*. After
Benjamin Peter Gloxin. Tender perennial
herb. For the common Gloxinia of
cultivation see *Sinningia speciosa*.
   perennis pe-*ren*-is. Perennial. Colombia to
   Peru.

Glyceria gli-*se*-ree-a classically gloo-*ke*-ree-a
Gramineae. From Gk. *glykis* (sweet) referring
to the edible seeds of one species. Perennial
Grass,.
   maxima mank-si-ma. Larger. Reed Grass.
   Europe, Asia.

Glyptostrobus glip-*to*-stro-bus *Cupressaceae*
From Gk, *glypto* (to carve) and *strobilus* (a
cone) referring to the pitted cone scales.
Deciduous conifer.
   lineatus See *Taxodium ascendans* 'Nutans'.

Goat's Beard see *Aruncus dioicus*
Goat's Rue see *Galega officinalis*
Godetia see *Clarkia*
Gold-rayed Lily see *Lilium auratum*
Golden Alyssum see *Aurinia saxatilis*
Golden Bell see *Forsythia suspensa*
Golden Club see *Orontium aquaticum*
Golden Column see *Trichocereus spachianus*
Golden Drop see *Onosma tauricum*
Golden Rod see *Solidago*
Golden Tom Thumb see *Parodia aureispina*
Golden Trumpet see *Allamanda cathartica*

Gomphrena gom-*free*-n *Amaranthaceae* L.
name for this or a similar plant. Anual herb.
   globosa glo-*bō*-sa. Spherical (the flower
   heads). Globe Amaranth. Old World
   tropics.

Goniolimon gōn-ee-ō-*lee*-mon
Plumbaginaceae. From Gk. *gonio* (angled) and
Limonium a related genus. Perennial herbs.
   callicomum ka-*li*-komum. (= *Limonium
   incanum*). With beautiful hair, referring to
   the flower heads. Siberia.
   tataricum ta-*tah*-ri-kum. (=*Limonium
   tataricum*). Of Tatary, C Asia, S Europe,
   N Africa, Caucasus, Russia.

Good King Henry see *Chenopodium bonus-
henricus*
Good Luck Plant see *Cordyline terminalis*

Goodyera gud-*yer*-ra *Orchidaceae*. After John
Goodyer (1592–1664). Hardy orchids.

Goodyera (continued)
   pubescens pew-*bes*-enz. Hairy, E N
   America.
   repens ree-penz. Creeping. Himalaya, E
   Asia, N America.
   tesselata tes-e-*lah*-ta. Chequered (the
   leaves). E N America.

Gooseberry see *Ribes uva-crispa*

Gordonia gor-*don*-ee-a *Theaceae*. After James
Gordon (died 1781), a nurseryman who
introduced *Ginkgo biloba*. Evergreen shrub or
small tree.
   alatamaha see *Franklinia alatamaha*
   axillaris aks-il-*lah*-ris. In the leaf axils (the
   flowers). China, Taiwan.

Gorse see *Ulex*
   Common see *U. europaeus*
   Dwarf see *U. minor*
   Spanish see *Genista hispanica*

Gossypium go-*sip*-ee-um *Malvaceae*. From
gossypion the L. name. Tender shrubs and
herbs.
   arboreum ar-bo-ree-um. Tree-like. Tree
   Cotton. Tropical Asia.
   herbaceum her-*bah*-kee-um. Herbaceous.
   Levant Cotton. Africa, W Asia, India.

Granadilla see *Passiflora edulis*
   Giant see *P. quadrangularis*.
   Red see *P. coccinea*
   Yellow see *P. laurifolia*
Grape, Northern Fox see *Vitis labrusca*
Grape Hyacinth see *Muscari*
   Oxford and Cambridge see *M. aucheri*
Grape Ivy see *Cissus rhombifolia*
   Miniature see *Cissus striata*
Grape Vine, Common see *Vitis vinifera*
Grapefruit see *Citrus* × *paradisi*

Graptopetalum grăp-tō-*pe*-ta-lum
Crassulaceae. From Gk. *graptos* (written
upon) and *petalon* (a petal) referring to the
markings on the petals. Tender succulents.
   amethystinum ă-me-*this*-ti-num. (=
   Pachyphytum amethystinum). Violet (the
   leaves). Mexico.
   pachyphyllum pă-kee-*fil*-lum. Thick-
   leaved. Mexico.
   paraguayense pă-ra-gwie-*en*-see. Of
   Paraguay. Ghost Plant, Mother of Pearl
   Plant. Cult.

Grass of Parnassus see *Parnassia palustris*
Greater Spearwort see *Rununculus lingua*

Grevillea gre-*vil*-ee-a *Proteaceae*. After
Charles Francis Greville (1749–1809).
Semi-hardy and tender, evergreen shrubs.
  robusta rō-*bus*-ta. Robust. Silky Oak. E
  Australia.
  rosmarinifolia rōs-ma-reen-i-*fo*-lee-a.
  Rosmarinus-leaved. New South Wales.
  sulphurea sul-*fu*-ree-a. Sulphur-yellow (the
  flowers). New South Wales.

Grey Sage Brush see *Atriplex canescens.*

Grindelia grin-*del*-ee-a *Compositae*. After
David H. Grindel (1776–1836), German
botanist. Evergreen shrub.
  chiloensis ki-lō-*en*-sis. Of Chiloe, where it
  doesn't grow. Argentina, Chile.

Griselinia gri-se-*leen*-ee-a *Cornaceae*. After
Francesco Griselini (1717–83), Venetian
naturalist. Semi-hardy, evergreen shrub or
tree.
  littoralis li-to-*rah*-lis. Growing near the sea
  shore (where it was originally collected).
  New Zealand.

Ground Ivy see *Glechoma hederacea*
Groundsel Tree see *Baccharis halimifolia*
Guava see *Psidium guajava*
Guelder Rose see *Viburnum opulus*
Guernsey Lily see *Nerine sarniensis*
Guinea Goldvine see *Hibbertia scandens.*

Gunnera gun-*e*-ra *Gunneraceae*
(*Haloragidaceae*). After Ernst Gunnerus
(1718–83), Norwegian bishop and botanist.
Perennial herbs.
  chilensis see *G. tinctoria*
  manicata măn-i-*kah*-ta. With long sleeves.
  Giant rhubarb Colombia.
  tinctoria tink-*to*-ree-a (= *G. chilensis*). Used
  in dyeing. Chile, Paraguay.

Guzmania guz-*mahn*-ee-a *Bromeliaceae*. After
Anastasio Guzman, 18th-century Spanish
naturalist. Tender, evergreen perennials.
  lingulata ling-gew-*lah*-ta. Tongue-like (the
  bracts). Tropical America.
  monostachia mon-ō-*stăk*-ee-a. With one
  spike. Tropical America.
  musaica mu-*sah*-i-ka. Like a mosaic (the
  leaves). Colombia.
  zahnii zahn-ee-ee. After Gottlieb Zahan
  who collected for Veitch in C America.
  Panama.

Gymnocalycium gim-nō-ka-*li*-kee-um
Cactaceae. From Gk. *gymnos* (naked) and

Gymnocalycium (continued)
calyx (a bud) referring to the naked flower
buds.
  anisitsii a-nee-*sit*-see-ee (= *G. damsii*).
  Derivation unclear. Paraquay.
  baldianum băl-dee-*ah*-num. (= *G.*
  venturianum). After Bald. Argentina.
  damsii see *G. anisitsii.*
  denudatum day-new-*dah*-tum. Naked, it is
  barely tubercled as most species. Spider
  Cactus, Brazil, Argentina.
  gibbosum gi-*bō*-sum. Swollen on one side.
  Argentina.
  mihanovichii mi-hahn-no-*vich*-ee-ee. After
  Mihanovich, a plant collector. Plain
  Cactus. Paraguay.
  multiflorum mul-tee-*flō*-rum. Many-
  flowered. Brazil, Argentina.
  platense pla-*ten*-see. From near the River
  Plate, Argentina.
  quehlianum kwel-ee-*ah*-num. After
  Leopold Quehl (1849–1923). Argentina
  saglionis săg-lee-*ō*-nis. After the
  Frenchman who was the first to grow it
  in Europe. N Argentina.

Gymnocarpium gim-nō-*kar*-pee-um.
Dryopteridaceae (Aspleniaceae) From Gk.
gymnos (naked) and *karpos* (a fruit), the sori
are not covered with an indusium. Ferns.
  dryopteris dree-*op*-te-ris. Oak fern. Oak
  Fern. N America, Europe, Asia.
  robertianum robert-ee-*ah*-num.
  Resembling *Geranium robertianum.*
  Limestone Polypody. N America, Europe.

Gymnocladus gim-*no*-kla-dus *Leguminosae*.
From Gk. *gymos (naked) and klados* (a branch)
referring to its deciduous nature. Deciduous
tree.
  dioica dee-ō-*ee*-ka. Dioecious. Kentucky
  Coffee Tree. E and C United States.

Gynura gin-*ew*-ra *Compositae*. From Gk. *gyne*
(female) and *oura* (a tail) referring to the
long stigma. Tender, scandent herb.
  aurantiaca ow-răn-tee-*ah*-ka. (= *G.*
  sarmentosa hort.). Orange (the flowers).
  Velvet Plant, Purple Passion Vine. Java.

Gypsophila gip-*sof*-i-la *Caryophyllaceae*. From
Gk. *gypsos* (gypsum) and *philos* (loving),
some species grow on lime. Annual and
perennial herbs.
  aretioides a-ray-tee-*oi*-deez. Like *Aretia*
  (now included in *Androsace*). N Iran.
  cerastioides ke-ŕs-tee-*oi*-deez. Like *Cerastium*.
  Himalaya.

Gypsophila (continued)
elegans ay-le-gahnz. Elegant. Caucasus, W
Asia.
paniculata pa-nik-ew-*la*-ta. With flowers in
panicles. Europe to C Asia.
repens ree-penz. Creeping. Europe.

H

Haageocereus hag-ee-ō-*kay*-ree-us *Cactaceae*.
After J. N. Haage (1826–78) and *Cereus* q.v.
Peru.
decumbens day-*kum*-benz. Prostrate.
versicolor ver-*si*-ko-lor. Variably coloured
(the spines).

Haberlea ha-*ber*-lee-a *Gesneriaceae*. After Carl
Constantin Haberle (1764–1832) Perennial
herbs.
ferdinandii-coburgii fer-di-*nahn*-dee-ee-kō-
burg-ee-ee. After King Ferdinand of
Bulgaria. C Bulgaria.
rhodopensis ro-do-*pen*-sis. Of the Rhodope
Mountains, Bulgaria. Bulgaria, N Greece.

Habranthus ha-*brǎnth*-us *Amaryllidaceae*.
From Gk. *habros* (graceful) and *anthos* (a
flower). Tender, bulbous herbs.
andersonii see H. tubispathus
robustus rō-*bus*-tus. (= *Zephyranthes
robusta*). Robust. Argentina.
tubispathus tew-bee-*spath*-us (=*H.
andersonii*). With a tubular spathe. S
America.

Hackberry see *Celtis occidentalis*

Hacquetia hǎ-*kay*-tee-a *Umbelliferae*. After
Balthasar Hacquet (1740–1815), Austrian
botanical writer. Perennial herb.
epipactis e-pi-*pǎk*-tis. Gk. name for a plant.
E Europe.

Haemanthus hiem-*ǎnth*-us *Amaryllidaceae*.
From Gk. *haima* (blood) and *anthos* (a
flower) referring to the colour of the flowers.
Tender bulbous herbs. Blood Lily, Red
Cape Tulip.
albiflos ǎl-bi-flōs. With white flowers.
White Paint Brush. S Africa.
coccineus kok-*kin*-ee-us. Scarlet (the
flowers). Ox-tongue lily. S. Africa.
katharinae see *Scadoxus multiflorus* ssp
katharinae
magnificus see *Scadoxus puniceus*
multiflorus see *Scadoxus multiflorus*

Haemanthus (continued)
puniceus see *Scadoxus puniceus*

Hair Grass see *Eleocharis acicularis*

Hakea hǎk-ee-a *Proteaceae*. After Baron
Christian Ludwig von Hake (1745–1818) a
German patron of botany. Evergreen hardy
and tender shrubs.
laurina low-*reen*-a. Like *Laurus*. Pincushion
Flower, Sea Urchin. W Australia.
lissosperma lis-ō-*sperm*-a. (= *H. sericea*
hort.). With smooth seeds. Tasmania, SE
Australia.
microcarpa mik-rō-*kar*-pa. With small fruits.
E Australia, Tasmania.

Hakonechloa ho-kōn-ee-*klō*-a *Gramineae*.
From *Hakone*, a region of Japan, and Gk.
chloa (a grass). Perennial grass.
macra mǎk-ra. Large. Japan.

Halesia haylz-ee-a *Styracaceae*. After Dr
Stephen Hales (1677–1761), English scientist
and inventor. Deciduous trees and shrubs.
Silver Bell, Snowdrop Tree. SE United
States.
carolina see *H. tetraptera*.
diptera dip-te-ra. Two-winged (the fruit).
magniflora mahg-ni-*flō*-ra. Large-
flowered.
monticola mon-*ti*-ko-la. Mountain-loving.
Mountain Snowdrop Tree.
tetraptera tet-*rap*-te-ra (= *H. carolina*) Four-
winged (the fruit). SE US.

× **Halimiocistus** ha-lim-ee-ō-*kis*-tus
Cistaceae. Intergeneric hybrids, from the
names of the parents. *Cistus* × *Halimium*.
Semi-hardy, evergreen shrubs.
'Ingwersenii' ing-gwa-*sen*-ee-ee. *Cistus
salviifolius* × *Halimium alyssoides*. After W.
E. Th. Ingwersen (1883–1960),
nurseryman and plant collector
specialising in alpines who discovered it in
Portugal about 1929.
× *sahucii* sa-hook-ee-ee. *Cistus salviifolius* ×
Halimium umbellatum. After M. Sahuc, a
member of the party that discovered it.
× *wintonensis* win-ton-*en*-sis. Of
Winchester, where it was raised by
Hillier's.

Halimium ha-*lim*-ee-um *Cistaceae*. From
Atriplex halimus, some species resemble it in
leaf. Semi-hardy, everygreen shrubs.
alyssoides ǎ-lis-*oi*-deez. Like *Alyssum*. SW
Europe.
halimifolium ha-lim-i-*fo*-lee-um. With

Halimium (continued)
leaves like *Atriplex halimus*.
Mediterranean region, N Africa.
lasianthum lă-see-*ănth*-um. With woolly
flowers. S Portugal, S Spain.
formosum for-*mō*-sum. (= *H. lasianthum*
hort.). Beautiful. S Portugal.
ocymoides ō-kim-*oi*-deez. Like *Ocimum*.
Spain, Portugal.
umbellatum um-bel-*ah*-tum. With flowers
in umbels.

Halimodendron ha-lim-ō-*den*-dron
Leguminosae (*Papilionoideae*). From Gk.
halimum (maritime) and *dendron* (a tree), it
grows in salt-rich soils. Deciduous shrub.
halodendron hă-lō-*den*-dron. Salt tree. Salt
Tree. C Asia to Mongolia.

Hamamelis hăm-a-*may*-lis *Hamamelidaceae*.
Gk. name for another plant. Decidious
shrubs. Witch Hazel.
× *intermedia* in-ter-*med*-ee-a. *H. japonica* ×
H. mollis. Intermediate (between the
parents).
japonica ja-*pon*-i-ka. Of Japan. Japanese
Witch Hazel.
mollis mol-is. Softly hairy (the young shoots
and leaves). Chinese Witch Hazel. W
China.
vernalis ver-*nah*-lis. Of spring (flowering).
SE United States.

Hamatocactus hamatacanthus see *Ferocactus
hamatacanthus*
setispinus see *Ferocactus setispinus*
Handkerchief Tree see *Davidia involucrata*

Hardenbergia hard-an-*berg*-ee-a *Leguminosae*
(*Papilionoideae*). After Franziska, Countess
von Hardenberg. Tender evergreen climber.
comptoniana komp-ton-ee-*ah*-na.
Compton, the family name of Lady
Northampton who grew it c. 1810. W
Australia.
violacea vee-o-*lah*-kee-a. Violet (the
flowers). E Australia, Tasmania.

Harebell see *Campanula rotundifolia*
Hare's-tail Grass see *Lagarus ovatus*
Harlequin Flower see *Sparaxis tricolor*
Harry Lauder's Walking Stick see *Corylus
avellana* 'Contorta'

Hatiora hă-tee-*o*-ra *Cactaceae*. After Thomas
Hariot (1560–1621), mathmatician and
cartographer. Originally called *Hariota*, a
name that had already been used. Brazil.
bambusioides see *H. salicornioides*.

Hatiora (continued)
gaetneri gairt-na-ree (= *Rhipsalidopsis
gaetneri*). After J. Gärtner (1739–91), a
Stuttgart physician. Easter Cactus.
rosea ro-see-a (= *Rhipsalidopsis rosea*).
Rose-coloured (the flowers).
salicornioides să-li-korn-ee-oi-deez. Like
Salicornia, a salt-marsh plant. Drunkard's
Dream. Bottle cactus.

Haworthia hay-*werth*-ee-a *Liliaceae*. After
Adrian Hardy Haworth (1768–1833),
entomologist and botanist who grew
succulents at Chelsea. Tender succulents. S
Africa.
attenuata a-ten-ew-*ah*-ta. Drawn out (the
leaves).
× *cuspidata* kus-pi-*dah*-ta. With a sharp,
short point (the leaves).
limifolia leem-i-*fo*-lee-a. With file-like
leaves. Fairies' Washboard.
margaritifera mar-gar-ri-*ti*-fe-ra. Pearl-
bearing (the leaves). Pearl Plant.
maughanii mawn-ee-ee. After Dr R.
Maughan Brown.
papillosa see *H. margaritifera*.
reinwardtii rien-*vart*-ee-ee. After
Reinwardt, see *Reinwardtia*.
truncata trun-*kah*-ta. Abruptly cut off (the
leaves.)
venosa vee-*nō*-sa. Conspicuously veined.
tesselata te-se-*lah*-ta. Chequered. War
Plant.

Hawthorn see *Crataegus*
  Common see *C. monogyna*
  Midland see *C. laevigata*
Hazel see *Corylus avellana*
  Turkish see *C. colurna*
Heart of Jesus see *Caladium*
Heartsease see *Viola tricolor*
Hearts on a String see *Ceropegia linearis var
woodii*.
Heath see *Erica*
  Cornish see *E. vagans*
  Cross-leaved see *E. tetralix*
  Spanish see *E. australis*
  Tree see *E. arborea*
Heather see *Calluna vulgaris*
  Bell see *Erica cinerea*

Hebe hay-bay. *Scrophulariaceae*. After Hebe,
goddess of youth. Evergreen shrubs. New
Zealand, apart from hybrids and cultivars.
albicans ăl-bi-kănz. Whitish (the leaves).
× *andersonii* ăn-der-*son*-ee-ee. After Isaac
Anderson-Henry (1800–84) who raised
it.
anomala see *H. odora*

Hebe (continued)

armstrongii arm-*strong*-ee-ee (*H. ochracea* misapplied). After J. B. Armstrong (1850–1926), botanist.

brachysiphon brǎ-kee-*see*-fon. (= *H. traversii* hort.). With a short tube (the corolla).

'Carnea' *kar*-nee-a. Flesh-coloured (the flowers).

cattaractae see *Parahebe cattaractae*

chathamica cha-*tǎm*-i-ka. Of the Chatham Islands.

cupressoides kew-pres-*oi*-deez. Like *Cupressus*.

darwiniana see *H. glaucophylla*

'Edinensis' ed-in-*en*-sis. Of Edinburgh, where it was raised.

× *franciscana* frǎn-sis-*kah*-na. *H. elliptica* × *H. speciosa*. (= *H. elliptica* hort.). Of San Francisco, it was described from plants in the Golden Gate Park.

glaucophylla glow-kō-*fil*-la. (= *H. darwiniana*). With glaucous leaves.

hulkeana hūlk-ee-*ah*-na. After Mr T. H. Hulke, who is said to have discovered it. New Zealand Lilac.

'Lindsayi' *lind*-zay-ee. After Mr R. Lindsay who raised it.

macrantha ma-*krǎnth*-a. Large-flowered.

ochracea ok-*rah*-kee-a. Ochre-coloured (the foliage).

odora ō-*dor*-a (= *H. anomala*). Fragrant.

pimeleoides pi-me-lay-*oi*-deez. Like *Pimelea*.

pinguifolia ping-gwi-*fo*-lee-a. With fat leaves.

rakaiensis rǎ-kie-*en*-sis. (= *H. subalpina* hort.). Of the Rakai Valley, Canterbury.

recurva re-*kur*-va. Curved back (the leaves).

salicifolia sǎ-li-ki-*fo*-lee-a. *Salix*-leaved. Also in Chile.

subalpina hort. see *H. rakaiensis*

traversii hort. see *H. brachysiphon*

Hedera he-*de*-ra *Araliaceae*. the L. name. Evergreen climbers. Ivy.

algeriensis ǎl-ge-ree-*en*-sis. (= *H. canariensis* hort.). Of Algeria.

'Margino-maculata' *mar*-gi-nō-mǎk-ew-*lah*-ta. With spotted margins (the leaves). Often seen labelled *H. helix* 'Marmorata'.

azorica a-*zo*-ri-ka. (= *H. canariensis azorica*). Of the Azores.

colchica kol-ki-ka. Of Colchis, the E coast of the Black Sea. Caspian region, Caucasus, Turkey.

'Dentata' den-*tah*-ta. Toothed (the leaves).

helix he-liks. Winding around, perhaps referring to it being wound around a staff

Hedera (continued)

carried by Bacchus or his attendants. Common Ivy, Europe, Caucasus.

'Hibernica' see *H. hibernica*

'Luzii' *luts*-ee-ee. After the German nursery of Ernst Luz. Often sold as 'Marmorata'.

'Pedata' pe-*dah*-ta. Like a bird's foot (the leaves).

poetica pō-*et*-i-ka. Of poets, who used it for wreaths. Poet's Ivy.

hibernica hi-*bern*-i-ka. (= *H. helix* 'Hibernica'). Irish. Irish Ivy. W Europe.

'Deltoidea' del-*toi*-dee-a. Deltoid i.e. shaped like the Gk. letter delta (Δ), the leaves.

'Digitata' di-gi-*tah*-ta. With finger-like lobes (the leaves).

Hedychium hay-*di*-kee-um *Zingiberaceae*. From Gk. *hedys* (sweet) and *chion* (snow) referring to the fragrant, white flowers of *H. coronarium*. Semi-hardy to tender, herbaceous perennials. Ginger lily.

coccineum kok-*kin*-ee-um. Scarlet. Scarlet Ginger Lily. NE India, Indo-China.

coronarium ko-*fo*-*nah*-ree-um. Used in garlands. Butterfly Lily, Garland Flower. Tropical Himalaya, SE Asia.

densiflorum dens-i-*flō*-rum. Densely flowered. Temperate. Himalaya, Assam.

flavescens flah-*ves*-enz. (= *H. flavum*). Yellow (the flowers). E Himalaya.

gardnerianum gard-na-ree-*ah*-num. After Edward Gardner (born 1784), political resident in Nepal. Kahili Ginger. Himalaya, Assam.

greenei green-ee-ee. After Mr H. F. Green in whose garden the originally discovered plants were grown. Bhutan, Assam.

Hedyotis see *Houstonia*.

Hedysarum hay-*dis*-a-rum *Leguminosae* (*Papilionoideae*) From Gk. *hedys* (sweet) referring to the fragrant flowers of *H. coronarium*. Perennial herb and shrub.

coronarium ko-ō-*nah*-ree-um. Used in garlands. French Honeysuckle. Europe.

multijugum mul-tee-*yoo*-gum. With many joined together, referring to the many leaflets. W Mongolia, Gansu, E Tibet

apiculatum a-pik-ew-*lah*-tum. With an abrupt, short point.

Helenium he-*le*-nee-um *Compositae*. From *helonion* the GK. name for another plant after Helen of Troy. Perennial herb.

Helenium (continued)
autumnale ow-tum-*nah*-lee. Of autumn
(flowering). Sneezweed. N America.

Helianthemum hay-lee-*ănth*-e-mum
Cistaceae. From Gk. *helios* (the sun) and
anthemon (a flower). Dwarf, evergreen shrubs.
Sun Rose, Rock Rose.
apenninum ă-pe-*neen*-um. Of the
Apennines. SW Europe, N Africa.
lunulatum loon-ew-*lah*-tum. Crescent-
shaped (the blotch at the base of the
petals). N Italy.
nummularium num-ew-*lah*-ree-um. (= *H.
chamaecistus. H. vulgare*). With coin-
shaped leaves (in the type specimen).
Europe, W Asia, Caucasus.
oelendicum ur-lănd-i-kum. Of Öland, SE
Sweden.
alpestre ăl-*pes*-tree. (= *H. alpestre*). Of
the lower mountains. C and S Europe.

Helianthus hay-lee-*ănth*-us *Compositae*. From
Gk. *helios* (the sun) and *anthos* (a flower).
Annual and perennial herbs.
annus ăn-ew-us. Annual. Sunflower. N
America, N Mexico.
decapetalus dek-a-*pe*-ta-lus. Ten-petalled. E
N America.
× *multiflorus* mul-tee-*flō*-rus. *H. annuus* ×
H. decapetalus. Many-flowered.
tuberosus tew-be-*rō*-sus. Tuberous (the
rhizome). Jerusalem Artichoke. E N
America.

× **Heliaporus** hay-lee-a-*po*-rus *Cactaceae*.
Intergeneric hybrid, from the names of the
parents. *Aporocatus × Heliocereus*.
smithii smith-ee-ee. *Aporocactus flagelliformis*
× *Heliocereus speciosus*. (= *Aporocactus ×
mallisonii*). After Smith.

Helichrysum hay-li-*kris*-um *Compositae*.
From Gk. *helios* (the sun) and *chryson*
(golden). Annual and perennial herbs and
shrubs.
angustifolium see *H. italicum*
bellidioides bel-i-dee-*oi*-deez. Like *Bellis*
New Zealand.
bracteatum brăk-tee-*ah*-tum. With
conspicuous bracts. Everlasting. Australia.
coralloides see *Ozothamnus coralloides*
frigidum fri-gi-dum. Of cold regions.
Corsica, Sardinia.
italicum ee-tă-li-kum. (= *H. angustifolium*).
Of Italy, S Europe.
serotinum se-*ro*-ti-num. (= *H. serotinum*).
Late flowering. Curry Plant. SW
Europe.

Helichrysum (continued)
ledifolium see *Ozothamnus ledifolius*
microphyllum hort. see *Plecostachys
serpyllifolia*
milfordiae mil-*ford*-ee-ie. After Mrs Helen
A. Milford (died 1940) who collected in
S Africa. S Africa.
orientale o-ree-en-*tah*-lee. Eastern. SE
Europe, W Asia.
petiolare pe-tee-o-*lah*-ree. (= *H. petiolatum*
hort.). With conspicuous petioles. S Africa
plicatum pli-*kah*-tum. Pleated (the leaves).
SE Europe.
rosmarinifolium see *Ozothamnus
rosmarinifolius*
selago see *Ozothamnus selago*
sibthorpii sib-*thorp*-ee-ee. (= *H. virgineum*).
After John Sibthorp (1758–96). Greece.
splendidum splen-di-dum. Splendid. S
Africa.

Helicodiceros muscivorum see *Dracunculus
muscivorus*

Helicotrichon he-lik-tō-*tri*-kon *Gramineae*
From Gk. *helix* (spiral) and *trichos* (a hair).
Perennial grass.
sempervirens sem-per-*vi*-rens. (= *Avena
candida. Avena sempervirens*). Evergreen.
SW Alps.

Helocereus hay-lee-ō-*kay*-ree-us *Cactaceae*.
From Gk. *helios* (the sun) and *Cereus* q.v.
cinnabarinus kin-a-bar-*reen*-us. Cinnabar red
(the flowers). Guatemala.
speciosus spek-ee-ō-sus. Showy. Sun Cactus.
Mexico.
amecamensis a-me-ka-*men*-sis. Of Ameca.
C Mexico.

Heliophila hay-lee-o-fi-la *Cruciferae*. From
Gk. *helios* (the sun) and *philos* (loving).
Annual herbs. S Africa.
leptophylla lep-tō-*fil*-la. Narrow-leaved.
linearis lin-ee-ah-ris. Narrow, with parallel
sides (the leaves).
longifolia long-gi-*fo*-lee-a. Long-leaved.

Heliopsis hay-lee-*op*-sis *Compositae* From
Gk. *helios* (the sun) and -*opsis* indicating
resemblance, referring to the flower heads.
Perennial herbs.
helianthoides hay-lee-ănth-*oi*-deez. Like
Helianthus. E. United States.
scabra skăb-ra. (= *H. scabra*). Rough (the
leaves).
'Patula' păt-ew-la. (= *H. patula*).
Spreading.

Heliopsis (continued)
'Zinniiflora' zin-ee-i-*flŏ*-ra. *Zinnia*-flowered.

Heliotropium hay-lee-o-*trŏ*-pee-um *Boraginaceae*. From Gk. *helios* (the sun) and *trope* (to turn) referring to the old belief that the flowerheads turned with the sun. Annual herb.
arborescens ar-bo-*res*-enz. (= *H. hybridum*. *H. peruvianum*). Becoming tree-like. Heliotrope, Cherry Pie. Peru.

Helipterum hay-*lip*-te-rum *Compositae*. From Gk. *helios* (the sun) and *pteron* (a wing) referring to their sun-loving nature and the feathery pappus bristles. Annual herbs. Straw flower. Australia.
albicans āl-bi-kǎnz. Whitish, the leaves and flower heads.
humboldtianum see *Pteropogon humboldtianum*.
manglesii see *Rhodanthe manglesii*
roseum see *Acroclinium roseum*.

Helleborus he-*le*-bo-rus *Ranunculaceae*. From *helleboros* the Gk. name for *H. orientale*. Perennial herbs. Hellebore.
argutifolius ar-gew-ti-*fo*-lee-us. (= *H. corsicus*. *H. lividus corsicus*). With sharply-toothed leaves. Corsica, Sardinia.
atrorubens aht-rō-*ru*-benz. Deep red (the flowers). Yugoslavia.
corsicus see *H. argutifolius*
foetidus foy-ti-dus. Stinking. Stinking Hellebore. SW Europe.
lividus lee-vi-dus. Lead-coloured. We Mediterranean islands.
corsicus see *H. argutifolius*
niger ni-ger. Black (the roots). Christmas Rose. E Alps, N Italy, Yugoslavia.
× *nigercors* ni-ger-kors. *H. argutifolius* × *H. niger*. From the names of the parents.
orientalis o-ree-en-*tah*-lis. Eastern. Lenten Rose. SE Europe, W Asia.
abchasicus ǎb-*kǎs*-i-kus. Of Abchasia, Caucasus.
purpurascens pur-pew-*rǎs*-enz. Purplish. E Europe.
× *sternii* stern-ee-ee. *H. argutifolius* × *H. lividus*. After Sir Frederick Stern (1884–1967) of Highdown.
viridis vi̱ri-dis. Green (the sepals). Green Hellebore. C and W Europe.

Helxine soleirolii see *Soleirolia soleirolii*

Hemerocallis hay-me-rō-*kǎ*-lis. *Liliaceae*

Hemerocallis (continued)
(*Hemerocallidaceae*) From Gk. *hemera* (day)– and *kallos* (beauty), the flowers only last for one day. Perennial herbs. Day Lily.
citrina ki-*tree*-na. Lemon-yellow. China.
dumortieri dew-mor-tee-*e*-ree. After B. C. Dumortier (1797–1828). Japan.
fulva ful-va. Tawny. S Europe to China.
middendorfii mid-an-*dorf*-ee-ee. After Alexander Theodor von Middendorf (1815–94), Russian traveller and plant collector. E Asia.
minor mi-nor. Smaller. NE Asia.
multiflora mul-tee-*flŏ*-ra. Many-flowered. China.
thunbergii thun-*berg*-ee-ee. After Thunberg, see *Thunbergia*. Japan.

Hemigraphis hay-mi-*grǎf*-is *Acanthaceae*. From Gk. *hemi* (half) and *graphis* (a brush) referring to the hairy filaments of the outer stamens. Tender herb.
alternata ǎl-ter-*nah*-ta. Alternate. Red Ivy. SE Asia.

Hemlock see *Tsuga*
  Eastern see *T. canadensis*
  Mountain see *T. mertensiana*
  Western see *T. heterophylla*
Hemp see *Cannabis sativa*
Hemp Agrimony see *Eupatorium cannabinum*
Hen and Chickens see *Sempervivum tectorum*
Henbane see *Hyoscyamus niger*

Hepatica he-*pǎ*-ti-ka *Ranunculaceae*. From Gk. *hepar* (the liver) referring to the shape and colour of the leaves. Perennial herbs.
americana a-me-ri-*kah*-na. American. E N America.
× *media* me-dee-a. *H. nobilis* × *H. transsilvanica*. Intermediate (between the parents).
nobilis nŏ-bi-lis. (= *H. triloba*) Notable. Europe, W Asia.
transsilvanica trahns-sil-*vah*-ni-ka. (= *H. angulosa*). Of Transylvania, Romania.

Heptapleurum arboricolum see *Schefflera arboricola*

Heracleum hay-ra-*klee*-um *Umbelliferae*. After Hercules (Herakles). Biennial herb.
mantegazzianum mǎn-tee-gǎts-ee-*ah*-num. After Paolo Mantegazzi (1831–1910). Giant Hogweed, Cartwheel Flower. Caucasus.

Herald's Trumpet see *Beaumontia grandiflora*

Herb Christopher see *Actaea spicata*

Hermodactylus her-mō-*dăk*-ti-lus *Iridaceae*.
From Hermes (Gk. name for Mercury) and
daktylos (a finger) referring to the finger-like
tubers. Perennial herb.
    tuberosus tew-be-*rō*-sus. (= *Iris tuberosa*)
    Tuberous. Snake's head Iris. S Europe, W
    Asia.

Herniaria her-nee-*ah*-ree-a *Caryophyllaceae*.
From *hernia* (a rupture) referring to
supposed medicinal properties. Perennial
herb.
    glabra glă-bra. Glabrous. Rupturewort.
    Europe, N Africa, W and C Asia

Hertia see *Othonna*.

Hesperis hes-pe-ris *Cruciferae*. From Gk.
hespera (the evening), the flowers are fragrant
in the evening. Biennial or perennial herb.
    matronalis mah-trō-*nah*-lis. Of matrons,
    from an old name, Mother of the Evening.
    Sweet Rocket, Dame's Violet. Europe, W
    and C Asia.

Heterocentron he-te-rō-*ken*-tron
Melastomataceae. From Gk. *heteros* (different)
and *kentron* (a spur) referring to the different
spurs of the anthers. Tender herbs and sub-
shrubs.
    elegans ay-le-ghanz. (= *Schizocentron
    elegans*). Spanish Shawl. Mexico. C
    America.
    macrostachyum măk-ro-*stăk*-ee-um. (= *H.
    roseum*). With large spikes. Mexico.
    subtriplinervium sub-trip-li-*nerv*-ee-um.
    Somewhat three-nerved. Mexico.

Heuchera hoy-ka-ra *Saxifragaceae*. After
Johann Heinrich von Heucher
(1677–1747). Perennial herb. Alum Root.
    sanguinea sang-*gwin*-ee-a. Blood-red (the
    flowers). SW United States. Mexico.

× *Heucherella* hoy-ke-*rel*-la *Saxifrageceae*.
Intergeneric hybrid, from the names of the
parents. *Heuchera* × *Tiarella*. Perennial herb.
    × *tiarelloides* tee-a-rel-*loi*-deez. *Heuchera*
    hybrid × *Tiarella cordiflora*. Like *Tiarella*.

Hibbertia hib-*bert*-ee-a *Dilleniaceae*. After
George HIbbert (1757–1837), merchant
who had a botanic garden at Clapham.
Tender, evergreen shrub.
    dentata den-*tah*-ta. Toothed (the leaves).
    SE Australia.

Hibbertia (continued)
    scandens skăn-denz. Climbing. Guinea
    Gold Vine. Snake Vine. Australia.

Hibiscus hi-*bis*-kus *Malvaceae*. Gk. name for
mallow. Tender annuals and perennials,
hardy and tender shrubs.
    × *archeri* ar-cha-ree. *H. rosa-sinensis* × *H.
    schizopetalus*. After the raiser, Mr A. S.
    Archer of Antigua.
    moscheutos mos-*kew*-tas. Musk-scented. SE
    United States.
    mutabilis mew-tah-bi-lis. Changeable (the
    flower colour). Cotton Rose. S China,
    Taiwan, Japan.
    rosa-sinensis ro-sa-si-*nen*-sis. As the
    common name, Rose of China. Origin
    uncertain.
    schizopetalus ski-zo-*pe*-ta-lus. With split
    petals. Japanese Hibiscus. E Africa.
    sino-syriacus see-nō-ree-*ah*-kus. The
    Chinese *H. syriacus* q.v. C China.
    syriacus si-ree-*ah*-kus. Of Syria where it
    was originally thought to be native.
    China, Taiwan.
    trionum tree-*ō*-num. Three-coloured.
    Flower of an Hour. Old World tropics.

Hickory see *Carya*
    Bitternut see *C. cordiformis*
    Mockernut see *C. tomentosa*
    Pignut see *C. glabra*
    Shagbark see *C. ovata*

Hieracium hee-e-*rah*-kee-um *Compositae*.
Gk. and L. name for these or similar plants.
Perennial herb.
    aurantiacum see *Pilosella aurantiaca*
    maculatum măk-ew-*lah*-tum. Spotted (the
    leaves). Europe.
    villosum vi-*lō*-sum. Softly hairy. Europe.
        waldsteinii văld-stien-ee-ee. After
        Waldstein, see *Waldsteinia*. Yugoslavia.

Hippeastrum hi-pee-*ăs*-trum *Amaryllidaceae*.
From Gk. *hippos* (a horse), the inflorescence
of *H. puniceum* was likened to a horse's head.
Tender bulbous herbs. Amaryllis.
    argentinum ar-jen-*teen*-um. (= *H.
    candidum*). Of Argentina.
    aulicum ow-li-kum. Of the court. Lily of
    the Palace. Brazil to Paraguay.
    × *johnsonii* jon-*son*-ee-ee. *H. reginae* × *H.
    vittatum*. After Mr Johnson, the raiser. St
    Joseph's Lily.
    puniceum pew-*ni*-kee-um. (= *H. equestre*).
    Reddish-purple (the flowers). Tropical
    America.

Hippeastrum (continued)
reticulatum ray-tik-ew-*lah*-tum. Net-veined (the perianth lobes). Brazil.
striatum stree-*ah*-tum. (= *H. rutilum*). Striped (the flowers). Brazil.
vittatum vi-tah-tum. Banded (the flowers). Peru.

Hippocrepis hi-pō-*kre*-pis *Leguminosae (Papilionoidae)*. From Gk. *hippos* (a horse) and *krepis* (a shoe) referring to the horse-shoe shaped segments of the pods.
comosa ko-*mō*-sa. Tufted. Horseshoe Vetch. Europe.

Hippophae hi-*po*-fa-ee *Elaeagnaceae*. The Gk. Name for another plant. Deciduous shrub or small tree.
rhamnoides răm-*noi*-deez. Like *Rhamnus*. Sea Buckthorn. Europe, Asia.

Hippuris hi-*pewr*-ris *Hippuridaceae*. From Gk. *hippos* (a horse) and *oura* (a tail). Aquatic perennial.
vulgaris vul-*gah*-ris. Common. Mare's Tail. Widely distributed.

Hoheria hō-*he*-ree-a *Malvaceae*. From *houhere* the Maori name for *h. populnea*. Semi-hardy, deciduous and evergreen, trees and shrubs. New Zealand.
angustifolia ang-gust-i-*fo*-lee-a. Narrow-leaved.
glabrata glăb-*rah*-ta. Rather glabrous.
lyallii lie-*ăl*-ee-ee. After David Lyall (1817–95), naval surgeon and naturalist, who collected the type specimen.
populnea pō-*pul*-nee-a. *Populus*-like (the leaves).
sexstylosa seks-sti-*lō*-sa. With six styles.

Holboellia hol-*burl*-ee-a *Lardizabalaceae*. After F. L. Holboell (1765–1829), a superintendant of the Copenhagen Botanic Garden. Evergreen climber.
coriacea ko-ree-*ah*-kee-a. Leathery (the leaflets). China.

Holcoglossum hol-cō-*gloss*-um *Orchidaceae*. From Gk. *holkos*, furrow, and *glossa*, tongue. Epiphytic orchids closely related to *Vanda*. SE Asia.
amesiana aymz-ee-*ah*-na (= *Vanda amesiana*). After Frederick Northrop Ames (1835–93) of Boston, an amateur orchid grower. India.

Holcus *hol*-kus *Gramineae*. The Gk. name for Sorghum. Perennial grass.

Holcus (continued)
mollis mol-lis. Softly hairy. Europe.

Holly see *Ilex*
  Blue see *I.* × *meserveae*
  Common see *I. aquifolium*
  Horned see *I. cornuta*
Hollyhock see *Alcea rosea*
  Fig-leaved see *Alcea ficifolia*

Holmskioldia holm-*shol*-dee-a *Verbenaceae*. After Theodor Holmskiold (1733–94), Danish botanist. Tender, evergreen shrub.
sanguinea sang-*gwin*-ee-a. Blood red (the flowers). Chinese-hat Plant. Himalaya.

Holodiscus ho-lō-*dis*-kus *Rosaceae*. From Gk. *holos* (entire) and *diskos* (a disc) referring to the unlobed disc. Deciduous shrub.
discolor dis-ko-lor. Two-coloured, the leaves are grey-hairy underneath. Ocean Spray. W N America.

Homalomena hŏm-a-lō-*mee*-na *Araceae*. From Gk. *homalos* meaning flat, and *nema*, thread, which refers to the stamens. Tropical evergreen perennials.
lindenii lin-*den*-ee-ee. After Lucien Linden, Belgian nurseryman. New Guinea.

Honeysuckle see *Lonicera*
  Common see *L. periclymenum*
  Perfoliate see *L. caprifolium*
  Trumpet see *L. sempervirens*
Hop see *Humulus lupulus*
Hop Hornbeam see *Ostrya carpinifolia*
Hop Tree see *Ptelea trifoliata*

Hordeum hor-dee-um *Gramineae*. The L. name for *H. vulgare* (barley). Perennial grass.
jubatum you-*bah*-tum. Maned (the inflorescence). Squirrel-tailed Grass. N temperate regions.

Horminum hor-*men*-um *Labiatae*. Gk. name for sage. Perennial herb.
pyrenaicum pi-ray-*nah*-i-kum. Of the Pyrenees. Alps, Pyrenees.

Hornbeam see *Carpinus*
  American see *C. caroliniana*
  Common see *C. betulus*
Hornwort see *Ceratophyllum*
Horse Briar see *Smilax rotundifolia*
Horse-chestnut see *Aesculus*
  Common see *A. hippocastanum*
  Indian see *A. indica*
  Japanese see *A. turbinata*
  Red see *A.* × *carnea*

Horse-radish see *Armoracia rusticana*
Horseshoe Vetch see *Hippocrepis comosa*

Hosta host-a Liliaceae (*Funkiaceae, Hostaceae*)
After Nicholas Tomas Host (1761–1834).
Herbaceous perennials. Plantain Lily.
albomarginata see *H. sieboldii*
crispula krisp-ew-la. Wavy-margined (the leaves). Cult.
decorata de–ko-*rah*-ta. Decorative. Origin uncertain.
elata see *H. montana.*
fortunei for-*tewn*-ee-ee. After Fortune, see *Fortunella.* Japan.
montana mon-*tah*-na (= *H. elata)*
Of mountains. Japan.
rectifolia rek-ti-*fo*-lee-a. With erect leaves. Japan.
sieboldiana see-bōld-ee-*ah*-na. After Siebold, see *Acanthopanax sieboldianus.* Japan.
seiboldii see-*bōld*-ee-ee. (= *H. albomarginata*). As above. Japan.
tardiflora tar-di-*flō*-ra. Late flowering. Japan.
undulata un-dew-*lah*-ta. Wavy-margined (the leaves). Cult.
  erromena e-*rō*-me-na. Vigorous.
ventricosa ven-tri-*kō*-sa. Swollen on one side (the corolla). E Asia.

Hottentot Fig see *Carpobrotus edulis*

Hottonia ho-*ton*-ee-a Primulaceae. After Peter Hotton (1648–1709), Dutch physician and botanist. Floating aquatic perennial herb.
palustris pa-*lus*-tris. Growing in marshes. Water Violet. Europe, Asia.

Hot Water Plant see *Achimenes*
Hound's Tongue see *Cynoglossum officinale*
Houseleek see *Sempervivum*
  Cobweb see *S. arachnoideum*
  Common see *S. tectorum*
  Hen and Chickens see *Jovibarba sobolifera*

Houstonia hoo-*stōn*-ee-a (= *Hedyotis)* Rubiaceae. For Dr. Wm. Houston (1695–1733), British botanist. Hardy perennial herbs. N America, Mexico.
caerulea) kie-*ru*-lee-a. Blue (the flowers). Bluets. EN America.
purpurea pur-*pewr*-ree-a. (= *Hedyotis purpurea*). Purple (the flowers). EN America.

Houttuynia hoo-*tie*-nee-a Saururaceae. After Martin Houttuyn (1720–94), Dutch naturalist. Herbaceous perennial.

Houttuynia (continued)
cordata kor-*dah*-ta. Heart-shaped (the leaves). Himalaya, SE Asia.

Howea how-ee-a Palmae. After Lord Howe, as was the island where they grow. Tender palms. Sentry Palm. Lord Howe Island (Australia).
belmoreana bel-mor-ree-*ah*-na. (= *Kentia belmoreana*). After De Belmore, a Governor of New South Wales. Curly Sentry Palm.
forsteriana for-sta-ree-*ah*-na. (= *Kentia forsteriana*). After William Forster, a senator of New South Wales. Paradise Palm.

Hoya hoy-a Asclepiadaceae. After Thomas Hoy (c. 1750–1822), gardener at Syon House. Tender, evergreen climbers.
australis ow-*strah*-lis. Southern. Australia.
bella be-la. Pretty. Miniature Wax Plant. India.
carnosa kar-*nō*-sa. Fleshy. Honey Plant. Wax Plant. SE Asia to Australia.
imperialis im-pe-ree-*ah*-lis. Showy. Borneo.

Huernia hoo-*ern*-ee-a Asclepiadaceae. After Justin Heurnius (1587–1652), Dutch missionary who collected in S Africa. Tender succulent.
keniensis ken-ee-*en*-sis. Of Kenya.

Humble Plant see *Mimosa pudica*
Humea elegans see *Calomeria amaranthoides*

Humulus hum-ew-lus Cannabaceae. L. version of *humela* the old German name. Perennial climber.
japonicus ja-*pon*-i-kus. Of Japan. E Asia.
lupulus lup-ew-lus. A small wolf, from an old name, willow-wolf, referring to its habit of climbing over willows. N temperate regions.

Hunnemannia hūn-ee-*măn*-ee-a Papaveraceae. After John Hunneman (c. 1760–1839), London bookseller and introducer of plants. Semi-hardy perennial or annual herb.
fumariifolia few-mah-ree-i-*fo*-lee-a. With leaves like *Fumaria.* Mexican Tulip Poppy. Mexico.

Huon Pine see *Lagarostrobus franklinii*

Hutchinsia see *Pritzelago*

Hyacinthoides hee-a-kinth-*oi*-deez Liliaceae (*Hyacinthaceae*). From *Hyacinthus* q.v. and

Hyacinthoides (continued)
Gk. *-oides* indicating resemblance. Bulbous
perennial herbs.
  hispanica his-*pah*-ni-ka. (= *Endymion
  hispanicus*. *Scilla hispanica*). Spanish.
  Spanish Bluebell. SW Europe, N Africa.
  non-scripta non-*skrip*-ta. (= *Endymion non-
  scriptus*. *Scilla non-scripta*). Unmarked.
  Bluebell. W Europe.

Hyacinthus hee-a-*kinth*-us *Liliaceae
(Hyacinthaceae)*. The Gk. name. Bulbous
herb.
  amethystinus see *Brimeura amethystina*
  azureus see *Muscari azureum*
  candicans see *Galtonia candicans*
  orientalis o-ree-en-*tah*-lis. Eastern. Dutch
  or Common Hyacinth. W Asia.

Hydrangea hi-*drang*-gee-a *Hydrangeaceae*.
From Gk. *hydor* (water) and *aggos* (a jar)
referring to the cup-shaped fruits. Deciduous
shrubs and climbers.
  arborescens ar-bo-*res*-enz. Becoming tree-
  like. E United States.
    discolor dis-ko-lor. (= *H. cinerea*). Two
    coloured, the leaves are grey-white
    beneath.
  aspera a-*spe*-ra. (= *H. villosa*). Rough (the
  leaves). Himalaya, E Asia.
  cinerea see *H. arborescens discolor*
  heteromalla het-e-*ro*-mall-la. Variably hairy.
  Himalaya, China.
    'Bretschneideri' bret-*shnie*-da-ree. After
    Dr Bretschneider who introduced it.
  integerrima see *H. serratifolia*
  involucrata in-vo-loo-*krah*-ta. With an
  involucre. Japan, Taiwan.
    'Hortensis' hor-*ten*-sis. Of gardens.
  macrophylla măk-rō-*fil*-la. Large-leaved.
  Japan (Cult.).
    Hortensia hor-*ten*-see-a. Possibly after
    Hortense, the daughter of the Prince
    of Nassau-Siegen.
    'Mariesii' ma-*reez*-ee-ee. After Maries
    who introduced it. see *Davillia mariesii*.
  paniculata pan-nik-ew-*lah*-ta. With flowers
  in panicles. Japan, China.
    'Praecox' *prie*-koks. Early flowering.
    'Tardiva' *tar*-di-va. Late flowering.
  petiolaris pe-tee-o-*lah*-ris. With
  conspicuous petioles. Japan, S Korea,
  Taiwan.
  quercifolia kwer-ki-*fo*-lee-a. *Quercus*-leaved.
  Oak-leaved Hydrangea. SE United
  States.
  sargentiana sar-jen-tee-*ah*-na. After
  Sargent, see *Prunus sargentii*. China.

Hydrangea (continued)
  serrata se-*rah*-ta. Saw-toothed (the leaves).
  Japan, S Korea.
  serratifolia se-rah-ti-*fo*-lee-a. (= *H.
  integerrima*). With saw-toothed leaves.
  Chile, Argentina.
  villosa see *H. aspera*.

Hydrilla hid-*r*il-la *Hydrocharitaceae*. From Gk.
hydor (water referring to its habit. Aquarium
plant.
  verticillata ver-ti-ki-*lah*-ta. Whorled (the
  leaves). Old World.

Hydrocharis hi-*dro*-ka-ris *Hydrocharitaceae*.
From Gk. *hydor* (water) and *charis* (grace).
Floating aquatic herb.
  morsus-ranae mor-sus-*rah*-nie. The
  common name. Frog's bit. Europe, Asia.

Hydrocleys hi-*dro*-klee-is *Limnocharitaceae*.
From Gk. *hydor* (water) and *kleis* (a key).
Tender aquatic herb.
  nymphoides nimf-*oi*-deez. Like *Nymphaea*.
  Water Poppy. Tropical S America.

Hygrophilia hi-*gro*-fi-la *Acanthaceae*. From
Gk. *hygros* (moist) and *philos* (loving), they
grow in wet places. Aquarium plants.
  difformis di-*for*-mis. (= *Synnema triflorum*).
  Of differing shapes (the leaves). SE Asia.
  polysperma po-lee-*sperm*-a. With many
  seeds. India, Bhutan.

Hylocereus hi-lō-*kay*-ree-us *Cactaceae*. From
Gk. *hyle* (a wood) and *Cereus* q.v. referring
to their epiphytic habit. Climbing cacti.
  ocamponis ō-cam-*pō*-nis (= *H. purpusii*).
  From Ocampo, Mexico.
  purpusii see *H. ocamponis*
  trigonus tri-*gō*-nus. Three angled (the
  stems). W Indies.
  undatus un-*dah*-tus. Wavy (the stem
  wings). Tropical America.

Hylotelephium hi-lō-te-*le*-fee-um
Crassulaceae. From Gk. *hylo*, woody, and
Telephium, the Gk. name for a plant.
Perennials with robust, sometimes woody
stems, formerly included in *Sedum*.
  cauticolum kaw-*ti*-ko-lum (= *Sedum
  cauticolum*). Growing on cliffs. N Japan.
  ewersii ay-*verz*-ee-ee (= *Sedum ewersii*).
  After Joseph Ewers (1781–1830), a German
  statesman. Himalaya, China.
  populifolium pō-pew-li-*fo*-lee-um (=
  Sedum populifolium). *Populus*-leaved.
  Siberia.
  seiboldii see-*bōld*-ee-ee (= *Sedum sieboldii*).

Hylotelephium (continued)
After Siebold who introduced it, see
Eleutherococcus sieboldii. Japan.
spectabile spek-*tah*-bi-lee (= *Sedum*
spectabile). Spectacular. China, Korea.
telephium te-le-fee-um (= *Sedum*
telephium). Gk. name of a plant. Europe
to Japan.

Hymenanthera hi-mayn-an-*the*-ra *Violaceae*.
From Gk. *hymen* (a membrane) and *anthera*
(an anther), the anthers are joined by a
membrane. Semi-hardy evergreen shrubs.
angustifolia ang-gust-i-*fo*-lee-a. Narrow-
leaved. New Zealand.
crassifolia krăs-i-*fo*-lee-a. Thick-leaved. SE
Australia, Tasmania, New Zealand.

Hymenocallis hi-mayn-ō-*kăl*-is
Amaryllidaceae. From Gk. *hymen* (a
membrane) and *kallos* (beauty), the stamens
are united by a membrane. Tender, bulbous
herbs. Spider Lily.
amancaes a-*măn*-kies. The Peruvian name.
Peru.
caribaea kă-ri-*bie*-a. Of the Caribbean. W
Indies.
× *festalis* fay-*stah*-lis. *H. longipetala* × *H.*
narcissiflora. Festive, gay.
harrisiana hă-ris-ee-*ah*-na. After Mr T.
Harris who imported it c. 1840. Mexico.
littoralis li-to-*rah*-lis. Of the shore. Tropical
America.
× *macrostephana* măk-rō-ste-*fah*-na. *H.*
narcissiflora × *H. speciosa*. Large-crowned.
narcissiflora nar-kis-i-*flō*-ra. (= *H. calathina*)
Narcisssus-flowered. Peru, Bolivia.

Hymenophyllum hi-mayn-ō-*fil*-lum
Hymenophyllaceae. From Gk. *hymen* (a
membrane) and *phyllon* (a leaf) referring to
the very thin fronds. Filmy Ferns. Widely
distributed.
tunbrigense tūn-brij-*en*-see. Of Tunbridge
Wells. Tunbridge Filmy Fern.
wilsonii wil-*son*-ee-ee. After Wilson.
Wilson's Filmy Fern.

Hyoscyamus hee-ō-*skee*-a-mus *Solanaceae*.
The Gk. name. Poisonous annual or
biennial herb.
niger ni-ger. Black, referring to the
poisonous properties, it was once
believed that any part of the body touched
with it would turn black and rot.
Henbane. Europe, W Asia, N Africa.

Hypericum hi-pe-*ree*-kum *Guttiferae*. From
Gk. *hyper* (above) and *eikon* (a picture), it

Hypericum (continued)
was hung above pictures to ward off evil
spirits. Perennial herbs and shrubs.
androsaemum ăn-dros-*ie*-mum. Gk. name
for a plant with red sap from *andros* (man)
and *haima* (blood). Tutsan. Europe, N
Africa.
beanii been-ee-ee. After Bean, see *Cytisus*
× *beanii*. China.
calycinum kăl-i-*kee*-num. With a
conspicuous calyx. Rose of Sharon,
Aaron's Beard. SE Europe, Turkey.
cerastioides ke-răs-tee-*oi*-deez. Like
Cerastium. SE Europe, NW Turkey.
coris ko-ris. From the resemblance of the
leaves to those of *Coris*. Alps, N Italy.
empetrifolium em-pet-ri-*fo*-lee-um.
Empetrum-leaved. Greece.
oliganthum o-lig-*ănth*-um. Few-
flowered. Crete.
forrestii fo-*rest*-ee-ee. After Forrest who
introduced it, see *Abies delavayi forrestii*.
Himalaya, China.
× *inodorum* on-o-*dō*-rum. *H. androsaemum*
× *H. hircinum*. Not scented. *H. hircinum*
smells of goats as the name suggests.
× *moserianum* mŏ-za-ree-*ah*-num. *H.*
calycinum × *H. patulum*. After Moser's
nursery, Versailles where it was raised c.
1887.
olympicum o-*lim*-pi-kum. (= *H.*
polyphyllum hort.). Of Mt Olympus,
applied to several mountains in SE Europe
and W Asia. SE Europe, Turkey.
polyphyllum hort. see *H. olympicum*
reptans rep-tănz. Creeping. Himalaya.
xylosteifolium zi-los-tee-i-*fo*-lee-um. With
leaves like *Lonicera xylosteum*. NW Turkey,
SW Georgia.

Hypoestes hi-pō-*es*-teez *Acanthaceae*. From
Gk. *hypo* (below) and *estia* (a house), the
bracts cover the calyx. Tender herbs.
Madagascar.
phyllostachya fil-lō-*stăk*-ee-a. (= *H.*
sanguinolenta hort.). With leafy spikes.
Polka-dot Plant, Baby's Tears, Freckle
Face.
taeniata tie-nee-*ah*-ta. From *Taenia* a genus
of tapeworms, referring to the worm-like
flowers.

Hypoxis hi-*poks*-is *Hypoxidaceae*. Gk. name
for a plant. Perennial herbs.
hirsuta hir-*soo*-ta. Hairy (the leaves). N
America.
hygrometrica hi-gro-*met*-ri-ka. Measuring
moisture, the flowers close in cloudy
weather. Australia.

Hypsela hip-*say*-la *Campanulaceae*. From Gk. *hypselos* (high), the following grows at high altitudes. Perennial herb.
  reniformis ray-ni-*form*-is. Kidney-shaped (the leaves). S America.

Hyssopus hi-*sŏp*-us *Labiatae*. Old name for another plant. Evergreen shrub.
  officinalis o-fi-ki-*nah*-lis. Sold as a herb. Hyssop. S Europe, W Asia.

I

Iberis i-*be*-ris *Cruciferae*. Gk. *iberis*, from Iberia. Annual herbs and sub-shrubs. Candytuft.
  amara a-*mah*-ra. Bitter. Rocket Candytuft. Europe.
  'Correifolia' ko-ree-i-*fo*-lee-a. With leaves like *Correa*.
  gibraltarica ji-brawl-*tah*-ri-ca. Of Gibraltar. Gibraltar Candytuft. S Spain, Morocco.
  saxatilis sǎks-*ah*-ti-lis. Growing on rocks. S Europe.
  sempervirens sem-per-*vi*-renz. Evergreen. S Europe, W Asia.
  umbellata um-bel-*ah*-ta. With flowers in umbels. Common Candytuft. Mediterranean region.

Ice Plant see *Mesembryanthemum crystallinum*.

Idesia i-*deez*-ee-a *Flacourtiaceae*. After E. I. Ides a Dutch traveller who visited China in the 18th century. Deciduous tree.
  polycarpa po-lee-*kar*-pa. With many fruits. Japan, China.

Ilex ee-leks *Aquifoliaceae*. The L. name for *Quercus ilex*. Evergreen trees and shrubs. Holly.
  × *altaclerensis* ǎl-ta-kle-*ren*-sis. *I. aquifolium* × *I. perado*. Of Highclere (Alta Clera), Berks. where it is thought to have been first raised.
    'Camelliifolia' ka-mel-ee-i-*fo*-lee-a. With leaves like *Camellia*.
    'Hendersonii' hen-der-*son*-ee-ee. After Mr Henderson, a friend of Mr Shepherd who supplied material of the Handsworth nursery of Fisher and Holmes.
    'Hodginsii' ho-*jinz*-ee-ee. After Edward Hodgins, the raiser.
    'Lawsoniana' law-son-ee-*ah*-na. After

Ilex (continued)
    the Lawson nursery of Edinburgh who distributed it.
  aquifolium ǎ-kwi-*fo*-lee-um. The L. name. Common Holly. Europe. W Asia.
  cornuta kor-*new*-ta. Horned, referring to the leaf spines. Horned Holly. China, Korea.
  crenata kray-*nah*-ta. With shallow, rounded teeth (the leaves). Japan, Korea.
    'Mariesii' ma-Ireez-ee-ee. After Maries who introduced it in about 1979, see *Davillia mariesii*.
  × *merseveae* mer-*serv*-ee-ie. *I. aquifolium* × *I. rugosa*. After Mrs Kathleen Meserve who raised it. Blue Holly.
  pernyi per-nee-ee. After the Abbé Paul Hubert Perny (1818–1907), French missionary in China who discovered it in 1858. C and W China.

Illicium il-*lik*-ee-um *Illiciaceae*. From L. *illicio* (to attract) referring to the fragrance. Evergreen trees and shrubs.
  anisatum ǎn-i-*sah*-tum. Anise-scented (the leaves). China, Japan.
  floridanum flor-ri-*dah*-num. Of Florida. S United States.
  henryi hen-ree-ee. After its discoverer, Augustine Henry (1857–1930), an Irish doctor who collected in China. W China.

Immortelle see *Xeranthemum annuum*

Impatiens im-*pǎt*-ee-enz *Balsaminaceae*. L. for impatient, referring to the explosive release of the seed when a ripe capsule is touched. Hardy and tender, annual and perennial herbs.
  balfourii bǎl-*for*-ree-ee. After Sir Isaac Bayley Balfour (1853–1922), Regius Keeper of the Edinburgh Botanic Garden. W Himalaya.
  balsamina bǎl;-sa-*meen*-a. Bearing balsam. SE Asia.
  capensis kap-*pen*-sis. (= *I. biflora*). Originally thought to be from the Cape of Good Hope. Jewel Weed. N America.
  glandulifera glǎnd-ew-*li*-fe-ra. (= *I. roylei*). Glandular. Himalayan Balsam, Policeman's Helmet. W Himalaya.
  hawkeri hawk-a-ree. (= *I. petersiana* hort.). After Lieutenant Hawker who sent plants from the South Sea Islands. New Guinea.
  noli-tangere no-lee-tang-*ge*-ree. Do not touch (the pods). Touch-me-not. Europe, Asia.
  walleriana wo-la-ree-*ah*-na. (= *I. holstii. I. sultanii*). After the Rev. Horace Waller

Impatiens (continued)
(1833–96), a missionary in C Africa. Busy
Lizzie. E Africa.

Incarvillea in-kar-*vil*-ee-a *Bignoniaceae*. After
Pierre d'Incarville (1706–57), French
missionary and plant collector in China.
Perennial herbs.
  delavayi del-a-*vay*-ee. After Delavay, see
Abies delavayi. China.
  mairei mair-ree-ee. After Edouard Ernest
Maire (vorn 1848), French missionary in
China. China.
    grandiflora gră̆nd-i-*flŏ*-ra. (= *I.
grandiflora*). Large-flowered. Himalaya,
W China.
  olgae ol-gie. After Olga Fedtschenko
(1845–1921). C Asia.

Incense Cedar see *Calocedrus decurrens*
Inch Plant see *Callisia, Tradescantia albiflora*
  Flowering see *Tradescentia cerinthoides*
  Striped see *Callisia elegans*
Indian Bean Tree see *Catalpa bignonioides*
Indian Currant see *Symphoricarpus rivularis*
Indian Hawthorn see *Rhaphiolepis indica*
Indian Physic see *Gillenia trifoliata*
Indian Shot see *Canna indica*
Indian Turnip see *Arisaema triphyllum*
Indigo, False see *Amorphia fruticosa, Baptisia
australis*
  Wild see *Baptisia tinctoria*

Indigofera in-di-*go*-fe-ra *Leguminosae
(Papilionoideae)*. From indigo and L. *fero* (to
bear, indigo is obtained from *I. tinctoria*.
Deciduous shrubs.
  decora de-*kŏ*-ra. Beautiful. China, Japan.
  heterantha he-te-ră̆nth-a. (= *I. gerardiana*).
With different flowers. NW Himalaya.

Indocalamus in-do-*kal*-a-mus *Graminae*.
From *indo*, Indian, and Gk. *kalamos*, a reed.
Small bamboo closely related to *Sasa*, in
which this species was once included.
China, Japan, Malaysia.
  tessellata te-se-*lah*-ta (= *Sasa tessellata*).
Tesselated, i.e. with minute squared
venation (the leaves). China.

Inula in-ew-la *Compositae*. The L. name for
I. helenium. Perennial herbs.
  acaulis a-*kaw*-lis. Stemless. W Asia.
  ensifolia ayns-i-*fo*-lee-a. With sword-
shaped leaves. Europe.
  helenium he-*len*-ee-um. From the
resemblance to *Helenium*. Elecampane. C
Asia.
  hookeri huk-a-ree. After Sir Joseph Dalton

Inula (continued)
Hooker (1817–1911). Himalaya, Burma,
W China.
  magnifica mahg-*ni*-fi-ka. Splendid.
Caucasus.
  oculus-christi ok-ew-lus-*kris*-tee. Literally,
the eye of Christ. E Europe.
  orientalis o-ree-en-*tah*-lis. Eastern.
Caucasus.
  royleana royl-ee-*ah*-na. After John Forbes
Royle (1798–1858) who collected in
India and the Himalaya. Himalaya.

Iochroma ee-ŏ-*krŏ*-ma *Solanaceae*. From Gk.
ion (violet) and *chroma* (colour) referring to
the colour of the flowers. Tender shrubs.
  coccineum kok-*kin*-ee-um. Scarlet (the
flowers). C America.
  cyaneum see-*ă̆n*-ee-um. (= *I. tubulosum*).
Blue (the flowers). S America.
  grandiflora gră̆nd-i-*flŏ*-ra. Large-flowered.
Ecuador, Peru.

Ionopsidium ee-on-op-*sid*-ee-um *Cruciferae*.
From Gk. *ion* (violet) and *-opsis* indicating
resemblance, from the resemblance to a
violet. Annual herb.
  acaule a-*kaw*-lee. Stemless. Violet Cress.
Portugal.

Ipheion i-*fay*-on *Liliaceae (Alliaceae)*.
Derivation obscure. Bulbous herb.
  uniflorum ew-ni-*flo*-rum. (= *Brodiaea
uniflora. Triteleia uniflora*). One flowered.
Argentina, Uruguay.

Ipomoea i-pom-*oy*-a. *Convolvulaceae* From
Gk. *ips* (a worm) and *homoios* (resembling).
Tender annual and perennial climbers.
Morning Glory.
  acuminata see *I. indica*
  alba ă̆l-ba. White (the flowers). Tropical
America.
  batatas ba-*tah*-tas. The Haitian name.
Sweet Potato Vine. S America.
  bonariensis bo-nah-ree-*en*-sis. Of Buenos
Aires. S America.
  coccinea kok-*kin*-ee-a. (= *Quamoclit
coccinea*). Scarlet (the flowers). E United
States.
  hederacea he-de-*rah*-a. Like *Hedra* (the
leaves). Tropical America.
  horsfalliae hors-*fă̆l*-ee-ie. After Mrs Charles
Horsfall who painted for the Botanical
Magazine. W Indies.
  indica in-di-ka (= *I. acuminata, I. learii*).
From India, the Orient. Blue Dawn
flower. Pantropical.
  learii see *I. indica*

Ipomoea (continued)
lobata lo-bah-ta (= *Mina lobata*) Lobed (the leaves). Mexico to S America.
nil nil. Violet (the flowers). Tropics.
pes-caprae pays-*kǎp*-rie. Like a goat's-foot (the leaves). Tropics.
purpurea pur-*pewr*-ree-a. Purple (the flowers). Tropical America.
quamoclit kwah-mo-klit. (= *Quamoclit pennata*). The Mexican name. Cypress Vine. Tropical America.
rubrocaerulea see *I. tricolor*
tricolor tri-ko-lor. (= *I. rubrocaerulea. I. violacea* hort.). Three-coloured (the flowers). Tropical America.
violacea hort. see *I. tricolor*

Ipomopsis i-pom-*op*-sis *Polemoniaceae*. From Gk. *ips* (a worm) and *-opsis* indicating resemblance. Annual herb.
rubra rub-ra. (= *Gilia rubra*). Red (the flowers). Standing Cypress. S United States.

Iresine ee-res-*ee*-nay *Amaranthaceae*. From Gk. *eiresione*, a branch wound round with wool. Tender perennials.
herbstii hairbst-ee-ee. After Hermann Carl Gottlieb Herbst (c. 1830–1904), Director of the Rio de Janiero Botanic Gardens, later a nurseryman of Richmond, Surrey. S America.
lindenii lin-*den*-ee-ee. After Linden, Belgian nurseryman. Ecuador.

Iris ee-ris *Iridaceae*. After the Gk. goddess of the rainbow. Rhizomatous or bulbous perennials.
bakeriana bay-ka-ree-*ah*-na. (= *I. bakeriana*). After J. G. Baker (1834–1920), Kew botanist and authority on bulbs. Iran.
bucharica bew-*kah*-ri-ka. Of Bokhara, Turkestan. C Asia.
bulleyana bul-ee-*ah*-na. After Arthur Kilpin Bulley (1861–1942), founder of Bees nursery, who employed Forrest and Kingdon-Ward as plant collectors. W China.
chamaeiris see *I. lutescens*
chrysographes kris-*o*-gra-feez. With gold lines (on the falls). W China.
clarkei *klark*-ee-ee. After Charles Baron Clarke (1832–1906) who collected in India. Himalaya.
cristata kris-*tah*-ta. Crested (the falls). E United States.
danfordiae dǎn-*ford*-ee-ie. After Mrs C. G. Danford who collected bulbs in W Asia, 1876–9. C Turkey.

Iris (continued)
douglasiana dūg-las-ee-*ah*-na. After David Douglas (1798–1834), who introduced many plants from WN America. W N America.
ensata ayn-*sah*-ta. Sword-like (the leaves). C to E Asia.
    'Kaempferi' *kemp*-fa-ree. After Englebert Kaempfer (1651–1716). Cult.
florentina see *I. germanica florentina*
foetidissima foy-ti-*dis*-i-ma. Very foetid. Stinking Iris, Gladdon. S and W Europe, N Africa.
forrestii fo-*rest*-ee-ee. After Forrest, see *Abies delavayi forrestii*. SW China.
fulva *ful*-va. Tawny (the flowers). Red Iris. E N America.
gatesii *gayts*-ee-ee. After the Rev. F. S. Gates of the American Mission in Armenia, c. 1888. W Asia.
germanica ger-*mahn*-i-ka. Of Germany. Cult.
    florentina flō-ren-*teen*-a. (= *I. florentina*). Of Florence. Orris. Cult.
gracilipes gra-*kil-i*-pays. Slender-stalked. Japan.
graebneriana grayb-na-ree-*ah*-na. After Karl Graebner (1871–1933). Turkestan.
graminea grah-*min*-ee-a. Grass-like (the leaves). C and S Europe, Caucasus.
histrio *his*-tree-ō. L. *histrio* (an actor) from the colouring of the flowers. W Asia.
    aintabensis ien-ta-*ben*-sis. Of Gaziantep (Aintab), S Turkey.
histrioides his-tree-*oi*-deez. Like *I. histrio*. C Turkey.
    'Major' *mah*-yor. Larger.
hoogiana hoog-ee-*ah*-na. After John Hoog of the ban Tubergen nursery. Turkestan.
innominata in-nom-i-*nah*-ta. Nameless. W United States.
japonica ja-*pon*-i-ka. Of Japan. Japan, China.
laevigata lie-vi-*gah*-ta. Smooth (the leaves). E Asia.
latifolia lah-tee-*fo*-lee-a. (= *I. xiphioides*). Broad-leaved. English Iris. Pyrenees, N Spain.
lutescens loo-*tes*-enz. (= *I. chamaeiris*). Yellowish (the flowers of some forms). SW Europe.
missouriensis mi-sur-ree-*en*-sis. Of the source of the Missouri River. C and W N America.
ochroleuca see *I. spuria ochroleuca*
orchioides or-kee-*oi*-deez. Orchid-like. C Asia.

Iris (continued)
orientalis o-ree-en-*tah*-lis (= *I. spuria* var.
ochroleuca). Eastern. Greece, W Asia.
pallida pá-li-da. Pale (the flowers).
Yugoslavia.
pseudacorus sood-*á*-ko-rus. False *Acorus*.
Yellow Flag. S and W Europe, N Africa.
pumila pew-mi-la. Dwarf. E Europe, W
Asia.
reticulata ray-tik-ew-*lah*-ta. Netted (the
bulb). W Asia.
ruthenica roo-*then*-i-ka. Of Ruthenia, SW
Russia. E Europe to China.
setosa say-*tō*-sa. Bristly (the apex of the
standards). NE Asia, Alaska.
sibirica si-*bi*-ri-ka. Siberian. C Europe to
Russia.
spuria spewr-ree-a. False. Europe, N Africa,
W Asia.
ochroleuca see *I. orientalis*
stolonifera sto-lōn-*i*-fe-ra. Bearing stolons.
C Asia.
stylosa see *I. unguicularis*
susiana soo-see-*ah*-na. Of Shush (Susa),
Iran. W Asia.
tectorum tek-*to*-rum. Growing on roofs.
Roof Iris, Wall Iris. China.
tenax ten-ahks. Tough (the leaves). W
United States.
tingitana ting-gi-*tah*-na. Of Tangier
(Tingi). Morocco.
tuberosa see *Hermodactylus tuberosus*
unguicularis un-gwik-ew-*lah*-ris. (= *I.
stylosa*). Clawed. SE Europe, N Africa, W
Asia.
verna ver-na. Of Spring. E United States.
versicolor ver-*si*-ko-lor. Variously coloured.
EN America.
xiphioides see *I. latifolia*
xiphium zi-fee-um. Gk. name for *Gladiolus*
from *xiphos* (a sword) referring to the
leaves. SW Europe, N Africa.

Ironweed see *Vernonia*
Iron Wood see *Ostrya virginiana*

Isatis ee-sa-tis *Cruciferae*. The Gk. name.
Biennial herb.
tinctoria tink-*to*-ree-a. Used in dyeing.
Woad. C and S Europe.

Itea ee-tee-a *Grossulariaceae* (*Iteaceae*) The Gk.
name for willow. Deciduous and evergreen
shrubs.
ilicifolia ee-lik-i-*fo*-lee-a. *Ilex*-leaved. W
China.
virginica vir-*jin*-i-ka. Of Virginia. Virginia
Willow. E United States.

Ivy see *Hedera*
Common see *H. helix*
Irish see *H. hibernica*
Poet's see *H. helix poetica*
Ivy-leaved Toadflax see *Cymbalaria muralis*

Ixia iks-ee-a *Iridaceae*. From Gk. *ixia* (bird
lime) referring to the sticky sap. Semi-hardy,
cormous perennial. Corn Lily.
viridiflora vi-ri-di-*flō*-ra. With green
flowers. S Africa.

Ixora iks-*o*-ra *Rubiaceae*. Portuguese version
of *Iswara* a Malabar deity to whom the
flowers were offered. Tender, evergeeen
shrubs.
chinensis chin-*en*-sis. Of China. SE Asia.
coccinea kok-*kin*-ee-a. Scarlet (the flowers).
Flame of the Woods. India.

J

Jacaranda jăk-a-*rănd*-a *Bignoniaceae*. The
Portuguese name, derived from the
Brazilian Indian name. Tender, semi-
evergreen tree.
mimosifolia mee-mo-si-*fo*-lee-a. (= *J.
ovalifolia. J. acutifolia* hort.). Mimosa-leaved.
NW Argentina.

Jack-in-the-Pulpit see *Arisaema triphyllum*
Jacobinia carnea see *Justicea carnea*
coccinea see *Pachystachys coccinea*
ghiesbreghtiana see *Justicea ghiesbreghtiana*
pauciflora see *Justicea rizzinii*
pohliana see *Justicea carnea*
Jacob's Coat see *Acalypha wilkesiana*
Jacob's Ladder see *Pedilanthes tithymaliodes
smalli, Polemonium caeruleum*
Jade Plant see *Crassula ovata*
Japanese Foam Flower see *Tanakaea
radicans*
Japan Pepper see *Zanthoxylum piperitum*

Jasione yă-see-*ō*-nay *Campanulaceae*. Gk.
name for another plant. Perennial herbs.
Sheep's Bit.
crispa kris-pa. Crisped (the stem hairs).
Pyrenees.
laevis lie-vis. (= *J. perennis*). Smooth. W
Europe.
montana mon-*tah*-na. (= *J. jankae*). Of
mountains. Europe.

Jasminum yăs-*meen*-um *Oleaceae*. From
yasmin the Persian name. Deciduous and

Jasminum (continued)
evergreen shrubs and climbers. Jasmine,
Jessamine.

 beesianum beez-ee-*ah*-num. After Bees
 nursery for whom Forrest introduced it in
 1906. China.
 humile hum-i-lee. Low growing. Himalaya,
 China.
  'Revolutum' re-vo-*loo*-tum. Rolled
  back (the leaf margins).
 mesnyi mez-nee-ee. After William Mesny
 (1842–1919), a Major-General in the
 Chinese National Army who collected in
 China. Primrose Jasmine. W China.
 nudiflorum new-di-*flō*-rum. Flowering
 naked, it flowers in winter when it is
 leafless. Winter Jasmine. China.
 officinale o-fik-i-*nah*-lee. Sold as a herb.
 Common Jasmine. Caucusus, Himalaya,
 China.
 parkeri park-a-ree. After R. N. Parker who
 discovered it in 1919 and later introduced
 it. Indian Himalaya.
 polyanthum po-lee-*ănth*-um. Many-
 flowered. China.
 × *stephanense* stef-an-*en*-see. *J. beesianum*
 × *J. officinale*. Of Saint Etienne, France
 where it was raised.

Jeffersonia jef-er-*sonee*-a *Berberidaceae*. After
Thomas Jefferson (1743–1826), 3rd
President of the United States and botanical
patron. Perennial herbs.
 diphylla di-*fil*-la. Two-leaved. E N
 America.
 dubia dub-ee-a. Doubtful. Vladivostok,
 Korea.

Jelly Beans see *Sedum pachyphyllum*
Jerusalem Cherry see *Solanum*
pseudocapsium
 False see *S. capsicastrum*
Jerusalem Cross see *Lychnis chalcedonica*
Jerusalem Sage see *Phlomis fruticosa*
Jerusalem Thorn see *Parkinsonia aculeata*
Jessamine see *Jasminum*
Jewel Weed see *Impatiens capensis*
Job's Tears see *Coix lacryma-jobi*
Joe-pye Weed see *Eupatorium maculatum, E.
purpureum*
Joseph's Coat see *Amaranthus tricolor*

Jovellana ho-vel-*lah*-na *Scrophulariaceae*. After
Gaspar Melchor de Jovellanos, 18th-century
patron of botany in Peru. Semi-hardy shrub.
 violaceae vee-o-*lah*-kee-a. Violet (the
 corolla). Chile.

Jovibarba yov-i-*bar*-ba *Crassulaceae*. From L.

Jovibarba (continued)
Jovis (Jupiter) and *barba* (a beard) presumably
referring to the fringed petals which
distinguishes the genus from *Sempervivum*.
Succulent herbs.
 sobolifera so-bo-*li*-fe-ra. (= *Sempervivum
 soboliferum*). Bearing offspring. Hen and
 Chickens Houseleek. Alps.

Joy Weed see *Alternanthera*
Judas Tree see *Ceris siliquastrum*

Juglans yoo-glahnz *Juglandaceae*. The L. name
for *J. regia* from *jovis* (of Jupiter) and *glans* (an
acorn or nut). Deciduous trees.
 ailantifolia ie-lăn-ti-*fo*-lee-a. *Ailanthus*-
 leaved. Japanese Walnut. Japan.
 cinerea kin-*e*-ree-a. Grey (the bark).
 Butternut. E N America.
 microcarpa mik-rō-*kar*-pa. With small fruit.
 Texan Walnut. SW United States, N
 Mexico.
 nigra nig-ra. Black (the bark). Black
 Walnut. E and C United States.
 regia ray-gee-a. Royal. Common Walnut.
 S Europe to China.

Juncus yung-kus *Juncaceae*. L. name for rush.
Perennial herb.
 effusus e-*few*-sus. Loose. Widely
 distributed. Common Rush.
  'Spiralis' spee-*rah*-lis. Spiral (the stems).
  Corkscrew Rush.

Juniperus you-*ni*-pe-rus *Cupressaceae*. The L.
name. Evergreen conifers. Juniper.
 chinensis chin-*en*-sis. Of China. China,
 Mongolia, Japan.
 communis kom-*ew*-nis. Common.
 Common Juniper. N temperate regions.
 conferta kōn-*fer*-ta. Crowded (the leaves).
 Japan.
 drupacea droo-*pah*-kee-a. With fleshy fruit.
 SE Europe, W Asia.
 horizontalis ho-ri-zon-*tah*-lis. Horizontal,
 the prostrate habit. N America.
  'Douglasii' dūg-*lăs*-ee-ee. After the
  Douglas nursery, Waukegon, Illinois.
  Waukegon Juniper.
  'Glauca' *glow*-ka. Glaucous (the foliage).
  'Plumosa' ploo-*mō*-sa. Feathery (the
  foliage).
  'Wiltonii' wil-*ton*-ee-ee. After South
  Wilton Nurseries, Wilton,
  Connecticut.
 × *media* me-dee-a. *J. chinensis* × *J. sabina*.
 Intermediate (between the parents).
  'Pfitzeriana' fits-a-ree-*ah*-na. After W.
  Pfitzer, a Stuttgart nurseryman.

Juniperus (continued)
procumbens prō-*kum*-benz. Prostrate. Japan.
recurva re-*kur*-va. Curved downwards (the shoots). Himalayan Juniper. Himalaya, W China.
  coxii koks-ee-ee. After E. H. M. Cox who, with Farrer, introduced it in 1920.
rigida ri-gi-da. Rigid (the leaves). Japan.
sabina sa-*been*-a. The L. name. Savin. C and S Europe, W Russia.
  'Tamariscifolia' tăm-a-risk-i-*fo*-lee-a *Tamarix*-leaved.
sargentii sar-*jent*-ee-ee. After Sargent who discovered and introduced it, see *Prunus sargentii*.
scopulorum skop-ew-*lo*-rum. Growing on cliffs. Rocky Mountain Juniper. W United States.
squamata skwah-*mah*-ta. Scaly (the bark). Himalaya, China.
  'Meyeri' may-a-ree. After F. N. Meyer who introduced it to the United States in 1910.
virginiana vir-jin-ee-*ah*-na. Of Virginia. Eastern Red Cedar. E and CN America.

Jupiter's Beard see *Anthyllis barba-jovis*

Justicea jūs-*tis*-ee-a *Acanthaceae*. After James Justice (1698–1763) Scottish botanist. Tender, evergreen shrubs.
brandegeana brănd-ee-gee-*ah*-na. (= *Beloperone guttata*). After Brandegee. Shrimp Plant. Mexico.
carnea kar-nee-a. (= *Jacobinia carnea*. *Jacobinia pohliana*). Flesh-coloured (the flowers). Brazilian Plume. S America.
ghiesbreghtiana geez-brekt-ee-*ah*-na. (*Jacobinia ghiesbreghtiana*). After Augustine Boniface Ghiesbreght (1810–93). Mexico.
rizzinii riz-*in*-ee-ee. (= *Jacobinia pauciflora*). After Carlos Toledo Rizzini. Brazil.

K

Kaffir Lily see *Clivia miniata, Schizostylis coccinea*
Kahili Ginger see *Hedychium gardnerianum*
Kaki see *Diopyros kaki*

Kalanchoe ka-*lăn*-kō-ee *Crassulaceae*. From the Chinese name of one species. Tender succulents.
beharensis bee-hah-*ren*-sis. Of Behara, Madagascar. Felt Brush. Velvet Leaf.

Kalanchoe (continued)
bentii bent-ee-ee. (= *K. teretifolia*). After Theodore Bent who collected in S Arabia about 1894. Arabia.
blossfeldiana blos-feld-ee-*ah*-na. After Robert Blossfeld, German nurseryman. Madagascar.
crenata kray-*nah*-ta. (= *Bryophyllum crenatum*). With shallow, rounded teeth (the leaves). Malaysia.
daigremontiana day-gre-mont-ee-*ah*-na. (= *Bryophyllum daigremontianum*). After Mme. and M. Daigremont. Devil's Backbone, Mother of Thousands. Madagascar.
delagonensis de-le-*gon*-en-sis = (*K. tubiflora*). Of Delagoa (Mozambique). Chandelier Plant. Madagascar, S Africa.
fedtschenkoi fet-*shenk*-ō-ee. After Boris Fedtschenko (1872–1933), Russian traveller. Madagascar.
flammea flăm-ee-a. Flame-coloured (the flowers). Somalia.
× *kewensis* kew-*en*-sis. K. *bentii* × K. *flammea*. Of Kew.
marmorata mar-mo-*rah*-ta. Marbled (the leaves). Pen Wiper. NE Africa.
millotii mi-*lot*-ee-ee. After Prof. Millot of the Natural History Museum, Paris. Madagascar.
pinnata pin-*ah*-ta. (= *Bryophyllum pinnatum*). Pinnate (the leaves). Air Plant. Origin uncertain.
pumila pew-mi-la. Dwarf. Madagascar.
teretifolia see *K. bentii*
tomentosa tō-men-*tō*-sa. Hairy. Panda Plant, Pussy Ears. Madagascar.
tubiflora see *K. delagonensis*
uniflora ew-ni-*flō*-ra. (= *Bryophyllum uniflorum*). One-flowered. Madagascar.

Kale, Ornamental see *Brassica oleracea Acephala*

Kalmia kăl-mee-a *Ericaceae*. After Pehr Kalm (1715–79), Finnish student of Linnaeus. Evergreen shrubs.
angustifolia ang-gust-i-*fo*-lee-a. Narrow-leaved. Sheep Laurel. EN America.
latifolia lah-tee-*fo*-lee-a. Broad-leaved. Calico Bush, Mountain Laurel. EN America.
polifolia pol-i-*fo*-lee-a. With leaves like *Teucrium polium*. N America.

Kalmiopsis kăl-mee-*op*-sis *Ericaceae*. From *Kalmia* q.v. and Gk. *-opsis* indicating resemblance. Evergreen shrub.
leachiana leech-ee-*ah*-na. After Mr and

Kalmiopsis (continued)
Mrs Leach who discovered it in 1930.
Oregon.

Kalopanax kăl-ō-păn-ăks *Araliaceae*. From
Gk. *kalos* (beautiful) and *Panax* a related
genus. see *Acanthopanax*. Deciduous tree.
  septemlobus sep-*tem*-lo-bus. (= *K. pictus*).
  Seven-lobed (the leaves). NE Asia.

Kangaroo Apple see *Solanum aviculare*
Kangaroo Vine see *Cissus antarctica*
Kangaroo's Paw see *Anigozanthus*
Katsura Tree see *Cercidiphyllum japonicum*
Kauri Pine see *Agathis australis*
Kentia see *Howea*
Kentucky Coffee Tree see *Gynmocladus
dioica*

Kerria ke-ree-a *Rosaceae*. After William Kerr
(died 1814), Kew gardner who introduced
K. japonica 'Pleniflora'. Deciduous shrub.
  japonica ja-*pon*-i-ka. Of Japan, where it is
  grown. China.
    'Pleniflora' play-ni-*flō*-ra. With double
    flowers.

King Cup see *Caltha palustris*
King of Bromeliads see *Vriesia hieroglyphica*
King of the Alps see *Eritrichium nanum*
King Plant see *Anoectochilus regalis*
King William Pine see *Athrotaxis
selaginoides*
Kingfisher Daisy see *Felicia bergeriana*
King's Spear see *Asphodeline lutea*

Kirengeshoma ki-reng-ge-*shō*-ma
Saxifragaceae. The Japanese name.
Herbaceous perennial.
  palmata pahl-*mah*-ta. Lobed like a hand
  (the leaves). Japan.

Kiwi Fruit see *Actinidia deliciosa*

Kleinia klie-nee-a *Compositae*, After Jacob
Theodore Klein (1685–1751), German
zoologist. Succulent perennials.
  fulgens ful-genz (= *senecio fulgens*). Shining.
  Tropical and South Africa.
  galpinii găl-*pin*-ee-ee (= *Senecio galpinii*).
  After Galpin. See *Kniphofia galpinii*.
  Transvaal.
  haworthii see *Senecio haworthii*
  neriifolia ner-ee-ee-*fō*-lee-a (= *Senecio
  neriifolius*). Oleander (*Nerium*) leaved.
  Canary Islands.
  repens see *Senecio serpens*
  tomentosa see *Senecio haworthii*

Kniphofia nee-*fof*-ee-a *Liliaceae*. After Johann
Hieronymus Kniphof (1704–63). Perennial
herbs. Red Hot Poker, Torch Lily.
  caulescens kaw-*les*-enz. With a stem. S
  Africa.
  'Erecta' e-*rek*-ta. Erect (the flowers).
  foliosa fo-lee-*ō*-sa. Leafy. Ethiopia.
  galpinii găl-*pin*-ee-ee. After Ernst E.
  Galpin, a plant collector of Barberton, SE
  Transvaal. S Africa.
  nelsonii see *K. triangularis*
  northiae north-ee-ie. After Miss Marianne
  North (1830–90), botanical artist. S
  Africa.
  triangularis tri-an-gew-*lah*-sis (= *K.
  nelsonii*). With three angles. S Africa.
  uvaria oo-*vah*-ree-a. Like a bunch of
  grapes. S Africa.

Kochia see *Bassia*

Koelreuteria kurl-roy-*te*-ree-a *Sapindaceae*.
After Joseph Gottleib Koelreuter
(1733–1806), German professor of botany.
Deciduous tree.
  paniculata pan-nik-ew-*lah*-ta. With flowers
  in panicles. China.

Kohleria kō-*le*-ree-a *Gesneriaceae*. After
Michael Kohler, 19th-century natural
history teacher in Zurich. Tender perennial
herbs and shrubs.
  amabilis a-*mah*-bi-lis. Beautiful. Colombia.
  bogotensis bo-go-*ten*-sis. Of Bogota.
  Colombia.
  digitaliflora di-gi-*tah*-li-flō-ra. With flowers
  like *Digitalis*. Colombia.
  eriantha e-ree-*ănth*-a. With woolly flowers.
  Colombia.
  lindeniana lin-den-ee-*ah*-na. After Linden,
  a Belgian nurseryman. Ecuador.
  tubiflora tew-bi-*flō*-ra. With tubular
  flowers. Columbia.

Kohl Rabi see *Brassica oleracea* Gongylodes

Kolkwitzia kol-*kwitz*-ee-a *Caprifoliaceae*.
After Richard Kolkwitz (born 1873),
German professor of botany. Deciduous
shrub.
  amabilis a-*mah*-bi-lis. Beautiful. Beauty
  Bush. China.

Kowhai see *Sophora tetraptera*
Kris Plant see *Homalomena lindenii*
Kumquat see *Fortunella japonica*

L

Labrador Tea see *Ledum groenlandicum*

+ *Laburnocytisus* la-burn-ō-*sit*-i-sus
Leguminosae (Papilionoideae) Graft hybrid, from
the names of the parents *Cytisus* × *Laburnum*.
Deciduous tree.
   adamii a-*dăm*-ee-ee. *Cytisus purpureus* +
   Laburnum anagyroides. After its raiser, Jean
   Louis Adam.

Laburnum la-*burn*-um *Leguminosae*
(Papilionoideae) The L. name. Deciduous
tree.
   alpinum ăl-*peen*-um. Alpine. Scotch
   Laburnum. Alps, Italy, Czechoslovakia,
   Yugoslavia.
   anagyroides ăn-a-gi-*roi*-deez. Like *Anagyris*.
   Common Laburnum. C and S Europe.
   × *watereri* waw-ta-ra-ree. *L. alpinum* × *L.*
   anagyroides. After the Waterer nursery
   where one form was raised.

Lace Flower see *Alsobia dianthiflora*
Lace Trumpets see *Sarracenia leucophylla*

Lachenalia lah-shen-*ahl*-ee-a *Liliaceae*
(Hyacinthaceae). After Werner de La Chenal
(1736–1800), professor of botany at Basle.
Tender, bulbous herbs. Cape Cowslip. S
Africa.
aloides ă-lō-*ee*-deez. Like *Aloe*.
   'Nelsonii' nel-*son*-ee-ee. After the Rev.
   John Gudgeon Nelson (1818–82) who
   raised it.
bulbifera bul-*bi*-fe-ra. Bulb-bearing.
mutabilis mew-*tah*-bi-lis. Changeable (the
   flower colour). Fairy Lachenalia.
orchioides or-kee-*oi*-deez. Orchid-like.
   glaucina glow-*keen*-a. Glaucous. Opal
   Lachenalia.

Lactuca lăk-*too*-ka *Compositae*. From L. *lac*
(milk) referring to the white sap. Annual
herb.
alpina see *Cicerbita alpina*
plumieri see *Cicerbita plumieri*
sativa sa-*tee*-va. Cultivated. Lettuce. Cult.

Ladies' Fingers see *Abelmoschus esculentus*
Lady of the Night see *Brassavola nodosa,*
Brunfelsia americana
Lady's Mantle see *Alchemilla*
   Alpine see *A. alpina*
Lady's Smock see *Cardamine pratensis*

Laelia lie-lee-a *Orchidaceae*. After Laelia, one
of the vestal virgins. Greenhouse orchids.

Laelia (continued)
   anceps ăn-keps. Two-edged (the pseudobulbs). Mexico.
   autumnalis ow-tum-*nah*-lis. Of autumn
   (flowering). Mexico.
   cinnabarina kin-a-ba-*reen*– a. Cinnabar-red.
   Brazil.
   grandis grănd-is. Large (the flowers).
   Brazil.
   harpophylla harp-ō-*fil*-la. With sickle-
   shaped leaves.
   pumila pew-mi-la. Dwarf. Brazil.
   purpurata pur-pew-*rah*-ta. Purple (the
   flowers). Brazil.
   speciosa spek-ee-ō-sa. Showy. Mexico.
   tenebrosa ten-e-*brō*-sa. Growing in shady
   places.
   xanthina zănth-*ee*-na. Yellow (the flowers).
   Brazil.

× *Laeliocattleya* lie-lee-ō-*kăt*-lee-a
Orchidaceae. Intergeneric hybrids, from the
names of the parents. *Cattleya* × *Laelia*.
Greenhouse orchids.

Lagarosiphon lă-ga-rō-*see*-fon
Hydrocharitaceae. From Gk. *lagaros* (narrow)
and *siphon* (a tube) referring to the narrow
perianth tube. Aquatic herb.
   major mah-yor (= *Elodea crispa*). Larger. S
   Africa.

Lagarostrobus lag-ah-rō-*strō*-bus
Podocarpaceae. From Gk. *lagaros*, lax, and
strobilos, cone; the female cones are loosely
constructed. Evergreen conifers. New
Zealand and Tasmania.
   franklinii frank-*lin*-ee-ee (= *Dacrydium*
   franklinii). After Sir John Franklin, Arctic
   explorer (1786–1847). Huon Pine.
   Tasmania.

Lagenaria lă-gen-*ah*-ree-a *Cucurbitaceae*.
From Gk. *lagenos* (a flask) referring to the
shape and use of the fruits. Annual herb.
   siceraria see-ke-*rah*-ree-a. From L. *sicera* (an
   intoxicating drink). Bottle Gourd. Old
   World tropics.

Lagerstroemia lah-ger-*strurm*-ee-a *Lythraceae*.
After Magnus von Lagerstrom (1696–1759),
Swedish merchant and friend of Linnaeus.
Semi-hardy, deciduous tree.
   indica in-di-ka. Indian. Crape Myrtle.
   China, Korea.

Lagurus la-*gew*-rus *Gramineae*. From Gk.
lagos (a hare) and *oura* (a tail) referring to
the inflorescence. Annual grass.

Lagurus (continued)
ovatus ō–*vah*–tus. Ovate (the inflorescence). Hare's Tail Grass. Mediterranean region.

Lamb's Lettuce see *Valerianella locusta*
Lamb's Tongue see *Stachys byzantina*

Lamiastrum galeobdolon see *Lamium galeodolon*

Lamium lă–mee–um *Labiatae*. The L. name. Perennial herbs. Dead Nettle.
galeobdolon gă–lee–*ob*–do–lon (= *Galeodolon luteum*, *Lamiastrum galeobdolon*). From the Latin name for a nettle-like plant. Yellow Archangel. Europe, W Asia.
garganicum gar–*gah*–ni–kum. Of Monte Gargano, S Italy. S Italy, SE Europe.
maculatum măk–ew–*lah*–tum. Spotted (the leaves). Spotted Dead Nettle. Europe, Asia.
orvala or–*vah*–la. From French *orvale* (= L. *aureis galli*) a kind of sage. S Europe.

Lampranthus lăm–*prănth*–us *Aizoaceae*. From Gk. *lampros* (shining) and *anthos* (a flower). Tender, succulent sub-shrubs. S Africa.
amoenus a–*moy*–nus. Pleasant.
aurantiacus ow–răn–tee–*ah*–kus. Orange.
aureus ow–ree–us. Golden.
blandus blănd–us. Charming.
brownii brown–ee–ee. After Robert Brown, naturalist and first Keeper of Botany at the British Museum.
coccineus kok–*kin*–ee–us. Scarlet.
conspicuus kon–*spik*–ew–us. Conspicuous.
multiradiatus mul–tee–ră–dee–*ah*–tus. (= L. *roseus*). With many rays.
spectabilis spek–*tah*–bi–lis. Spectacular.
zeyheri zay–ha–ree. After Carl L. P. Zeyher who collected in S Africa.

Lantana lăn–*tah*–na *Verbenaceae*. L. name for *Viburnum*, from the similar inflorescence. Tender shrubs.
camara ka–*mah*–ra. The S American name. Yellow Sage. Tropical America.
montevidensis mon–tee–vid–*en*–sis. Of Montevideo. S America.

Lapageria lah–pa–*zhe*–ree–a *Liliaceae* (*Philesiaceae*) After Josephine de la Pagerie, wife of Napoleon Bonaparte. Evergreen climber.
rosea ros–ee–a. Rose-coloured (the flowers). Chile, Argentina.

Lapeirousia lah–pay–*rooz*–ee–a *Iridaceae*. After

Lapeirousia (continued)
Baron Phillipe de la Peyrouse. (1744–1818), French botanist. Semi-hardy, cormous herb.
laxa = *Anomatheca laxa*

Larch see *Larix*
  Common, European see *L. decidua*
  Dunkeld see *L.* × *marschlinsii*
  Golden see *Pseudolarix amabilis*
  Japanese see *L. kaempferi*

Larix lă–riks *Pinaeceae*. The L. name. Deciduous conifers. Larch.
decidua de–kid–ew–a. (= L. *europaea*). Deciduous. Common Larch, European Larch. Europe.
× *eurolepsis* see *L.* × *marschlinsii*
kaempferi kemp–fa–ree. (= L. *leptolepis*). After Engelbert Kaempfer (1651–1716), German physcian and botanist. Japanese Larch. Japan.
× *marschlinsii* marsh–*lin*–see–ee (= L. × *eurolepis*). L. *decidua* × L. *kaempferi*. After Marschlin. Dunkeld Larch.

Larkspur see *Consolida*

Lathyrus lă–thi–rus *Leguminosae*. Gk. name for the pea. Annual and perennial herbs.
cirrhosus ki–*rō*–sus. Climbing by tendrils. Pyrenees, Cevennes.
'Grandiflora' grănd–i–*flō*–ra. Large flowered. S Europe.
latifolius lah–tee–*fo*–lee–us. Broad-leaved. Europe.
magellanicus see *L. nervosus*
nervosus ner–vō–sus (= L. *magellanica*). With conspicuous veins. Lord Anson's Blue Pea. S America.
odoratus o–dō–*rah*–tus. Scented. Sweet Pea. S Italy, Sicily.
rotundifolius ro–tund–i–*fo*–lee–us. With round leaves (leaflets). E Europe, W Asia.
splendens splen–denz. Splendid. S California, N Mexico.
tingitanus ting–gi–*tahn*–us. Of Tangier (Tingi). S Europe, N Africa.
tuberosus tew–be–*rō*–sus. Tuberous. Europe, W Asia.
vernus ver–nus. Of spring (flowering). Europe.

Laurus low–rus *Lauraceae*. L. name for *L. nobilis*. Evergreen trees or shrubs.
azorica a–zo–ri–ka. Of the Azores, Canary Islands, Azores.
nobilis nō–bi–lis. Notable. Bay Laurel. Mediterranean region.

Laurustinus see *Viburnum tinus*

Lavandula la-*văn*-dew-la *Labiatae*. From L. *lavo* (to wash) from its use in soaps. Evergreen shrubs. Lavender.
angustifolia ang-gust-i-*fo*-lee-a. (= *L. spica*). Narrow-leaved. Common Lavender. W Mediterranean region.
dentata den-*tah*-ta. Toothed (the leaves). Mediterranean region, Spain.
× *intermedia* in-ter-*med*-ee-a. *L. angustifolia* × *L. latifolia*. Intermediate (between the parents). Many cultivated lavenders belong here.
lanata lah-*nah*-ta. Woolly (the leaves and shoots). Spain.
stoechas stoy-kas. Of the Stoechades (now the Isles d'Hyères) off the S coast of France near Toulon. French Lavender. SW Europe.

Lavatera lah-va-*te*-ra *Malvaceae*. After the Lavater brothers, 16th-century Zurich naturalists. Annual herbs and shrubs.
arborea ar-*bo*-ree-a. Tree-like. Tree Mallow. S and W Europe.
maritima ma-*ri*-ti-ma. Growing near the sea. W Mediterranean region, NW Africa.
olbia ol-bee-a. Olbia, the L. name of several towns including one in S France and one in Sardinia. W Mediterranean coasts.
trimestris tri-*mays*-tris. Of three months (the flowering period). Mediterranean region.

Lavender see *Lavandula*
    Common see *L. angustifolia*
    French see *L. stoechas*
Lavender Cotton see *Santolina chamaecyparissus*

Layia lay-ee-a *Compositae*. After G. Tradescant Lay (died 1845). Annual Herb.
platyglossa plăt-ee-*glos*-a. (= *L. elegans*). Broad-tongued. Tidy Tips. California.

Ledebouria lay-da-*bour*-ree-a *Liliaceae* (*Hyacinthaceae*). After Carl Friedrich von Ledebour (1785–1851). Tender, bulbous herbs. S Africa.
ovalifolia ō-vah-li-*fo*-lee-a. (= *Scilla ovalifolia*). With oval leaves.
socialis so-kee-*ah*-lis. (= *Scilla violacea*). Growing in colonies. Silver Squill.

Ledum *lay*-dum *Ericaceae*. From *ledon* the Gk. name for *Cistus*. Evergreen shrubs.
groenlandicum grurn-*lănd*-i-kum. Of

Ledum (continued)
Greenland. Labrador Tea. N America, Greenland.
palustre pa-*lus*-tree. Growing in marshes. Marsh Ledum. Arctic and alpine N hemisphere.

Leek see *Allium porrum*

Legousia le-*goos*-ee-a *Campanulaceae*. Derivation not traced. Annual herb.
speculum-veneris spek-ew-lum-*ven*-e-ris. (= *Specularia speculum-veneris*). As the common name. Venus's Looking Glass. C and S Europe.

Leiophyllum lay-ō-*fil*-lum *Ericaceae*. From Gk. *leios* (smooth) and *phyllon* (a leaf) referring to the glossy leaves. Evergreen shrub.
buxifolium buks-i-*fo*-lee-um. *Buxus*-leaved. Sand Myrtle. E United States.

Lemaireocereus la-mair-ree-ō-*kay*-ree-us *Cactaceae*. After Charles Lemaire (1801–71), French cactus specialist, and *Cereus* q.v.
marginatus see *Pachycereus marginatus*
pruinosus see *Stenocereus pruinosus*
thurberi see *Stenocereus thurberi*
treleasei see *Stenocereus treleasei*

Lemboglossum lem-bō-*glos*-um. *Orchidaceae*. From Gk. *lembos*, a boat, and *glossum*, lip; the lip is boat-shaped. Epiphytic orchids previously included in *Odontoglossum*. C America, Mexico
bictoniense bik-ton-ee-*en*-see (= *Odontoglossum bictoniense*). Of Bicton, Devon. Mexico, C America.
cervantesii ser-văn-*tes*-ee-ee (= *Odontoglossum cervantesii*). After Prof. Vincentio Cervantes. Mexico, Guatemala.
cordatum cor-*dah*-tum (= *Odontoglossum cordatum*) Heart-shaped (the lip). Mexico, Guatemala.
rossii ros-ee-ee (= *Odontoglossum rossii*). After John Ross, who collected orchids in Mexico in the 19th century. Mexico, Guatemala.
uroskinneri ew-rō-*ski*-na-ree (= *Odontoglossum uroskinneri*). After Skinner, see *Cattleya skinneri*.

Lemon see *Citrus limon*
Lemon Balm see *Melissa officinalis*
Lemon Mint see *Monarda citriodora*
Lemon-scented Gum see *Eucalyptus citriodora*
Lemon verbena see *Aloysia triphylla*

Lenten Rose see *Helleborus orientalis*

Leonotis le-ō-*nō*-tis *Labiatae*. From Gk. *leon* (a lion) and *otis* (an ear), the corolla has been likened to a lion's ear. Tender shrub.
   leonurus lee-ō-*new*-rus. Like a lion's tail (the flowering shoots). Lion's Ear. S Africa.

Leontopodium lee-on-to-*pod*-ee-um *Compositae*. Classical name for another plant from Gk. *leon* (a lion) and *podion* (a foot). Perennial herb.
   alpinum ăl-*peen*-um. Alpine. Edelweiss. Europe (mountains).

Leopard Flower see *Belamcanda chinensis*
Leopard's Bane see *Arnica montana, Doronicum*

Lepismium lep-is-*mee*-um *Cactaceae*. From Gk. *lepis*, scale; a reference to the rudimentary scale-like leaves. Shrubby cacti. Brazil, Bolivia, Argentina.
   ianothele ee-ahn-ō-thay-lay (= *Pfeiffera ianothele*). With purple nipples.
   houlletianum hoo-lay-tee-*ah*-num (= *Rhipsalis houlletianum*). After M. Houllet, assistant curator of the Jardin des Plantes. Paris. Snowdrop Cactus.
   warmingiana var-ming-gee-*ah*-na (= *Rhipsalis warmingiana*). After Prof. Johannes Eugenius Bülow Warming (1841–1924), a Danish botanist. Brazil.

Leptinella lep-tin-*el*-a *Compositae*. Gk. *lept*, meaning slender, thin or narrow, with *ella* used to form a diminutive. The foliage is very fine, soft and fern-like. Tufted perennial herbs, usually aromatic, and creeping.
   atrata ah-*trah*-ta (= *Cotula atrata*). Black (the flowers) New Zealand.
   potentillina po-ten-til-*een*-na (= *Cotula potentillina*). Like *Potentilla*. New Zealand.
   squalida skwah-li-dah (= *Cotula squalida*). Dirty. New Zealand.

Leptospermum lep-to-*sperm*-um *Myrtaceae*. From Gk. *leptos* (slender) and *sperma* (a seed) referring to the narrow seeds. Semi-hardy, evergreen shrubs.
   humifusum see *L. rupestre*
   lanigerum lah-*ni*-ge-rum. (= *L. cunninghamii*). Woolly (the leaves and shoots). Australia, Tasmania.
   rupestre roo-*pes*-tree (= *L. humifusum*). Of rocky places. Tasmania.
   scoparium skō-*pah*-ree-um. Broom-like.

Leptospermum (continued)
   Manuka, Tea Tree. Australia, New Zealand.
   'Chapmanii' chăp-*măn*-ee-ee. After Sir Frederick Chapman who found it in 1889.
   'Keatleyi' *keet*-lee-ee. After Captain Keatley who found it.
   'Nicholsii' ni-*kolz*-ee-ee. After Mr William Nichols who supplied a nursery with the seed from which it was raised.

Lespedeza les-pe-*dee*-za *Leguminosae* (*Papilionoideae*) After Vincente Manuel de Céspedes, governor of Florida in about 1790. Deciduous shrubs. Bush Clover.
   bicolor bi-ko-lor. Two coloured. N China, Japan.
   thunbergii thun-*berg*-ee-ee. After Thunberg, see *Thunbergia*. N China, Japan.

Lettuce see *Lactuca sativa*

Leucadendron loo-ka-*den*-dron *Proteaceae*. From Gk. *leukos* (white) and *dendron* (a tree) referring to the silvery foliage. Tender tree.
   argenteum ar-*gen*-tee-um. Silvery. Silver Tree. S Africa.

Leucanthemopsis lew-kan-theem-op-sis *Compositae*. From *Leucanthemum* and Gk. *opsis*, meaning appearance. Dwarf perennials formerly included in *Chrysanthemum*. Europe, N. Africa.
   alpina al-*peen*-a (= *Chrysanthemum alpinum*) Alpine. Alps.

Leucanthemum lew-kan-*thee*-mum. *Compositae*. From Gk. *leucos*, white, and *anthemon*, flower. Annual and perennial herbs, some formerly included in *Chrysanthemum*.
   maximum *max*-i-mum (= *Chrysanthemum maximum*). Largest. Pyrenees.
   × *superbum* soo-*perb*-um (= *Chrysanthemum maximum* hort., *Chrysanthemum* × *superbum*). *Leucanthemum lacustre* × *L. maximum*. Superb. Shasta Daisy. Garden origin.
   vulgare vul-*gah*-ree. Common. Ox-eye daisy; Moon daisy. Temperate Eurasia.

Leuchtenbergia loyk-tan-*berg*-ee-a *Cactaceae*. After Prince Maximilian E. Y. N. von Beauharnais (1817–52), Duke of Leuchtenberg, Germany.

Leuchtenbergia (continued)
principis pring-ki-pis. Distinguished.
Mexico.

Leucocoryne loo-kō-*ko*-ri-nay *Liliaceae*
(*Alliaceae*) From Gk. *leukos* (white) and
coryne (a club). Bulbous herb.
ixioides iks-ee-*oi*-deez. Like *Ixia*. Chile.

Leucojum loo-*kō*-yum *Amaryllidaceae*. Gk.
name for a plant from *leukon* (white) and
ion (violet). Bulbous herbs. Snowflake.
aestivum ies-ti-vum. Of summer. Summer
Snowflake. Europe, W Asia.
autumnale ow-tum-*nah*-lee. Of autumn. W
Mediterranean region, N Africa.
roseum ros-ee-um. Rose coloured (the
flowers). Corsica, Sardinia.
vernum ver-num. Of spring. Spring
Snowflake. Europe.

Leucothoe loo-*ko*-thō-ee *Ericaceae*. After
Leucothoe of Gk. mythology, daughter of
Orchamus, king of Babylon and Eurynome,
who is said to have been changed into a
shrub by her lover, Apollo, after being buried
alive by his rival, Clytia. Evergreen and
deciduous shrubs.
davisiae day-*vis*-ee-ie. After Miss N. J.
Davis. California, Oregon.
fontanesiana font-a-neez-ee-*ah*-na. After
Desfontaines, see *Desfontainea*. SE United
States.
grayana gray-*ah*-na. After Asa Gray
(1810–88). Japan.
keiskei kies-kee-ee. After Keisuke Ito
(1803–1901). Japanese physician and
botanist. Japan.

Levisticum le-*vis*-ti-kum *Umbelliferae*.
Probably derived from *Ligusticum*, a related
genus. Perennial herb.
officinale o-fi-ki-*nah*-lee. Sold as a herb.
Lovage. S Europe.

Lewisia loo-*is*-ee-a *Portulacaceae*. After
Captain Meriwether Lewis (1774–1809) of
the Lewis and Clark expedition 1806–7.
Perennial herbs.
columbiana ko-lūm-bee-*ah*-na. Of British
Columbia. W N America.
cotyledon ki-ti-*lay*-don. From the
resemblance to *Cotyledon*. California,
Oregon.
howellii how-*el*-ee-ee. After Thomas
Jefferson Howell (1842–1912).
nevadensis ne-va-*den*-sis. Of the Sierra
Nevada, California. W United States.

Lewisia (continued)
pygmaea pig-*mie*-a. Dwarf. W United
States.
rediviva re-di-*vee*-va. Broought back to life,
the original herbarium specimen
continued to grow after pressing. W
United States.
tweedyi twee-dee-ee. After Tweedy. NW
United States.

Leycesteria les-*e*-ree-a *Caprifoliaceae*. After
William Leycester. Chief Justice in Bengal
during the early 19th century. Deciduous
shrub.
formosa for-*mō*-sa. Beautiful. Himalaya, W
China.

Leymus ley-mus *Graminae*. An anagram of
Elymus q.v., in which this genus was once
included. Perennial grasses.
arenaria å-ray-*nah*-ree-us (= *Elymus
arenaria*). Growing in sandy places; it is a
coloniser of sand dunes. Lyme Grass.
Europe, Asia.

Liatris lee-*aht*-ris *Compositae*. Derivation
obscure. Perennial herbs. Gay Feather.
graminifolia grah-min-i-*fo*-lee-a. Grass-
leaved. SE United States.
pycnostachya pik-nō-*stāk*-ee-a. With dense
spikes. Button Snake Root. SE United
States.
spicata spee-*kah*-ta. (= *L. callilepis*
hort.). With flowers in a spike. E United States.

Libertia lee-*bert*-ee-a *Iridaceae*. After Marie
A. Libert (1782–1863), Belgian botanist.
Evergreen perennial herbs.
formosa for-*mō*-sa. Beautiful. Chile.
grandiflora grānd-i-*flō*-ra. Large flowered.
New Zealand.
ixioides iks-ee-*oi*-deez. Like *Ixia*. New
Zealand.
pulchella pul-*kel*-la. Pretty. New Zealand,
Australia, Tasmania.

Libocedrus chilensis see *Austrocedrus chilensis*
decurrens see *Calocedrus decurrens*

Ligularia lig-ew-*lah*-ree-a *Compositae*. From
L. *ligula* (a strap) referring to the strap-like
ray florets. Perennial herbs.
dentata den-*tah*-ta. (= *L. clivorum. Senecio
clivorum*). Toothed (the leaves). China,
Japan.
× *hessei* hes-ee-ee. *L. dentata* × *L.
wilsoniana*. After Hesse.
hodgsonii hoj-*son*-ee-ee. After C. P.

Ligularia (continued)
Hodgson, Consul in Japan c. 1840, who discovered it. Japan.
przewalskii sha-*văl*-skee-ee (= *Senecio przewalskii*). After its discoverer, the Russian explorer Nicholai M. Przewalski. N China.
stenocephala sten-ŏ-*kef*-a-la. Narrow-headed. E Asia.
tangutica see *Senecio tanguticus*
tussilaginea see *Farfugium japonicum*
veitchiana veech-ee-*ah*-na. After the Veitch nursery. China.
wilsoniana wil-son-ee-*ah*-na. After Ernest Wilson, see *Magnolia wilsonii*. China.

Ligustrum li-*gus*-trum Oleaceae. The L. name. Evergreen and semi-evergreen shrubs and trees. Privet.
delavayanum del-a-vay-*ah*-num. After Delavay who introduced it, see *Abies delavayi*. China.
henryi hen-ree-ee. After Henry who discovered it, see *Illicium henryi*. China.
japonicum ja-*pon*-i-kum. Of Japan. N China, Korea, Japan.
lucidum loo-ki-dum. Glossy (the leaves). China.
obtusifolium ob-tew-si-*fo*-lee-um. Blunt-leaved. Japan.
ovalifolium ŏ-vah-li-*fo*-lee-um. Oval-leaved. Japan.
quihoui kee-*hoo*-ee. After M. Quihou. China.
sinense si-*nem*-see. Of China.
vulgare vul-*gah*-ree. Common. Common Privet. Europe, N Africa, SW Asia.

Lilac see *Syringa*
Common see *S. vulgaris*
Persian see *S. × persica*
Rouen see *S. × chinensis*

Lilium lee-lee-um Liliaceae. The L. name. Bulbous perennial herbs. Lily.
auratum ow-*rah*-tum. Marked with gold (the flowers). Gold-rayed Lily. Japan.
brownii brown-ee-ee. After Mr F. E. Brown of Slough in whose nursery it first flowered, in 1837, S China.
bulbiferum bul-*bi*-fe-rum. Bearing bulbs (in the upper leaf axils). C Europe.
canadense kăn-a-*den*-see. Of Canada or NE North America. Canada Lily. E N America.
candidum kăn-di-dum. White (the flowers). Madonna Lily. SE Europe to Lebanon.
cernuum kern-ew-um. Nodding (the flowers). NE Asia.

Lilium (continued)
chalcedonicum kăl-kay-dŏ-ni-kum. Of Chalcedon (now Kadikoy) nr. Istanbul. Greece, Albania.
davidii dă-*vid*-ee-ee. After David, see *Davidia*. W China.
dauricum dow-rik-um (= L. *pennsylvanicum*). Of Daur. NE Asia.
formosanum for-mŏ-*sah*-num. Of Taiwan (Formosa).
hansonii hăn-*son*-ee-ee. After Peter Hanson of New York, artist and grower of lilies.
henryi hen-ree-ee. After Henry, see *Illicium henryi*. C China.
japonicum ja-*pon*-i-kum. Of Japan.
lancifolium lăn-ki-*fo*-lee-um. (= L. *tigrinum*). With lance-shaped leaves. Tiger Lily. Cult., probably a hybrid.
leichtlinii liekt-*lin*-ee-ee. After Max Leichtlin. Japan.
longiflorum long-gi-*flŏ*-rum. With long flowers. Easter Lily. Japan.
mackliniae ma-*klin*-ee-ie. Discovered by Kingdon Ward and named after his second wife, Jean Macklin. Manipur.
martagon mar-ta-gon. A Turkish word used for a sort of turban and this plant. Turk's-cap Lily. S Europe to Siberia.
monadelphum mon-a-*delf*-um. (= L. *szovitsianum*). From Gk. *monos* (one and *adelphos* (a brother), referring to the united stamens. Crimea, Caucasus.
nepalense ne-pa-*len*-see. Of Nepal. Himalaya.
pardalinum par-da-*leen*-um. Spotted like a leopard. Panther Lily. W N America.
pensylvanicum see L. *dauricum*.
philippinense fil-i-peen-*en*-see. Of the Philippines.
pumilum pew-mi-lum. Dwarf. Siberia, E China.
pyrenaicum pi-ray-*nah*-i-kum. Of the Pyrenees.
regale ray-*gah*-lee. Royal. W China.
rubellum rub-*el*-lum. Reddish (the flowers). Japan.
sargentiae sar-*jent*-ee-ie. After Charles Sargent's wife. Japan.
speciosum spek-ee-ŏ-sum. Showy. S Japan.
superbum soo-*per*-bum. Superb. E United States.
szovitsianum see L. *monadelphum*
× *testaceum* test-*ah*-kee-um. L. *candidum* × L. *chalcedonicum*. Brick-coloured.
tigrinum see L. *lancifolium*
tsingtauense tsing-tow-*en*-see. Of Qingdao (Tsingtao), E China. Korea, NE China.

Lilium (continued)
  wardii ward-ee-ee. After Kingdon Ward,
  see *Cassiope wardii*. Tibet.

Lily of China see *Rohdea japonica*
Lily of the Palace see *Hippeastrum aulicum*
Lily of the Valley see *Convallaria majalis*
Lily of the Valley Tree see *Clethra arborea*
Lily Tree see *Magnolia denudata*
Lime see *Citrus aurantiifolia, Tilia*
  Broad-leaved see *Tilia platyphyllos*
  Common see *T* × *vulgaris*
  European White see *T. tomentosa*
  Red-twigged see *T. platyphyllos* 'Rubra'
  Small-leaved see *T. cordata*
  Weeping Silver see *T.* 'Petiolaris'

Limnanthes lim-*nănth*-eez *Limnanthaceae*.
From Gk. *limne* (a marsh) and *anthos* (a
flower). Annual herb.
  douglasii dūg-*lăs*-ee-ee. After David
  Douglas, see *Iris douglasii*. Poached Egg
  Flower, California, Oregon.

Limnophila lim-*no*-fil-a *Scrophulariaceae*.
From Gk. *limne* (a marsh) and *phileo* (to
love). Aquarium plants. SE Asia.
  heterophylla he-te-rō-*fil*-la. With variable
  leaves, the leaves above and below the
  water are different.
  sessiliflora se-si-li-*flō*-ra. With un-stalked
  flowers.

Limonium lee-*mō*-nee-um *Plumbaginaceae*.
From Gk. *leimon* (a meadow), referring to
their habitat. Annual, biennial and perennial
herbs.
  bellidifolium bel-i-di-*fo*-lee-um. With
  leaves like *Bellis*. The true species is rarely
  grown. Europe, Asia (coasts).
  bonduellei bon-*dwel*-ee-ee. After M.
  Bonduelle, French army surgeon (c
  1813–70). Algerian Statice. Algeria.
  cosyrense ko-si-*ren*-see. Of Pantellaria
  (Cossyra) an island between Sicily and
  Tunisia. S Europe.
  incanum see *Goniolimon callicomum*
  latifolium lah-tee-*fo*-lee-um. Broad-leaved.
  E Europe.
  sinuatum sin-ew-*ah*-tum. Wavy-edged (the
  leaves). Mediterranean region, W Asia.
  spicatum see *Psylliostachys spicata*
  suworowii see *Psylliostachys suworowii*
  tataricum see *Goniolimon tataricum*

Linanthus leen-*ănth*-us *Polemoniaceae*. From
Gk. *linon* (flax) and *anthos* (a flower). Annual
herb.

Linanthus (continued)
  androsaceus ăn-dros-*ah*-kee-us. Like
  Androsace. California.
    luteus loo-tee-us (= *Gilia lutea*). Yellow
    (the flowers).

Linaria leen-*ah*-ree-a *Scrophulariaceae*. From
Gk. *linon* (flax) referring to the similar
leaves. Annual and perennial herbs. Toadflax.
  alpina ăl-*peen*-a. Alpine. Alpine Toadflax.
  Alps.
  cymbalaria see *Cymbalaria muralis*
  genistifolia gen-i-sti-*fo*-lee-a. *Genista*-
  leaved. E Europe.
    dalmatica dăl-*mă*-ti-ka. (= *L. dalmatica.
    L. macedonica*). Of Dalmatia. SE
    Europe.
  maroccana mă-ro-*kah*-na. Of Morocco.
  Bunny Rabbits.
  origanifolia see *Chaenorhinum origanifolium*
  purpurea pur-*pewr*-ree-a. Purple (the
  flowers). Italy, Sicily.
  reticulata ray-tik-ew-*lah*-ta. Net-veined
  (the corolla). N Africa.
  triornithophora tree-or-ni-*tho*-fo-ra. From
  Gk. *tri* (three), *ornis* (a bird) and *phorea*
  (to bear). The flowers are in threes and
  resemble three birds with long tails.
  Three Birds Flying. Spain, Portugal.
  vulgaris vul-*gah*-ris. Common. Common
  Toadflax. Europe.

Lindera lin-*de*-ra *Lauraceae*. After Johann
Linder (1676–1723), Swedish Botanist.
Deciduous shrubs or trees.
  benzoin ben-*zō*-in. From the Arabic name
  for another plant. Spice Bush. E N
  America.
  obtusiloba ob-tew-si-*lō*-ba. Bluntly-lobed
  (the leaves). E Asia.

Ling see *Calluna vulgaris*

Linnaea lin-*ie*-a *Caprifoliaceae*. After Linnaeus
(1707–78), who popularised the binomial
system of naming plants, i.e. using one name
for the genus and one for the species.
Evergreen sub-shrub.
  borealis bo-ree-*ah*-lis. Northern. Twin
  Flower. Northerly latitudes of the N
  hemisphere.

Linum leen-um *Linaceae*. The L. name for
flax. Annual and perennial herbs and shrubs.
  alpinum see *L. perenne alpinum*
  arboreum ar-bo-ree-um. Tree-like. Tree
  Flax. Greece, Turkey.
  austriacum ow-stree-*ah*-kum. Austrian. C
  and S Europe.

Linum (continued)
campanulatum kăm-păn-ew-*lah*-tum. Bell-shaped (the flowers). W Mediterranean region.
flavum flah-vum. Yellow (the flowers). Golden Flax. C and SE Europe.
grandiflorum grănd-i-*flō*-rum. Large-flowered. N Africa.
monogynum mon-*o*-gi-num. With one pistil. New Zealand.
narbonense nar-bon-*en*-see. Of Narbonne. SW Europe.
perenne pe-*ren*-ee. Perennial. Europe.
   alpinum ăl-*peen*-um. (= *L. alpinum*). Alpine.
salsoloides see *L. suffruticosum salsoloides*
suffruticosum su-froo-ti-*kō*-sum. Sub-shrubby. Spain.
   salsoloides sal-so-*loi*-deez. (= *L. salsoloides*). Like *Salsola*. SW Europe.
usitatissimum ew-see-tah-*tis*-i-mum. Most useful. Flax. Cult.

Lion's Ear see *Leonotis leonurus*
Lippia citriodora see *Aloysia triphylla*
   nodifera see *Phyla nodifera*

Lipstick Vine see *Aeschynanthus radicans*

Liquidambar li-kwid-*ăm*-bar *Hamamelidaceae*. From. L. *liquidus* (liquid) and *ambar* (amber), referring to the resin obtained from the bark. Deciduous trees.
formosana for-mō-*sah*-na. Of Taiwan (Formosa). SE Asia.
   Monticola mon-*ti*-ko-la. Growing on mountains, China.
styraciflua sti-ra-*ki*-floo-a. Flowing with styrax (storax, an aromatic balsam used in medicine and perfumery). Sweet Gum. E United States, Mexico, Guatemala.

Liriodendron li-ree-ō-*den*-dron *Magnoliaceae*. From Gk. *leiron* (a lily) and *dendron* (a tree). Deciduous trees.
chinense chin-*en*-see. Of China. Chinese Tulip Tree.
tulipifera tew-lip-*i*-fe-ra. Tulip-bearing, referring to the tulip-like flowers. Tulip Tree. EN America.

Liriope lee-*ree*-o-pay *Liliaceae (Convallariaceae)*. After Liriope a fountain nymph and the mother of Narcissus. Evergreen, perennial herbs.
graminifolia grah-min-i-*fo*-lee-a. Grass-leaved. China, Vietnam.
muscari mus-*kah*-ree. From the resemblance to *Muscari*. Japan, China.

Liriope (continued)
spicata spee-*kah*-ta. With flowers in spikes. China, Vietnam.

Lithocarpus lith-ō-*kar*-pus. *Fagaceae*. From Gk. *lithos* (stone) and *karpos* (a fruit) referring to the hard acorns. Evergreen trees.
densiflorus dens-i-*flō*-rus. Densely flowered. Tanbark Oak, California, Oregon.
edulis ed-*ew*-lis. Edible. Japan.
henryi hen-ree-ee. After Henry, see *Illicium henryi*. Japan.

Lithodora lith-ō-*do*-ra *Boraginaceae*. From Gk. *lithos* (a stone) and *dorea* (a gift). Evergreen shrubs.
diffusa di-*few*-sa. (= *Lithospermum diffusum*). Spreading. SW Europe.
oleifolia o-lee-i-*fo*-lee-a. (= *Lithospermum oleifolium*). With leaves like *Olea*. Spain.
rosmarinifolia rōs-ma-reen-i-*fo*-lee-a. (= *Lithospermum rosmarinifolium*). *Rosmarinus*-leaved. S Italy, Sicily.

Lithops lith-ops *Aizoaceae*. From Gk. *lithos* (stone) and *ops* (appearance). Tender succulents that resemble stones. Living Stones, S and SW Africa.
bromfieldii brom-*feeld*-ee-ee. After Mr H. Bromfield.
comptonii komp-*ton*-ee-ee. After Dr R. H. Compton, Director of Kirstenbosch Botanic Garden, S Africa.
fulleri *ful*-a-ree. After Mr E. R. Fuller, Postmaster at Prieska, Cape Province.
hookeri huk-er-ee (= *L. turbiniformis*). After Hooker. NW Cape.
karasmontana kă-ras-mon-*tah*-na. Of the Little Karasberg Mountains, Namibia.
   bella bel-la. Pretty.
   erniana ern-ee-*ah*-na. After F. Erni.
   mickbergensis mik-berg-*en*-sis. (= *L. mickbergensis*). Of Mickberg, Namibia.
lesliei lez-lee-ee. After its discoverer, Mr T. N. Leslie who collected in the Transvaal.
marmorata mar-mo-*rah*-ta. Marbled.
mickbergensis see *L. kaasmontana mickbergensis*
olivacea o-li-*vah*-kee-a. Olive green.
optica op-ti-ka. Seeing, referring to the translucent, eye-like window in the leaves.
pseudotruncatella soo-dō-trunk-a-*tel*-la. False *L. truncatella*.
   mundtii munt-ee-ee. After Mundt of Windhuk.
salicola sa-*li*-ko-la. Growing in salty places.
schwantesii shvănt-es-ee-ee. After Martin

Lithops (continued)
Heinrich Gustav Schwantes (1881–1960),
botanist. Namibia.
  marthae mar-thie. After the discoverer's
  wife, Martha Erni.
  turbiniformis see *L. hookeri*

Lithospermum diffusum see *Lithodora diffusa*
  doerfleri see *Moltkia doerfleri*
  oleifolium see *Lithodora oleifolia*
  purpurocaeruleum see *Buglossoides
  purpurocaerulea*
  rosmarinifolium see *Lithodora rosmarinifolia*
Little Candles see *Mammillaria prolifera*

Littonia li-*ton*-ee-a *Liliaceae* (*Colchicaceae*).
After Samuel Litton (1781–1847), professor
of botany at Dublin. Tender perennial herb.
  modesta mo-*des*-ta. Modest. S Africa.

Living Stones see *Lithops*
Livingstone Daisy see *Dorotheanthus
bellidiformis*

Livistona li-vi-*ston*-a *Palmae*. After Patrick
Murray, Baron of Livingston (died 1671)
whose plant collection helped to found
Edinburgh Botanic Garden in 1670. Tender
palms.
  australis ow-*strah*-lis. Southern. Australian
  Fan Palm. E Australia.
  chinensis chin-*en*-sis. Of China. Chinese
  Fan Palm. S Japan, China.

Lobelia lō-*bel*-ee-a *Lobeliaceae*. After Mathias
de l'Obel (1538–1616), Flemish botanist.
Annual and perennial herbs.
  cardinalis kar-di-*nah*-lis. Scarlet (the
  flowers). Cardinal Flower. E United
  States.
  erinus e-ri-nus. Gk. name of a plant.
  Edging Lobelia. S Africa.
  siphilitica si-fi-li-ti-ka. A reference to
  supposed medicinal properties. E United
  States.
  splendens *splen*-denz. (= *L. fulgens*).
  Splendid. Mexico.
  tenuior ten-*ew*-ee-or. Slender. W Australia.

Lobivia lo-*biv*-ee-a *Cactaceae*. An anagram of
Bolivia where some species grow. A now
obsolete genus.
  allegraiana see *Echinopsis hertrichiana*
  bruchii see *Echinopsis bruchii*
  famatimensis see *Rebutia famatimensis*
  densispina see *Echinopsis kuehnrichii*
  ferox see *Echinopsis ferox*
  hertichiana see *Echinopsis hertrichiana*
  jajoina see *Echinopsis marsoneri*

Lobivia (continued)
  pygmaea see *Rebutia pygmaea*

Lobularia lob-ew-*lah*-ree-a *Cruciferae*. From
L. *lobulus* (a small pod) referring to the fruit.
Annual or perennial herb.
  maritima ma-*ri*-ti-ma. (= *Alyssum
  maritimum*). Growing near the sea. Sweet
  Alyssum. S Europe.

Locust see *Robinia pseudacacia*
  Caspian see *R. caspica*
  Honey see *Gleditsia triacanthos*
Loganberry see *Rubus loganobaccus*

Loiseleuria lwū-ze-*lur*-ree-a *Ericaceae*. After
Jean Louis Auguste Loiseleur-
Deslongchamps (1774–1849). Prostrate,
evergreen shrub.
  procumbens prō-*kum*-benz. Prostrate.
  Alpine Azalea. Arctic-alpine N
  hemisphere.

Lollipop Plant see *Pachystachys lutea*

Lomatia lō-*mah*-tee-a *Proteaceae*. From Gk.
loma (a border), the seeds are edged with a
wing. Semi-hardy, evergreen shrubs.
  ferruginea fe-roo-*gin*-ee-a. Rusty (the hairs
  of the shoots). Chile, Argentina.
  myricoides mi-ri-*koi*-deez. Like *Myrica*. SE
  Australia.
  tinctoria tink-*to*-ree-a. Used in dyeing.
  Tasmania.

London Pride see *Saxifraga × urbium*

Lonicera lon-i-*se*-ra *Caprifoliaceae*. After
Adam Lonitzer (1528–86), German naturalist.
Deciduous and evergreen shrubs and
climbers. Honeysuckle.
  × *americana* a-me-ri-*kah*-na *L. caprifolium*
  × *L. etrusca*. Of America where it was once
  thought to be native.
  × *brownii* brown-ee-ee. *L. hirsuta* × *L.
  sempervirens*. After Brown.
    'Fuchsioides' fuks-ee-*oi*-deez. *Fuchsia*-
    like.
    'Plantierensis' plon-tee-e-*ren*-sis. From
    the Simon-Louis nursery at Plantières
    near Metz. France.
  caprifolium kăp-ri-*fo*-lee-um. Old generic
  name from L. *capra* (a goat) and *folium* (a
  leaf). Perfoliate Honeysuckle. Europe,
  Caucasus, Turkey.
  etrusca e-*troos*-ka. Of Tuscany.
  Mediterranean region.
  fragrantissima frah-gran-*tis*-i-ma. Very
  fragrant. China.

Lonicera (continued)
× *heckrottii* hek-*rot*-ee-ee. *L.* × *americana* ×
L. sempervirens. After Heckrott.
henryi hen-ree-ee. After Henry, see
Illicium henryi. W China.
hildebrandiana hil-da-brănd-ee-*ah*-na.
After Mr A. H. Hildebrand (1852–1918)
who collected in India. Burma, Thailand,
China.
involucrata in-vo-loo-*krah*-ta. With an
involucre, the conspicuous bracts around
the flowers which enlarge in fruit. W N
America, Mexico, S Canada.
japonica ja-*pon*-i-ka. Of Japan. Japan,
China, Korea.
   'Aureoreticulata' ow-ree-ō-ray-tik-ew-
   lah-ta. Veined with gold (the leaves).
   'Halliana' hawl-ee-*ah*-na. After Dr
   George Hall who introduced it to the
   United States in 1862.
   repens ree-penz. Creeping.
nitida ni-ti-da. Glossy (the leaves). China.
   'Fertilis' *fer*-ti-lis. Fertile, i.e. bearing
   fruit.
periclymenum pe-ree-*klim*-en-um. From
periklymenon the Gk, name for
honeysuckle. Common Honeysuckle,
Woodbine. europe, N Africa.
   'Belgica' *bel*-gi-ka. Belgian.
   'Serotina' se-*ro*-ti-na. Late flowering.
pileata pi-lee-*ah*-ta. Capped (the fruit).
China.
sempervirens sem-per-*vi*-renz. Evergreen.
Trumpet Honeysucke. S and E United
States.
standishii stăn-*dish*-ee-ee. After John
Standish (1814–75), of the Standish and
Noble nursery, for whom Fortune
introduced it. China.
syringantha si-ring-*gănth*-a. With flowers
like *Syringa*. China, Tibet.
tatarica ta-*tah*-ri-ka. Of Tatary, C Asia.
× *tellmanniana* tell-măn-ee-*ah*-na. *L.*
tragophylla × *L. sempervirens* 'Superba'.
After Tellmann.
tragophylla tră-go-*fil*-la. From Gk. *tragos* (a
goat) and *phyllon* (a leaf) it is related to *L.*
caprifolium q.v. China.

Loofah Gourd see *Luffa cylindrica*
Loosestrife see *Lysimachia*
   Purple see *Lythrum salicaria*
   Yellow see *Lysimachia vulgaris*

Lophomyrtus lō-fō-*mur*-tus *Myrtaceae*. From
Gk. *lophos*, a crest, and *Myrtus* q.v. shrubs
and small trees New Zealand.
   bullata bu-*lah*-ta (= *Myrtus bullata*). With
   puckered leaves.

Lophophora lo-fo-*fo*-ra *Cactaceae*. From Gk.
lophos (a crest) and *phoreo* (to bear) referring
to the tufts of woolly hairs on the aeroles.
   williamsii wil-*yămz*-ee-ee. After Williams.
   Dumpling Cactus, Mescal Button,
   Peyote. Texas, N Mexico.

Loquat see *Eriobotrya japonica*
Lords and Ladies see *Arum maculatum*

Lotus lō-tus *Leguminosae* (*Papilionoideae*). A
Gk. name used for several plants. Perennial
herb.
   corniculatus kor-nik-ew-*lah*-tus. With small
   horns. Bird's foot Trefoil. Europe.

Lotus see *Nelumbo*
   American see *N. lutea*
   Blue Egyptian see *Nymphaea caerulea*
   Indian or Sacred see *Nelumbo nucifera*
   White Egyptian see *Nymphaea lotus*
Lovage see *Levisticum officinale*
Love Grass see *Eragrostis elegans*
   Japanese see *E. amabilis*
Love-in-a-Mist see *Nigella damascena*
Love-in-a-Puff see *Cardiospermum*
halicacabum
Love-lies-Bleeding see *Amaranthus caudatus*

Luculia lu-*kew*-lee-a *Rubiacea*. From *lukuli*
swa, the native name of *L. gratissima*. Tender,
evergreen shrubs.
   grandifolia grănd-i-*fo*-lee-a. Large-leaved.
   Bhutan.
   gratissima grah-*tis*-i-ma. Very pleasing.
   Himalaya, China.
   pinceana pin-see-*ah*-na. After Robert
   Taylor Pince (c 1804–71), Exeter
   nurseryman and partner of Lucombe.
   Nepal, Assam.

Luffa luf-a *Cucurbitaceae*. From the Arabic
name. Annual climber.
   aegyptiaca see *L. cylindrica*
   cylindrica si-*lin*-dri-ka (= *L. aegyptiaca*).
   Cylindrical (the fruit). Tropical Asia and
   Africa.

Luma loo-ma *Myrtaceae*. From the native
name. Shrubs and small trees. Argentina,
Chile.
   apiculata a-pik-yew-*lay*-ta (= *Myrtus*
   apiculatus). With a small broad point at
   the tip (the leaves).

Lunaria loon-*ah*-ree-a *Cruciferae*. From L.
luna (the moon) referring to the pods.
Annual and perennial herbs.

Lunaria (continued)
annua ăn-ew-a. Annual. Honesty. SE
Europe.
rediviva re-di-*veev*-a. Coming back to life,
i.e. perennial. Europe.

Lungwort see *Pulmonaria*

Lupinus lu-*peen*-us *Leguminosae*
(*Papilionoideae*). The L. name. Annual and
perennial herbs. Lupin.
arboreus ar-*bo*-ree-us. Tree-like. California.
chamissonis sha-mi-*sō*-nis. After Ludolf
Adalbert von Chamisso (1781–1838).
California.
densiflorus dens-i-*flō*-rus. Densely-
flowered. California.
excubitus eks-*kew*-bi-tus. Presumably
likening its habit to that of a sentinel
(*excubitor*). S California.
hartwegii hart-*weg*-ee-ee. After Karl
Theodore Hartweg (1812–71). Mexico.
luteus loo-tee-us. Yellow. Mediterranean
region.
mutabilis mew-*tah*-bi-lis. Changeable (the
flower colour). Andes.
polyphyllus po-lee-*fil*-lus. With many leaves
(leaflets). WN America.
pubescens pew-*bes*-enz. Hairy. Mexico, C
America.
subcarnosus sub-kar-*nō*-sus. Somewhat
fleshy. Texas.
texensis teks-*en*-sis. Of Texas. Texas
Bluebonnet.

Luronium loo-*ron*-ee-um *Alismataceae*. From
Gk. *luron*, another name for *Alisma*. Aquatic
perennial herb.
natans nă-tănz. (= *Alisma natans*). Floating.
Europe.

Luzula luz-ew-la *Juncaceae*. From Italian
lucciola (a firefly). Perennial herbs. Wood
Rush.
nivea ni-vee-a. Snow-white (the flowers).
W and C Europe (mountains). Snow
Rush.
sylvatica see *L. maxima*
maxima max-i-ma (= *L. sylvatica*). Largest.
S, W & C Europe.

Lycaste li-*kăs*-tay *Orchidaceae*. After Lycaste,
daughter of Priam, the last King of Troy.
Greenhouse orchids.
aromatica ă-rō-*mă*-ti-ka. Fragrant. Mexico,
C America.
brevispatha brev-ee-*spăth*-a. With a short
spathe. C America.

Lycaste (continued)
ciliata sil-ee-*ay*-ta (= *L. fimbriata*). Fringed
with hair (the petal margin). Peru.
cruenta kroo-*en*-ta. Blood red (a blotch on
the lip). Mexico, C America.
deppei dep-ee-ee. After Ferdinand Deppe
(died 1861). Mexico, C America.
fimbriata see *L. ciliata*
macrophylla măk-rō-*fil*-la. Large-leaved. C
and S America.
schilleriana shil-a-ree-*ah*-na. After Herr
Schiller of Hamburg who owned a large
orchid collection. Columbia.
skinneri skin-ee-ee (= *L. virginalis*). After
Skinner. Guatemala, C America.
virginalis see *L. skinneri*

Lychnis lik-nis *Caryophyllaceae*. The classical
name, from Gk. *lychnos* (a lamp). Perennial
herbs.
alpina ăl-*peen*-a. (= *Viscaria alpina*). Alpine.
Alpine Campion. Europe, Siberia, N
America.
× *arkwrightii* ark-*riet*-ee-ee. L. *chalcedonica*
× L. × *haageana*. After Arkwright.
chalcedonica kăl-kay-*dōn*-i-ka. Of Kadikoy
(Chalcedon) near Istanbul. Maltese Cross,
Jerusalem Cross. Russia.
coeli-rosea see *Silene coeli-rosea*
coronaria ko-rō-*nah*-ree-a. Used in
garlands. Rose Campion. SE Europe, N
Africa to C Asia.
coronata ko-rō-*nah*-ta. Crowned. China.
flos-jovis flōs-*yov*-is. Flower of Jupiter.
Alps.
× *haageana* hah-gee-*ah*-na. L. *coronata* × L.
fulgens. After F. A. Haage.
lagascae see *Petrocoptis glaucifolia*
viscaria vis-*kah*-ree-a. Sticky (the stems).
German Catchfly. Europe to E Asia.

Lycium lik-ee-um *Solanaceae*. From the Gk.
name for a thorny tree. Deciduous shrub.
barbarum bar-ba-rum. (= *L. chinense. L.
halimifolium*). Foreign. Box Thorn.
China.

Lycopersicum li-kō-*per*-si-kum *Solanaceae*.
Gk. name for another plant from *lykos* (a wolf)
and *persicon* (a peach). Annual herb.
esculentum es-kew-*lent*-um. Edible.
Tomato. Andes.

Lycoris li-ko-ris *Amaryllidaceae*. After Lycoris,
a beautiful Roman actress and mistress of
Marc Antony. Tender, bulbous herbs.
africana see *L. aurea*
aurea ow-ree-a (= *L. africana*). Golden (the
flowers). Golden Spider Lily. SE Asia.

Lycoris (continued)
incarnata in-kar-*nah*-ta. Flesh pink (the flowers). C China.
radiata răd-ee-*ah*-ta. Radiating (the long-exserted stamens). Red Spider Lily. China, Japan.
sanguinea sang-*gwin*-ee-a. Blood red (the flowers). Japan.
squamigera skwah-*mi*-ge-ra. Scaly (? the bulb). Resurrection Lily. Japan.

Lyme Grass see *Leymus arenarius*

Lyonia lie-*on*-ee-a *Ericaceae*. After John Lyon (c 1765–1814) a Scottish gardener who travelled and collected in the SE United States. Deciduous shrubs.
ligustrina li-gus-*tree*-na. Like *Ligustrum*. EN America.
mariana mă-ree-*ah*-na. Of Maryland. E and S United States.

Lyonothamnus lie-on-ō-*thăm*-nus *Rosaceae*. After W. S. Lyon (1851–1916) who discovered it in 1884 and Gk. *thamnos* (a shrub). Semi-hardy, evergreen tree.
floribundus flō-ri-*bun*-dus. Profusely flowering. Santa Catalina Island (California).
asplenifolius a-splay-ni-*fo*-lee-us. With leaves like *Asplenium*. S California (islands).

Lysichiton li-si-*ki*-ton or loo-si-*ki*-ton. Araceae. From Gk. *lysis* (releasing) and *chiton* (a cloak) referring to the shedding of the large spathe. Perennial herbs.
americanum a-me-ri-*kah*-num. American. Skunk Cabbage. WN America.
camtschatcense kamt-shăt-*ken*-see. Of Kamchatka. Kamchatka to N Japan.

Lysimachia li-si-*măk*-ee-a or loo-si-*măk*-ee-a *Primulaceae*. After King Lysimachos of Thrace (c 360–281 B.C.) who is said to have pacified a bull with a piece of loosestrife. In Gk. his name means ending strife. Perennial herbs.
ciliata ki-lee-*ah*-ta. Fringed with hairs (the petals). N America.
clethroides kleth-*roi*-deez. Like *Clethra*. E Asia.
ephemerum e-*fem*-e-rum. From *ephemeron* the L. name of a plant. SW Europe.
leschenaultii lay-shen-*olt*-ee-ee. After Louis Theodore Leschenault de la Tour (1773–1826), French botanist. India.
nemorum ne-mo-rum. Growing in woods. Europe, Caucasus.

Lysimachia (continued)
nummularia num-ew-*lah*-ree-a. With coin-shaped leaves. Creeping Jenny. Europe, Caucasus.
punctata punk-*tah*-ta. Dotted (the undersides of the leaves). Europe, W Asia.
thyrsiflora thurs-i-*flō*-ra. With flowers in a thyrse (L. *thyrsus*, a staff) a type of inflorescence. Europe, Asia, N America.
vulgaris vul-*gah*-ris. Common. Yellow Loosestrife. Europe, Asia.

Lythrum *lith*-rum *Lythraceae*. From Gk. *lythron* (blood) referring to the colour of the flowers. Herbaceous perennials.
salicaria săl-i-*kah*-ree-a. Like *Salix*. Purple Loosestrife. Europe, Asia.
virgatum vir-*gah*-tum. Wand-like. E Europe, W Asia.

M

Maackia mahk-ee-a *Leguminosae (Papilionoideae)*. After Richard Maack (1825–86), Russian naturalist. Deciduous tree.
amurensis ăm-ew-*ren*-sis. Of the Amur region, NE Asia. Manchuria.

Macleaya ma-*klat*-a *Papaveraceae*. After Alexander Macleay (1767–1848). Herbaceous perennials.
cordata kor-*dah*-ta. Heart-shaped (the leaves). China, Japan.
microcarpa mik-rō-*kar*-pa. With small fruits. Plume Poppy. C China.

Maclura ma-*kloo*-ra *Moraceae*. After William Maclure (1763–1840), American geologist. Deciduous tree.
pomifera pom-*i*-fe-ra. Apple-bearing, referring to the fruit. Osage Orange. S and C United States.

Madagascar Jasmine see *Stephanotis floribunda*
Madagascar Lace Plant see *Aponogeton madagascariensis*
Madonna Lily see *Lilium candidum*
Madrona see *Arbutus menziesii*

Magnolia măg-*nol*-ee-a *Magnoliaceae*. After Pierre Magnol (1638–1715), French professor of botany. Deciduous and evergreen, shrubs and trees.
acuminata a-kew-min-*ah*-ta. With a long

Magnolia (continued)
point (the leaves). Cucumber Tree. EN
America.
campbellii kăm-*belee*-ee. After Dr
Archibald Campbell (1805–74), Political
Resident in Darjeeling. Himalaya.
   mollicomata mol-i-kom-*ah*-ta. Softly
hairy.
conspicua see *M. denudata*
delavayi del-a-*vay*-ee. After Delavay who
discovered it, see *Abies delavayi*. China.
denudata day-new-*dah*-ta. (= *M. conspicua.
M. heptapeta*). Naked, the flowers open
before the leaves emerge. Yulan, Lily Tree.
China.
fraseri fray-za-ree. After John Fraser
(1750–1811), a collector of American
plants who introduced it to England. Ear-
leaved Umbrella Tree. SE United States.
grandiflora grănd-i-*flō*-ra. Large-flowered.
Bull Bay. S United States.
heptapeta see *M. denudata*
hypoleuca hi-pō-*loo*-ka. White beneath (the
leaves). Japan.
kobus kō-bus. From *kobushi* the Japanese
name. Japan.
liliiflora lee-lee-i-*flō*-ra. (= *M. quinquepeta*).
With flowers like *Lilium*. China.
   'Nigra' *nig*-ra. Black, the very dark
purple flowers.
× *loebneri lurb*-na-ree. *M. kobus* × *M.
stellata*. After Max Löbner, German
botanist.
macrophylla măk-rō-*fil*-la. Large-leaved. SE
United States.
officinalis o-fi-ki-*nah*-lis. Sold as a herb, the
bark and flower buds are used medicinally
in China. China.
   biloba bi-*lō*-ba. Two-lobed (the leaves).
quinquepeta see *M. liliiflora*
salicifolia să-li-ki-fo-lee-a. *Salix*-leaved.
Japan.
sieboldii see-*bōld*-ee-ee. After Siebold, see
Acanthopanax sieboldii. S Japan, Korea.
   sinensis si-*nen*-sis. Of China.
× *soulangiana* soo-lon-zhee-*ah*-na. *M.
denudata* × *M. liliiflora*. After Etienne
Soulange-Bodin, French cavalry officer
and Director of the Royal Institute of
Horticulture who raised it.
   'Brozzonii' bro-*zon*-ee-ee. After
Camillo Brozzoni in whose garden it
was raised.
   'Lennei' *len*-ee-ee. After Peter Joseph
Lenné (1789–1866), German botanist.
sprengeri spreng-a-ree. After Carl L.
Sprenger, German nurseryman. China.
stellata ste-*lah*-ta. Star-like (the flowers).
Star Magnolia. Japan.

Magnolia (continued)
× *thompsoniana* tomp-son-ee-*ah*-na. *M.
tripetala* × *M. virginiana*. After Archibald
Thompson a nurseryman of Mile End,
London who raised it in 1808.
   tripetala tri-*pe*-ta-la. With three petals.
Umbrella Tree. EN America.
× *veitchii veech*-ee-ee. *M. campbellii* × *M.
denudata*. After Peter C. M. Veitch
(1850–1929) who raised it in 1907.
virginiana vir-jin-ee-*ah*-na. Of Virginia.
Sweet Bay. E United States.
wilsonii wil-*son*-ee-ee. After its discoverer
and introducer Ernest Henry Wilson
(1876–1936), one of the most prolific of
all plant collectors. Originally sent to
China to introduce *Davidia*, he made
many expeditions there and introduced
numerous fine plants. China.

Mahonia ma-hon-ee-a *Berberidaceae*. After
Bernard McMahon (died 1816), American
horticulturist. Evergreen shrubs.
aquifolium ă-kwi-*fo*-lee-um. The L. name
for holly. Oregon Grape. WN America.
bealei beel-ee-ee. After T. C. Beale who
grew plants collected by Fortune in his
garden in Shanghai. China.
fortunei for-*tewn*-ee-ee. After Fortune who
introduced it in 1846, see *Fortunella*.
China.
japonica ja-*pon*-i-ka. Of Japan where it is
cultivated. ? China.
lomariifolia lō-mah-ree-i-*fo*-lee-a. With
leaves like *Lomaria* (now *Blechnum*).
Burma, W China.
× *media* me-dee-a. *M. japonica* × *M.
lomariifolia*. Intermediate (between the
parents).
pinnata pin-*nah*-ta. Pinnate (the leaves).
California.
pumila pew-mi-la. Dwarf. California,
Oregon.
trifoliolata tri-fo-lee-ō-*lah*-ta. With three
leaflets. S Texas.
   glauca glow-ka. Glaucous. Texas, N
Mexico.

Maianthemum mah-*uănth*-e-mum *Liliaceae*.
From Gk. *maios* (May) and *anthemon*
(blossom), referring to its flowering time.
Perennial herbs. May Lily.
bifolium bi-*fo*-lee-um. Two leaved. Europe,
Asia.
canadense kăn-a-*den*-see. Of Canada or NE
North America. N America.

Maidenhair Fern see *Adiantum*
Australian see *A. formosum*

Maidenhair Fern (continued)
  Common see *A. capillus-veneris*
  Delta see *A. raddianum*
  Giant see *A. trapeziforme*
  Kashmir see *A. venustum*
  N. American see *A. pedatum*
  Rose-fronded see *A. pedatum*
  'Japonicum'
  Rough see *A hispidulum*
  Trailing see *A. caudatum*
Maidenhair Tree see *Ginkgo biloba*
Maize see *Zea mays*

Malcolmia măl-*kol*-mee-a *Cruciferae.* After
William Malcolm (died 1798) and his son
William (c. 1768–1835), London
nurserymen. Annual herb.
  maritima ma-*ri*-ti-ma. Growing near the
  sea. Virginia Stock. Greece, Albania.

Mallow see *Malva*
  Hairy see *Anisodontea scabrosa*
  Marsh see *Althaea officinalis*
  Musk see *Malva moschata*
  Poppy see *Callirhoe*
  Tree see *Lavatera arborea*

Malope ma-*lō*-pee *Malvaceae.* Greek name for
a related plant. Annual herb.
  trifida *tri*-fi-da. Three-lobed (the leaves).
  Spain, N Africa.

Maltese Cross see *Lychnis chalcedonica*

Malus mah-lus *Rosaceae.* L. name for apple.
Deciduous trees. Apple, Crab.
  × *atrosanguinea* aht-rō-sang-*gwin*-ee-a. *M.
  halliana* × *M. sieboldii*. Deep red (the
  flowers).
  baccata bah-*kah*-ta. Bearing berries.
  Siberian Crab. NE Asia.
  mandshurica mănd-*shu*-ri-ka. Of
  Manchuria. Manchurian Crab.
  coronaria ko-rō-*nah*-ree-a. Used in
  garlands. EN America.
    'Charlotte' shar-*lot*-ie. After Mrs
    Charlotte de Wolf who found it in
    1902.
  × *eleyi* e-lee-ee. *M. niedzwetzkyana* × *M.
  spectabilis.* After Charles Eley who raised
  it.
  floribunda flō-ri-*bun*-da. Profusely
  flowering. Cult.
  halliana hawl-ee-*ah*-na. After Dr G. R.
  Hall who introduced this and the
  following to N America in about 1863.
  Cult.
    'Parkmanii' park-*măn*-ee-ee. After

Malus (continued)
    Francis Parkman, a historian and friend
    of Dr Hall.
  'Hillieri' *hil*-ee-a-ree. After Hillier's who
  raised it about 1928.
  hupehensis hew-pee-*hen*-sis. Of Hubei
  (Hupeh). C and W China.
  'Lemoinei' la-*mwăn*-ee-ee. After Lemoine
  who raised it.
  × *purpurea* pur-*pewr*-ree-a. M. ×
  atrosanguinea × *M. niedzwatzkyana.* Purple
  (the foliage).
  × *robusta* rō-*bust*-a. *M. baccata* × *M.
  prunifolia.* Robust.
  sargentii sar-*jent*-ee-ee. After Sargent who
  discovered and introduced it, see *Prunus
  sargentii.* Japan.
  spectabilis spek-*tah*-bi-lis. Spectacular. Cult.
  tschonoskii chon-*os*-kee-ee. After
  Tschonoski, a Japanese plant collector.
  Japan.

Malva mă;-va *Malvaceae.* L. name for mallow.
Biennial or perennial herbs. Mallow.
  alcea ăl-kee-a. From Gk. *alsaia* a sort of
  mallow. Europe.
    'Fastigiata' fa-stig-ee-*ah*-ta. Erect-
    branched.
  moschata mos-*kah*-ta. Musk-scented. Musk
  Mallow. Europe, N Africa.
  sylvestris sil-*ves*-tris. Of woods. Europe, N
  Africa, Asia.

Malvastrum campanulatum see *Sphaeralcea
purpurata*
  coccineum see *Sphaeralcea coccinea*
  scabrosum see *Anisodontea scabrosa*

Malvaviscus măl-va-*vis*-kus *Malvaceae.* From
Malva q.v. and L. *viscus* (glue) referring to
the pulp around the seeds. Tender shrub.
  arboreus ar-*bo*-ree-us. (= *M. mollis*). Tree-
  like. S America.

Mammillaria măm-i-*lah*-ree-a *Cactaceae.*
From L. *mammilla* (a nipple) referring to the
shape of the tubercles. Unless otherwise
stated C Mexico.
  aurihamata ow-ree-hah-*mah*-ta. With
  golden hooks, referring to the yellow,
  hooked spines.
  bocasana bō-ka-*sah*-na. Of the Sierra de
  Bocas. Powder Puff, Snowball Cactus.
  camptotricha kămp-tō-*tree*-ka. From Gk.
  kamptos (curved) and *trichos* (hair) referring
  to the slender, curved spines.
  candida *kăn*-di-da. White (the spines).
  celsiana see *M. muehlenpfordtii*
  compressa kom-*pres*-a. Compressed.

Mammillaria (continued)
decipiens day-*kip*-ee-enz. Deceptive. N Mexico.
densispina dens-i-*speen*-a. Densely spiny.
elegans see *M. haageana*
elongata ay-long-*gah*-ta. Elongated. Gold Lace Cactus.
erythrosperma e-rith-rō-*sperm*-a. With red seeds.
gigantea gi-*găn*-tee-a. Very large.
glochidiata glo-kid-ee-*ah*-ta. With barbed bristles.
haageana hahg-ee-*ah*-na. (= *M. elegans*) After F. A. Haage.
hahniana hahn-ee-*ah*-na. After Hahn. Old Woman Cactus.
heyderi hay-da-ree. After Heyder (1804–84), German cactus grower. Coral Cactus. Texas.
kunzeana kunts-ee-*ah*-na. After Gustav Kunze (1793–1851).
magnimamma mahg-ni-*măm*-a. With large tubercles.
microhelia mik-rō-*hay*-lee-a. A small sun, perhaps referring to the yellow spine clusters.
microheliopsis see *M. microhelia.*
muehlenpfordtii – mew-len-*fort*-ee-ee (= *M. celsiana*). After F. Muehlenpfordt (fl 1922), botanist. C Mexico.
parkinsonii par-kin-*son*-ee-ee. After John Parkinson (c 1772–1847), Consul-General in Mexico in 1838.
polythele po-li-*thay*-lay (= *M. tetracantha*). With many nipples. C Mexico.
plumosa ploo-*mō*-sa. Feathery (the spines). Feather Cactus. N Mexico.
prolifera prō-*li*-fe-ra. Proliferous. Little Candles, Silver Cluster Cactus. W Indies.
pygmaea pig-*mie*-a. Dwarf.
schelhasii shel-*hah*-see-ee. After Schelhas.
schiedeana shee-dee-*ah*-na. After Christian J. W. Schiede (died 1836) who collected in Mexico.
spinosissima speen-ō-*sis*-i-ma. Very spiny.
tetracantha see *M. polythele*
uncinata un-kee-*nah*-ta. Hooked (the spines).
zeilmanniana ziel-măn-ee-*ah*-na. After Zeilmann. Rose Pincushion.

Mandarin Lime see *Citrus* × *limonia*
Mandarin Orange see *Citrus reticulata*

Mandevilla măn-da-*vil*-a *Apocynaceae*. After Henry John Mandeville (1773–1861) a minister in Buenos Aires who introduced *M. suaveolens*. Evergreen climbers.

Mandevilla (continued)
boliviensis bo-liv-ee-*en*-sis. (= *Dipladenia boliviensis*). Of Bolivia. Bolivia, Ecuador.
laxa lak-sa (= *M. suaveolens*). Loose or open. Chilean Jasmine. Argentina.
sanderi sahn-da-ree. (= *Dipladenia sanderi*). After the Sander nursery of St Albans and Bruges. Brazil.
splendens splen-denz. (= *Dipladenia splendens*). Splendid. Brazil.

Mandragora măn-*drăg*-o-ra *Solanaceae*. The Gk. name. Perennial herb.
officinarum o-fi-ki-*nah*-rum. Sold as a herb. Mandrake. S Europe.

Manettia ma-*net*-ee-a *Rubiaceae*. After Saveria Manetti (1723–85). Tender, evergreen climber.
inflata see *M. luteo-rubra*
luteo-rubra loo-tee-ō-*roo*-bra (= *M. inflata*). Yellow and red (the flowers). Brazilian fire cracker. Paraguay, Uruguay.

Manuka see *Leptospermum scoparium*
Manzanita see *Arctostaphylos manzanita*
Maple see *Acer*
  Amur see *A. tatarica* ssp. *ginnala*
  Ash-leaved see *A. negundo*
  Field see *A. campestre*
  Hedge see *A. campestre*
  Hornbeam see *A. carpinifolium*
  Japanese see *A. palmatum*
  Montpelier see *A. monspessulanum*
  Nikko see *A. maximowiczianum*
  Norway see *A. platanoides*
  Oregon see *A. macrophyllum*
  Red see *A. rubrum*
  Silver see *A. saccharinum*
  Sugar see *A. saccharum*
  Vine see *A. circinatum*

Maranta ma-*răn*-ta *Marantaceae*. After Bartolommeo Maranti, 16th-century Venetian botanist. Tender, evergreen herbs.
arundinacea a-run-di-*nah*-kee-a. Reed-like. Arrowroot. Tropical America.
bicolor bi-ko-lor. Two-coloured (the leaves). Brazil, Guiana.
leuconeura loo-kō-*newr*-ra. (= *M. leuconeura massangeana*). White-veined (the leaves). Prayer Plant. Brazil.
  erythroneura e-rith-rō-*newr*-ra. Red-veined (the leaves).
  kerchoveana ker-chov-ee-*ah*-na. After Kerchove. Rabbit's Foot.

Marble Plant see *Neoregelia marmorata*
Mare's Tail see *Hippuris vulgaris*

Marguerite see *Leucanthemum vulgare*
   White see *Argyranthemum frutescens*

Margyricarpus mar-gi-*kar*-pus *Rosaceae.*
From Gk. *margarites* (a pearl) and *karpos* (a
fruit) referring to the white fruits. Dwarf,
evergreen shrub.
   pinnatus pin-*nah*-tus. (= *M. setosus*).
   Pinnate (the leaves). Pearl Fruit. S
   America.

Marigold see *Tagetes*
   African see *T. erecta*
   French see *T. patula*
   Pot see *Calendula officinalis*
   Signet see *T. tenuifolia*
Mariposa Lily see *Calochortus*
Marjoram, Pot see *Origanum onites*
   Sweet see *O. marjorana*
   Wild see *O. vulgare*
Marmalade Bush see *Streptosolen jamesonii*
Marrow see *Cucurbita pepo*
Marsdenia erecta see *Cionura erecta*
Marsh marigold see *Caltha palustris*
Martynia louisianica see *Proboscidea louisianica*
Marvel of Peru see *Mirabilis jalapa*

Masdevallia măs-da-*vah*-lee-a *Orchidaceae.*
After José Masdevall (died 1801), Spanish
botanist. Greenhouse orchids.
   amabilis a-*mah*-bi-lis. Beautiful. Peru.
   bella see *Dracula bella*
   caudata kaw-*dah*-ta. With a slender tail (the
   sepals). Venezuela, Colombia.
   chimaera see *Dracula chimaera.*
   coccinea kok-*kin*-ee-a. Scarlet. Colombia,
   Peru.
   militaris see *M. ignea*
   muscosa see *Porroglossum muscosa*
   simula see *Dryadella simula*
   tovarensis tō-vah-*ren*-sis. Of Tovar,
   Venezuela. Venezuela, Colombia.

Mask Flower see *Alonsoa*
Masterwort see *Astrantia*
Matricaria eximia hort. see *Tanacetum
parthenium*

Matteuccia ma-*too*-kee-a *Dryopteridaceae*
(*Athyriaceae*). After Carlo Matteucci, 19th-
century Italian physicist. Fern.
   struthiopteris stroo-thee-*op*-te-ris. (= *M.
   germanica*). Presumably from Gk.
   struthokamelos (an ostrich) and *pteris* (a
   fern), the fronds resemble ostrich
   feathers. Ostrich-feather Fern. Europe,
   Asia.

Matthiola mă-tee-ō-la *Cruciferae.* After

Matthiola (continued)
Pierandrea Mattioli (1500–77), Italian
botanist. Annual or biennial herbs. Stock.
   longipetala long-gi-*pe*-ta-la. Long-petalled.
   S Ukraine.
      bicornis bi-*kor*-nis. (= *M. bicornis*). Two-
      horned (the fruits). Night-scented
      Stock. Greece, Aegean region.
   incana in-*kah*-na. Grey. Brompton Stock.
   S and W Europe (coasts).
      Annua ăn-ew-a. Annual. Ten Week
      Stock.

Matucana mat-oo-*kah*-na *Cactaceae.* Named
for the village Matucana, in Peru.
   aurantiaca ow-ran-tee-*ah*-kă (= *Borzicactus
   aurantiacus*). Orange (the flowers). N
   Peru.
   haynei hayn-ee-ee. (= *Borzicactus haynei*).
   After Frederich Hayne (1763–1832). N
   Peru.

Maurandya maw-*răn*-dee-a *Scrophulariaceae.*
After Catherine Maurandy, wife of Prof. A.
J. Maurandy. Tender perennials. Mexico.
   barclayana see *Asarina barclayana*
   erubescens see *Asarina erubescens*
   scandens see *Asarina scandens.*

Maxillaria măks-i-*lah*-ree-a *Orchidaceae.*
From L. *maxilla* (a jaw), in some species the
column and lip resemble the jaw of an insect.
Greenhouse orchids.
   grandiflora grănd-i-*flō*-ra. Large-flowered.
   N Andes.
   longisepala long-gi-*sep*-a-la. With long
   sepals. Venezuela.
   luteoalba loo-tee-ō-*ăl*-ba. Yellow-white
   (the flowers). Costa Rica to Ecuador.
   picta pik-ta. Painted (the flowers).
   Colombia, Brazil.
   praestans prie-stahnz. Distinguished.
   Mexico, Guatemala.
   rufescens roo-*fes*-enz. Reddish (the
   flowers). C and S America.
   sanderiana sahn-da-ree-*ah*-na. After the
   Sander orchid nursery. Ecuador, Peru.
   tenuifolia ten-ew-i-*fo*-lee-a. Slender-
   leaved. C America.
   variabilis vă-ree-*ah*-bi-lis. Variable. C
   America.
   venusta ven-*us*-ta. Charming. S America.

May Apple see *Passiflora incarnata*
May Lily see *Maianthemum*
Maypop see *Passiflora incarnata*

Maytenus may-ten-us *Celastraceae.* From

Maytenus (continued)
maiten the native name of *M. boaria.*
Evergreen tree.
> *boaria* bō-ah-ree-a. Of cattle, which eat the leaves. Chile, Argentina.

Mazus mǎ-zus *Scrophulariaceae.* From Gk.
mazos (a teat) referring to the swellings in the throat of the corolla. Perennial herbs.
> *pumilio* pew-*mi*-lee-o. Dwarf. New Zealand, Australia.
> *radicans* rah-di-kǎnz. With rooting stems. New Zealand.
> *reptans* rep-tǎnz. Creeping. Himalaya.

Meadow Rue see *Thalictrum*
Meadowsweet see *Filipendula ulmaria*

Meconopsis may-kō-*nop*-sis *Papaveraceae.*
From Gk. *mekon* (a poppy) and *-opsis* indicating resemblance. Perennial or monocarpic herbs.
> *baileyi* see *M. betonicifolia*
> *betonicifolia* be-ton-i-ki-*fo*-lee-a. (= *M. baileyi*). *Betonica*-leaved. Himalayan Blue Poppy. China.
> *cambrica* kǎm-bri-ka. Welsh. Welsh Poppy. W Europe.
> *chelidoniifolia* ke-li-dōn-ee-i-*fo*-lee-a. *Chelidonium*-leaved. W China.
> *grandis* grǎnd-is. Large. Himalaya, W China.
> *horridula* ho-*rid*-ew-la. Somewhat spiny. Himalaya, W China.
> *integrifolia* in-teg-ri-*fo*-lee-a. With entire leaves. E Himalaya, W China.
> *napaulensis* nǎ-paw-*len*-sis. Of Nepal. Himalaya, W China.
> *quintuplinervia* kwin-tup-li-*ner*-vee-a. Five-veined (the leaves). Harebell Poppy. Tibet.
> *regia* ray-gee-a. Royal. Nepal.
> *simplicifolia* sim-pli-ki-*fo*-lee-a. With simple leaves. Himalaya, Tibet.
> *superba* soo-*per*-ba. Superb. Tibet.
> *villosa* vi-*lō*-sa. Softly hairy. Himalaya.

Medinilla me-di-*ni*-la *Melastomataceae.* After José de Medinilla, Governor of the Marianna Islands in 1820. Tender, evergreen shrub.
> *magnifica* mahg-*ni*-fi-ka. Magnificent. Philippines.

Medlar see *Mespilus germanica*
Medusa's Head see *Euphorbia caput-medusae*

Melaleuca me-la-*loo*-ka *Myrtaceae.* From Gk. *melas* (black) and *leukos* (white) referring to

Melaleuca (continued)
the black trunk and white shoots of many species. Semi-hardy and tender, evergreen shrubs.
> *armillaris* arm-i-*lah*-ris. Encircled, the inflorescence encircles the shoot. SE Australia.
> *elliptica* e-*lip*-ti-ka. Elliptic (the leaves). W Australia.
> *hypericifolia* hi-pe-ree-ki-*fo*-lee-a. *Hypericum*-leaved. New South Wales.
> *wilsonii* wil-*son*-ee-ee. After Mr Charles Wilson who discovered it. SE Australia.

Melia mee-lee-a *Meliaceae.* Gk. name for the ash, from the similar leaves. Semi-hardy, deciduous tree.
> *azedarach* a-*zed*-a-rǎk. From the native name. Bead Tree. Iran, Himalaya, China.

Melianthus me-lee-ǎnth-us *Melianthaceae.* From Gk. *meli* (honey) and *anthos* (a flower) referring to the abundant nectar. Semi-hardy sub-shrub.
> *major* mah-yor. Larger. Honey Flower. S Africa.

Melissa me-*lis*-a *Labiatae.* Gk. name for honeybees which are attracted to the flowers. Herbaceous perennial.
> *officinalis* o-fi-ki-*nah*-lis. Sold as a herb. Lemon Balm. S Europe.

Melittis me-*li*-tis *Labiatae.* Derivation as for *Melissa.* Perennial herb.
> *melissophyllum* me-lis-ō-*fil*-lum. With leaves like *Melissa.* Europe.

Melocactus may-lō-*kǎk*-tus *Cactaceae.* From L. *melopepo* (an apple-shaped melon) and *Cactus* q.v.
> *broadwayi* brord-way-ee. After W. R. Broadway who discovered it in 1914. W Indies.

Melon see *Cucumis melo*

Mentha men-tha *Labiatae.* The L. name. Perennial herbs. Mint. Most hybrids occur naturally.
> *aquatica* a-*kwah*-ti-ka. Growing in or near water. Water Mint. Europe, N Africa, Asia.
> × *gracilis* gra-kil-is (= *M.* × *gentilis*) *M. arvensis* × *M.* × *spicata.* Related to. Ginger Mint.
> *longifolia* long-gi-*fo*-lee-a. Long-leaved. Horse Mint. Europe.
> × *piperita* pi-pe-*ree*-ta. *M. aquatica* × *M.* ×

Mentha (continued)
spicata. Like pepper. Black Peppermint. In the wild with the parents.
officinalis o-fi-ki-*nah*-lis. Sold as a herb. White Peppermint.
citrata ki-*trah*-ta. Lemon-scented. Eau de Cologne Mint.
'Crispa' *kris*-pa. Crisped (the leaves). Curly Mint.
pulegium poo-*leg*-ee-um. The L. name. Pennyroyal. Europe, W Asia.
requienii rek-wee-*en*-ee-ee. After Esprit Requien (1788–1851) who studied the flora of S France and Corsica. Corsican Mint. Italy, Corsica, Sardinia.
spicata spee-kah-ta. With flowers in spikes. Spearmint, Garden Mint.
suaveolens swah-*vee*-o-lenz. (= *M.* × *rotundifolia* hort.). Sweetly scented. Round-leaved Mint. S and W Europe.
× *villosa* vi-*lŏ*-sa. *M.* × *spicata* × *M. suaveolens*. Softly hairy.
   alopecuroides ă-lŏ-pek-ew-*roi*-deez. Like *Alopecurus*. Bowles' Mint.

Mentzelia ment-*zel*-ee-a *Loasaceae*. After Christian Mentzel (1622–1701), German botanist. Annual herb.
lindleyi lind-lee-ee. (= *Bartonia aurea*). After John Lindley (1799–1865), professor of botany at London University. Blazing Star. C California.

Menyanthes may-nee-*ănth*-eez *Menyanthaceae*. From *menanthos* (moonflower) the Gk. name for *Nymphoides peltata*, a related plant. Aquatic or bog garden perennial herb.
trifoliata tri-fo-lee-*ah*-ta. With three leaves (leaflets). Buck Bean, Bog Bean. N hemisphere.

Menziesia men-*zeez*-ee-a *Ericaceae*. After Archibald Menzies (1754–1842), naval surgeon and botanist who collected in W N America. Deciduous shrubs.
ciliicalyx ki-lee-i-*kă*-liks. With the calyx fringed with hairs. Japan.
   purpurea pur-*pewr*-ree-a. Purple (the flowers).

Merendera me-ren-*de*-ra *Liliaceae*. From *quita meriendas*, the Spanish name for a *Colchicum*. Cormous herbs.
filifolia fee-li-*fo*-lee-a. With thread-like leaves. SW Europe.
montana mon-*tah*-na (= *M. pyrenaica*) Of mountains. Iberia, C Pyrenees.
pyrenaica see *M. montana*

Merendera (continued)
robusta rŏ-*bus*-ta. Robust. N Afghanistan, Russia.
sobolifera so-bo-*li*-fe-ra. Bearing offspring. Balkans, Romania.
trigyna tri-gi-na. With three pistils. Caucasus, Turkey, N Iran.

Merrybells see *Uvularia*

Mertensia mer-*tenz*-ee-a *Boraginaceae*. After Franz Karl Mertens (1764–1831), German botanist. Perennial herbs.
ciliata ki-lee-*ah*-ta. Fringed with hairs (the leaves). W United States.
echioides e-kee-*oi*-deez. Like *Echium*. Himalaya.
longiflora long-gi-*flŏ*-ra. Long-flowered. W United States.
maritima ma-*ri*-ti-ma. Growing near the sea. N Europe (coasts).
primuloides preem-ew-*loi*-deez. Like *Primula*. Himalaya.
sibirica si-*bi*-ri-ka. Of Siberia.
virginica vir-*jin*-i-ka. Of Virginia. E United States.

Mescal Button see *Lophophora williamsii*

Mesembryanthemum mes-em-bree-*ănth*-e-mum *Aizoaceae*. Previously spelled *Mesembrianthemum* meaning flowering at mid-day, the current spelling refers to the position of the ovary. Succulent annual.
cordifolium see *Aptenia cordifolia*
criniflorum see *Dorotheanthus bellidiformis*
crystallinum kris-ta-*leen*-um. (= *Cryophytum crystallinum*). Crystalline, the appearance of the leaves. Ice Plant. S Africa.
tricolor see *Dorotheanthus tricolor*

Mespilus *mes*-pi-lus *Rosaceae*. The L. name. Deciduous tree.
germanica ger-*mah*-ni-ka. Of Germany. Medlar. Europe, SW Asia.

Metapanax met-a-*păn*-ăks *Araliaceae* From Gk. *meta*, between, and *Panax*, a related genus; the species in this genus has been moved around various genera in the group. Evergreen shrubs and trees. China and North Vietnam.
davidii dă-vid-ee-ee (= *Pseudopanax davidii*). After David, who introduced it, see *Davidia*. China.

Metasequoia me-ta-se-*kwoy*-a *Taxodiaceae*. From Gk. *meta* (changed) *Sequoia* q.v. to

Metasequoia (continued)
which it is related. Deciduous conifer only
discovered in 1941.
  glyptostroboides glip-to-stro-*boi*-deez. Like
  Glyptostrobus. Dawn Redwood. China.

Metrosideros may-trō-si-*day*-ros *Myrtaceae.*
From Gk. *metra* (heart-wood) and *sideros*
(iron) referring to the very hard wood. Semi-
hardy, evergreen trees and shrubs.
  excelsa eks-*kel*-sa. (= *M. tomentosa*). Tall.
  New Zealand.
  kermadecensis kerm-a-dek-*en*-sis. Of the
  Kermadec Islands.
  robustus rō-*bus*-tus. Robust. New Zealand.
  umbellatus um-bel-*ah*-tus. (= *M. lucida*).
  The flowers appear to be in umbels. New
  Zealand.

Mexican Orange Blossom see *Choisya
ternata*
Mexican Sunflower see *Tithonia rotundifolia*
Mezereon see *Daphne mezereum*
Michaelmas Daisy see *Aster novi-belgii*

Michauxia mee-*shō*-ee-a *Campanulaceae.*
After André Michaux (1746–1803), French
traveller and plant collector. Biennial herbs.
  campanuloides kǎm-pǎn-ew-*loi*-dees. Like
  Campanula. E Mediterranean region.
  tchihatcheffi chee-ha-*chef*-ee-ee. After
  Count Pierre de Tchihatcheff (1808–90),
  Russian traveller and writer. Turkey.

Microbiota mik-rō-bee-*a*-ta *Cupressaceae.*
From Gk. *micros* (small) and *Biota (Thuja).*
Dwarf, evergreen conifer.
  decussata day-kus-*ah*-ta. With the leaves in
  pairs, one pair at right angles to the next.
  SE Siberia.

Microcachrys mik-rō-*kǎk*-ris *Podocarpaceae.*
From Gk. *mikros* (small) and *kachrys* (a cone).
Dwarf, evergreen conifer.
  tetragona tet-ra-*gōn*-a. Four-angled (the
  shoots).

Microcoelum mik-rō-*koy*-lum *Palmae.* From
Gk. *mikros* (small) and *koilos* (a hollow)
referring to a small hollow in the endosperm.
Tender Palm.
  weddellianum we-del-ee-*ah*-num. (= *Cocos
  weddelliana*). After Dr H. A. Weddell who
  collected in S America in the 19th century.
  Weddell Palm. Brazil.

Microglossa albescens see *Aster albescens*
Mignonette see *Reseda odorata*

Mila *mee*-la *Cactaceae.* An anagram of Lima,
capital of Peru.
  caespitosa kie-spi-*tō*-sa. Tufted. Peru.

Milium *mi*-lee-um *Gramineae.* L. name for
millet. Perennial grass.
  effusum e-*few*-sum. Spreading. Europe,
  Asia.
    'Aureum' *ow*-ree-um. Golden. Bowles'
    Golden Grass.

Milkweed see *Asclepias*

Miltonia mil-*ton*-ee-a *Orchidaceae.* After
Charles Fitzwilliam, Viscount Milton
(1786–1857), horticultural patron.
Greenhouse orchids. Pansy Orchid.
  candida kǎn-di-da. White (the lip). Brazil.
  clowesii klowz-ee-ee. After Clowes, see
  Anguloa clowesii. Brazil.
  endresii see *Miltoniopsis warscewiczii*
  flavescens flah-*ves*-enz. Yellowish. Paraguay.
  phalaenopsis see *Miltoniopsis phalaenopsis*
  regnellii reg-*nel*-ee-ee. After Mr Regnell
  who introduced it. Brazil.
  roezlii see *Miltoniopsis roezlii*
  spectabilis spek-*tah*-bi-lis. Spectacular.
  Brazil.
  vexillaria veks-i-*lah*-ree-a. Standard-
  bearing. Colombia.
  warscewiczii see *Miltoniopsis warscewiczii*

Miltoniopsis mil-*ton*-ee-op-sis *Orchidaceae.*
From *Miltonia* q.v., and *opsis*, resembling.
Epiphytic and lithophytic orchids. C & S
America.
  phalaeonopsis fâ-lie-*nop*-sis. (= *Miltonia
  phalaeonopsis*). Moth-like (the flowers).
  roezlii rurz-lee-ee (= *Miltonia roezlii*). After
  Benedict Roezl (1824–85), a plant
  collector. Colombia.
  warscewiczii var-sha-*vich*-ee (= *Miltonia
  warscewiczii*). After Warscewicz, see
  Alonsoa warscewiczii. Colombia, Peru.

Mimosa mee-*mos*-a *Leguminosae.* From Gk.
mimos (a mimic) referring to the sensitive
leaves. Tender shrub.
  pudica pu-*dee*-ka. Shy. Humble Plant,
  Sensitive Plant. Tropical America.

Mimosa see *Acacia dealbata*
  Pink see *Albizia julibrissin*

Mimulus mee-*mew*-lus *Scrophulariaceae.* A
diminutive of L. *mimus* (a mimic), the
flowers resemble a monkey's face. Perennial
herbs and shrubs. Monkey Flower.
  aurantiacus ow-rǎn-tee-*ah*-kus. (= *M.*

Mimulus (continued)
glutinosus. Diplacus aurantiacus). Orange.
Oregon, California.
× *burnetii* bur-*net*-ee-ee. *M. cupreus* × *M.
luteus*. After Dr Burnet of Aberdeen who
raised it about 1901.
cardinalis kar-di-*nah*-lis. Scarlet. Cardinal
Monkey Flower. W N America.
cupreus kew-pree-us. Coppery (the
flowers). S Chile.
glutinosus see *M. aurantiacus*
guttatus gu-*tah*-tus. Spotted (the flowers).
W N America, Mexico.
× *hybridus* hib-ri-dus. *M. guttatus* × *M.
luteus*. (= *M. tigrinus* hort.) Hybrid.
lewisii loo-*is*-ee-ee. After Lewis, see
Lewisia. W N America.
luteus loo-tee-us. Yellow. Chile.
moschatus mos-*kah*-tus. Musk-scented. N
America.
primuloides preem-ew-*loi*-deez. Like
Primula. W United States.
puniceus pew-*ni*-kee-us. (= *Diplacus
puniceus*). Reddish-purple. S California,
N Mexico.
ringens ring-genz. Gaping (the corolla). N
America.
tigrinus hort. see *M.* × *hybridus*
variegatus vă-ree-a-*gah*-tus. Variegated (the
flowers). Chile.

Mina mee-na *Convolulaceae*. After Joseph
Mina of Mexico. Tender climber.
lobata see *Ipomoea lobata*

Mind-your-own-Business see *Soleirolia
soleirolii*
Mint see *Mentha*
 Bowles' see *M.* × *villosa alopecuroides*
 Corsican see *M. requienii*
 Curly see *M.* × *piperita* 'Crispa'
 Eau de Cologne see *M.* × *piperita citrata*
 Garden see *M.* × *spicata*
 Ginger see *M.* × *gracilis*
 Horse see *M. longifolia*
 Round-leaved see *M. suaveolens*
 Water see *M. aquatica*
Mint Bush see *Prostanthera*

Minuartia min-*wah*-tee-a *Caryophyllaceae*.
After Juan Minuart (1693–1768) of
Barcelona. Perennial herbs.
laricifolia lă-ri-ki-fo-lee-a. (= *Arenaria
laricifolia*). With leaves like *Larix*. Europe.
verna ver-na. Of spring (flowering).
Europe.

Mirabilis mee-*rah*-bi-lis *Nyctaginaceae*. L. for
wonderful. Annual herbs.

Mirabilis (continued)
jalapa ha-lah-pa. Of Jalapa (Xalapa),
Mexico. Four o'Clock Plant, Marvel of
Peru. Tropical America.
longiflora long-gi-*flō*-ra. Long-flowered.
SW United States, Mexico.
multiflora mul-tee-*flō*-ra. Many-flowered.
SW United States.

Miscanthus mis-*kắnth*-us *Gramineae*. From
Gk. *miskos* (a stem) and *anthos* (a flower)
referring to the stalked spikelets. Perennial
grasses.
sacchariflorus să-ka-ri-*flō*-rus. With flowers
like *Saccharum* (sugar cane). Asia.
sinensis si-nen-sis. Of China. E Asia.

Mistflower see *Conoclinium coelestinum*
Mistletoe see *Viscum album*

Mitchella mi-*chel*-la *Rubiaceae*. After Dr John
Mitchell (1711–68), Virginian physician and
botanist. Dwarf, evergreen shrub.
repens ree-penz. Creeping. Partridge Berry.
E and C North America.

Mitella mi-*tel*-la *Saxifragaceae*. A diminutive
of Gk. *mitra* (a cap) referring to the fruit.
Perennial herbs. Bishop's Cap.
breweri broo-a-ree. After Brewer, see *Picea
breweriana*. W N America.
caulescens kaw-*les*-enz. With a stem. W N
America.
diphylla di-*fil*-la. Two-leaved. E N
America.

Mitraria mi-*trah*-ree-a *Gesneriaceae*. From
Gk. *mitra* (a cap) referring to the fruit. Semi-
hardy, evergreen climber.
coccinea kok-*kin*-ee-a. Scarlet (the flowers).
Chile, Argentina.

Mock Orange see *Philadelphus coronarius*
Mole Plant see *Euphorbia lathyris*

Molinia mo-*leen*-ee-a *Gramineae*. After Juan
Ignacio Molina (1740–1829), Chilean
botanist. Perennial grass.
caerulea kie-*ru*-lee-a. Blue. Purple Moor
Grass. Europe.

Moltkia molt-kee-a *Boraginaceae*. After Count
Joachim Gadske Moltke (1746–1818), Danish
statesman and naturalist. Herbs and shrubs.
doefleri durf-la-ree. (= *Lithospermum
doefleri*). After J. D. Doerfler. Albania.
× *intermedia* in-ter-*med*-ee-a. *M. petraea
× M. suffruticosa*. Intermediate (between
the parents).

Moltkia (continued)
  petraea pe–*trie*-a. Growing on rocks. SE
  Europe.
  suffruticosa su–froo–ti–*kō*-sa. Sub-shrubby.
  Italy.

Moluccella mo–lu–*kel*-la *Labiatae*. Derivation
obscure, possibly from Molucca. Annual
herb.
  laevis lie-vis. Smooth. Bells of Ireland,
  Shell Flower. W Asia.

Monarch of the East see *Sauromatum
venosum*
Monarch of the Veldt see *Arctotis fastuosa*

Monarda mo–*nar*-da *Labiatae*. After Nicholas
Monardes (1493–1588), Spanish botanist
and physician. Annual, biennial and
perennial herbs.
  citriodora kit–ree–o–*dō*-ra. Lemon-scented.
  Lemon Mint. S United States, Mexico.
  didyma di-di-ma. In pairs (the stamens or
  the leaves). Oswego Tea, Sweet
  Bergamot. E United States.
  fistulosa fist–ew–*lō*-sa. Tubular. E N
  America.
  media me–dee-a. Intermediate. E N
  America.

Monkey Flower see *Mimulus*
Monkey Puzzle see *Araucaria araucana*
Monkshood see *Aconitum*

Monstera mon–*ste*-ra *Araceae*. Derivation
obscure, possibly from the monstrous
appearance of the leaves.
  deliciosa day–li–kee–*ō*-sa. Delicious (the
  fruit). Swiss Cheese Plant. Mexico, C
  America.

Montbretia see *Crocosmia* × *crocosmiiflorua*

Montia mon–*tee*-a *Portulacaceae*. After
Giuseppe Monti (1682–1760), Italian
professor of botany. Annual herbs.
  perfoliata per–fo–lee–*ah*-ta. With leaves
  joined around the stem. W N America.
  sibirica si–*bi*-ri-ka. Siberian. W N
  America.

Monvillea mon–*vil*-ee-a *Cactaceae*. After
Monville, a 19th-century, French cactus
specialist. An obsolete name.
  cavendishii see *Cereus saxicola*
  spegazzinii see *Cereus spegazzinii*

Moonstones see *Pachyphytum oviferum*
  Sticky see *P. glutinicaule*

Moosewood see *Acer pensylvanicum*
Mop-headed Acacia see *Robinia pseudacacia*
'Umbraculifera'

Moraea mo–*rie*-a *Iridaceae*. After Robert
More (1703–80), amateur botanist.
Cormous perennials. S Africa.
  iridioides see *Dietes iridioides*
  ramosissima rah–mō–*si*-si-ma. Much
  branched.
  spathacea spa–*thah*-kee-a. Spathe-like.
  tricuspidata tr–kus–pi–*dah*-ta. Three-
  pointed.

Morina mo–*reen*-a *Dipsacaceae*. After Louis
Pierre Morin (1635–1715), French botanist.
Perennial herb.
  longifolia long–gi–fo–lee-a. Long-leaved.
  Whorl Flower. Himalaya.

Morisia mō–*ris*-ee-a *Cruciferae*. After
Giuseppe Giacinto Moris (1769–1869),
Italian botanist. Perennial herb.
  monanthos mon–*ánth*-os. One-flowered.
  Corsica, Sardinia.

Morning Glory see *Ipomoea*

Morus mō-rus *Moraceae*. The L. name for *M.
nigra*. Deciduous trees. Mulberry.
  alba ál-ba. White (the fruits). White
  Mulberry. China.
  nigra nig-ra. Black (the ripe fruit).
  Common Mulberry, Black Mulberry. W
  Asia.

Mosaic Plant see *Fittonia verschaffeltii
argyroneura*
Mossfern see *Selaginella pallescens*
Mother-in-law's Tongue see *Sansevieria
trifasciata*
Mother of Pearl Plant see *Graptophyllum
paraguayense*
Mother of Thousands see *Kalanchoe
daigremontiana, Saxifraga stolonifera*
Mount Atlas Daisy see *Anacyclus pyrethrum*
var. *depressus*
Mount Wellington Peppermint see
Eucalyptus coccifera
Mountain Ash see *Sorbus aucuparia*
Mountain Avens see *Dryas octopetala*
Mountain Laurel see *Kalmia latifolia*
Mountain Pepper see *Drimys lanceolata*
Mountain Tobacco see *Arnica montana*
Mourning Widow see *Geranium phaeum*
Mouse-tail Plant see *Arisarum proboscideum*
Moutan see *Paeonia suffruticosa*
Mrs Robb's Bonnet see *Euphorbia robbiae*

Muehlenbeckia moo-lan-*bek*-ee-a
Polygonaceae. After Henri Gustave
Muehlenbeck (1789–1845), French
physician and botanist. Deciduous, twining
shrubs.
   axillaris ăks-i-*lah*-ris. Axillary, the flowers
   are borne in the leaf axils. New Zealand,
   Tasmania, Australia.
   complexa kom-*pleks*-a. Embraced, the
   perianth swells, enclosing the fruit. New
   Zealand.
      triloba tri-lo-*bah*. Three-lobed (the
      leaves).

Mulberry see *Morsu*
   Common or Black see *M. nigra*
   Paper see *Broussonetia papyrifera*
   White see *M. alba*
Mullein see *Verbascum*
   Cretan see *V. creticum*
   Dark see *V. nigrum*
   Moth see *V. blattaria*
   Nettle-leaved see *V. chaixii*
   Purple see *V. phoeniceum*

Muscari mus-*kah*-ree *Liliaceae (Hyacinthaceae).*
The Turkish name. Bulbous herbs. Grape
Hyacinth.
   armeniacum ar-men-ee-*ah*-kum. Of
   Armenia. SE Europe, Caucasus, Turkey.
   aucheri ow-ka-ree. After P. M. R. Aucher-
   Eloy (1792–1838). Turkey.
   azureum a-*zew*-ree-um. Sky-blue (the
   flowers). Caucasus, NW Turkey.
   botryoides bot-ree-*oi*-deez. Like a bunch of
   grapes. Europe.
   comosum ko-*mō*-sum. With a tuft (of sterile
   flowers). Tassel Hyacinth. Europe, W
   Asia.
      'Plumosum' ploo-*mō*-sum. Feathery
      (the inflorescence).
   latifolium lah-tee-*fo*-lee-um. Broad-leaved.
   NW Turkey.
   macrocarpum măk-rō-*kar*-pum. With large
   fruit. Greece, W Turkey.
   moschatum mos-*kah*-tum. Musk-scented.
   Musk Hyacinth. W Asia.
   neglectum ne-*glek*-tum. (= *M. racemosum*).
   Overlooked. Europe, N Africa, W Asia,
   Caucasus.
   paradoxum see *Bellevallia pycnantha*
   racemosum see *M. neglectum*
   tubergenianum see *M. aucheri*

Mutisia mew-*tis*-ee-a *Compositae.* After José
Celestino Mutis (1732–1808), Spanish
botanist who studied the S American flora.
Semi-hardy evergreen climbers. Climbing
Gazania.

Mutisia (continued)
   clematis kle-ma-tis. Climbing. Colombia.
   decurrens day-*ku*-renz. With the base of the
   leaf gradually merging with the stem.
   Chile, Argentina.
   ilicifolia ee-li-ki-*fo*-lee-a. *Ilex*-leaved.
   Chile.
   oligodon o-*li*-go-don. Few-toothed. Chile,
   Argentina.

Myosotidium mee-os-ō-*tid*-ee-um
Boraginaceae. From *Myosotis* q.v. to which it
is related. Evergreen perennial herb.
   hortensia hor-*tens*-ee-a. Of gardens (from
   where it was originally described).
   Chatham Island Forget-me-not. Chatham
   Islands.

Myosotis mee-os-ō-tis *Boraginaceae.* The Gk.
name for another plant, from *mus* (a mouse)
and *otos* (an ear) referring to the leaves.
Annual, biennial and perennial herbs.
Forget-me-not.
   alpestris ăl-*pes*-tris. (= *M. rupicola*). Of
   lower mountains. Europe.
   azorica a-*zo*-ri-ka. Of the Azores.
   caespitosa kie-spi-*tō*-sa. Tufted. Europe.
   scorpioides skor-pee-*oi*-deez. (= *M.
   palustris*). Like a scorpion's tail (the
   inflorescence). Europe, Asia.
   sylvatica sil-*vă*-ti-ka. Of woods. Europe,
   Asia.

Myrica mi-ree-ka *Myricaceae.* From *myrike*
the Gk. name for *Tamarix.* Deciduous and
evergreen shrubs.
   californica kăl-i-*forn*-i-ka. Of California.
   Californian Bayberry.
   cerifera kay-*ri*-fe-ra. Wax-bearing (the
   fruits). Wax Myrtle. SE United States.
   gale gah-lee. From old English *gagel*. Sweet
   Gale, Bog Myrtle. Northerly N
   hemisphere.
   pensylvanica pen-sil-*vahn*-i-ka. Of
   Pennsylvania. Bayberry. E N America.

Myricaria mi-ree-*kah*-ree-a. *Tamaricaceae.*
From Gk. *myrike (Tamarix)* which it
resembles. Deciduous shrub.
   germanica ger-*mahn*-i-ka. (= *Tamarix
   germanica*). Of Germany. Europe to the
   Himalaya.

Myriophyllum mi-ree-ō-*fil*-lum
Haloragidaceae. From Gk. *myrios* (many) and
phyllon (a leaf) referring to the finely divided
leaves. Aquatic herbs. Water Milfoil.
   aquaticum a-*kwah*-ti-kum. (= *M.

Myriophyllum (continued)
proserpinacoides). Growing in water. S
America.
heterophyllum he-to-rō-*fil*-lum. With
variable leaves. E North America.
spicatum spee-*kah*-tum. With flowers in
spikes. Temperate N hemisphere.
verticillatum ver-ti-ki-*lah*-tum. Whorled
(the leaves). Temperate N hemisphere.

Myrrhis mi-ris *Umbelliferae.* The Gk. name
for a plant. Perennial herb.
odorata o-dō-*rah*-ta. Scented. Sweet Cicely.
Europe.

Myrsine mur-si-nay *Myrsinaceae.* Gk. name
for myrtle. Evergreen shrub.
africana ă-ri-*kah*-na. African. African
Boxwood. Africa, Himalaya, China.

Myrtillocactus mur-ti-lō-*kăk*-tus *Cactaceae*
From L. *myrtillus* (a small myrtle) and *Cactus*
q.v. referring to the myrtle-like fruits.
Mexico.
geometrizans gee-ō-*met*-ri-zănz. Regularly
marked.
schenkii shenk-ee-ee. After Professor H.
Schenk, Director of the Darmstadt Botanic
Garden.

Myrtle see *Myrtus*
Common see *M. communis*
Tarentum see *M. communis* 'Tarentina'
Myrobalan see *Prunus cerasifera*

Myrtus mur-tus *Myrtaceae.* The Gk. and L.
name. Semi-hardy, evergreen trees and
shrubs. Myrtle.
bullata see *Lophomyrtus bullata*
communis kom-*ew*-nis. Common.
Common Myrtle. W Asia.
'Tarentina' tă-ren-*teen*-a. Of Taranto, S
Italy. Tarentum Myrtle.
luma see *Luma apiculata*
nummularia num-ew-*lah*-ree-a. With coin-
shaped leaves. S South America, Falkland
Islands.
ugni see *Ugni molinae*

N

Namaqualand Daisy see *Arctotis fastuosa*

Nandina năn-*deen*-a *Berberidaceae.* From
nanten the Japanese name. Evergreen shrub.
domestica do-*mes*-ti-ka. Cultivated. China.

Narcissus nar-*kis*-us *Amaryllidaceae.* After
Narcissus of Gk. mythology who, it is said,
was turned into this plant after killing himself
because he couldn't reach the person he saw
reflected in a pool. Bulbous perennials.
asturiensis a-stu-ree-*en*-sis. Of Asturia,
Spain. N Spain, N Portugal.
bulbocodium bul-bō-*kō*-dee-um. From Gk.
bulbos (a bulb) and *kodiom* (wool). Hoop-
petticoat Daffodil. SW Europe, N Africa.
citrinus ki-*tree*-nus. Lemon-yellow.
conspicuus kon-*spik*-ew-us.
Conspicuous.
romeiuxii see *N. romieuxii*
tenuifolus ten-ew-i-*fo*-lee-us. Slender-
leaved.
canaliculatus hort. see *N. tazetta italicus*
cantabricus kăn-*tăb*-ri-kus. Of Cantabria,
Spain. S Spain, N Africa.
monophyllus mo-nō-*fil*-lus. One-leaved.
N Africa.
cyclamineus sik-la-*min*-ee-us. Like
Cyclamen. NW Spain, NW Portugal.
jonquilla yong-*kwil*-la. From *junquillo* the
Spanish name, from *Juncus,* referring to
the slender leaves. Spain, Portugal.
juncifolius hort. see *N. requienii*
minor mi-nor. (= *N. nanus*). Smaller.
Pyrenees, N Spain.
nanus see *N. minor.*
× *odorus* o-dō-rus. *N. jonquilla* × *N.
pseudonarcissus.* Scented. Campernelle.
papyraceus pă-pi-*rah*-kee-us. Paper-like (the
white flowers). Mediterranean region,
SW Europe.
poeticus pō-*e*-ti-kus. Of poets. Poet's
Narcissus. C and S Europe.
recurvus re-*kur*-vus. Curved back (the
perianth segments). Pheasant's-eye
Daffodil.
pseudonarcissus soo-dō-nar-*kis*-us. False
Narcissus. Wild Daffodil. W Europe.
requienii rek-wee-*en*-ee-ee. (= *N.
juncifolius* hort.). After Requien, see *Mentha
requienii.* S France, Spain.
romieuxii rom-*ew*-ee-ee. (= *N. bulbocodium
romieuxii*). After Romieux of Geneva who
grew it. Morocco.
rupicola roo-*pi*-ko-la. Growing on rocks.
Spain, Portugal.
watieri wo-tee-*e*-ree. After M. Water,
Inspector of Woods and Forests in
Morocco c 1920. Atlas Mountains.
tazetta ta-*ze*-ta. Italian name meaning a
small cup. Mediterranean region.
italicus ee-*tă*-li-kus. (= *N. canaliculatus*
hort.). Italian. N and E Mediterranean
region.
triandrus tree-*ăn*-drus. With three stamens

Narcissus (continued)
(three are larger than the other three).
Angel's Tears. Portugal, Spain.
  concolor see *N. triandrus pallidulus*
  pallidulus pa-*lid*-ew-lus. (= *N. triandrus
  concolor*). Rather pale. Golden Angels
  Tears.
  viridiflorus vi-ri-di-*flō*-rus. With green
  flowers. SW Spain, Morocco.

Nasturtium nas-*tur*-tee-um *Cruciferae. From*
L. nasus tortus (a twisted nose) referring to
the smell of the leaves. Aquatic perennial
herb.
  officinale o-fi-ki-*nah*-lee. Sold as a herb.
  Watercress. Europe.

Nasturtium see *Tropaeolum majus*
Native's Comb see *Pachycereus
pectenaboriginum*
Navelwort see *Omphalodes*
  Venus's see *O. linifolia*
Neanthe bella see *Chamaedorea elegans*

Nectaroscordum nek-ta-rō-*skor*-dum *Liliaceae*
(*Alliaceae*). From Gk. *nektar* (nectar) and
skordon (garlic). Perennial herb.
  siculum sik-ew-lum. (= *Allium siculum*). Of
  Sicily. S Europe, Turkey.

Neillia neel-ee-a *Rosaceae*. After Patrick Neill
(1776–1851), Scottish naturalist. Deciduous
shrubs.
  sinensis si-*nen*-sis. Of China.
  thibetica ti-*be*-ti-ka. (= *N. longeracemosa*). Of
  Tibet. China.

Nelumbo ne-*lum*-bō *Nymphaeaceae*. The
Sinhalese name. Aquatic perennial herbs.
Lotus.
  lutea loo-tee-a. Yellow (the flowers).
  American Lotus. E North America.
  nucifera new-*ki*-fe-ra. Nut-bearing. Indian
  Lotus, Sacred Lotus. S Asia to Australia.

Nemesia ne-*me*-see-a *Scrophulariaceae*. From
nemesion the Gk. name for a similar plant.
Annual herbs. S Africa.
  strumosa stroo-*mō*-sa. With cushion-like
  swellings.
  versicolor ver-*si*-ko-lor. Variously coloured.
  compacta kom-păk-ta. Compact.

Nemophila ne-*mo*-fi-la *Hydrophyllaceae*. From
Gk. *nemos* (a glade) and *phileo* (to love), they
grow in shady places. Annual herbs. Baby
Blue Eyes. California.
  maculata măk-ew-*lah*-ta. Spotted (the
  corolla).

Nemophila (continued)
  menziesii men-zeez-ee-ee. (= *N. insignis*).
  After Menzies, see *Menziesia*.

Neobuxbaumia nee-ō-bux-*bow*-mee-a
Cactaceae. After F. Buxbaum (1900–79),
Austrian specialist in Cacti. *Gk. neo*, new,
distinguishes the genus from *Buxbaumia*, a
genus of mosses. Huge columnar or tree-like
cacti.
  euphorbioides u-*for*-bee-oy-dees (=
  Cephalocereus euphorbioides). Euphorbia-
  like. E Mexico.
  polylopha po-lee-*lō*-pha (= *Cephalocereus
  polylopha*). Many crested, referring to the
  numerous ribs. E Mexico.

Neolitsea nee-ō-*lit*-see-a *Laureaceae*. From
Gk. *neos* (new) and *Litsea*, a related genus
(from the Japanese name). Semi-hardy,
evergreen shrub or tree.
  sericea say-*ri*-kee-a. Silky (the young
  growths), Japan, Korea, China.

Neolloydia nee-ō-*loyd*-ee-a *Cactaceae.*/ After
Francis Ernst Lloyd (1868–1947), American
botanist, the name *Lloydia* having already
been used for a genus of bulbous herbs.
  grandiflora see *N. conoidea*
  conoidea kon-ō-*i*-dee-a. (= *N. texensis*).
  Cone-like. Texas, N Mexico.
  macdowellii see *Thelocactus macdowellii*

Neoporteria nee-ō-por-*te*-tree-a *Cactaceae*.
After Carlos Porter, a Chilean entomologist.
Chile.
  fusca fus-ka. Dark (the spines).
  subgibbosa sub-gi-*bō*-sa. Somewhat swollen
  on one side.
  villosa vil-*lō*-sa. Softly hairy.

Neoregelia nee-ō-ray-*gel*-ee-a *Bromeliaceae*.
After Eduard Albert von Regel (1815–92).
Tender, evergreen herbs. Brazil.
  carolinae kǎ-ro-*leen*-ie. (= *N. marechallii*).
  After Carolina. Blushing Bromeliad.
  'Tricolor' *tri*-ko-lor. Three-coloured
  (the leaves).
  marmorata mar-mo-*rah*-ta. Marbled (the
  leaves). Marble Plant.
  sarmentosa sar-men-*tō*-sa. Creeping.
  spectabilis spek-*tah*-bi-lis. Spectacular.
  Finger-nail Plant.

Nepenthes nay-*pen*-theez *Nepenthaceae*. Gk.
name of a plant. tender herbs. Pitcher Plant.
  gracilis grǎ-ki-lis. Graceful. SE Asis.
  khasiana kah-zee-*ah*-na. Of the Khasi
  Hills, Assam.

Nepenthes (continued)
maxima mahk-si-ma. Larger. Borneo to
New Guinea.
mirabilis mee-*rah*-bi-lis. Wonderful. SE
Asia.
rafflesiana răf-alz-ee-*ah*-na. After Sir
Thomas Stamford Raffles (1781–1826),
scientific patron and founder of Singapore.
SE Asia.
sanguinea sang-*gwin*-ee-a. Blood-red (the
pitchers). Malaysia.
ventricosa ven-tri-*kō*-sa. Swollen on one
side (the pitchers). Philippines.

Nepeta ne-pe-ta *Labiatae*. The L. name.
Perennial herbs.
cataria ka-*tah*-ree-a. Of cats, which are
attracted to it. Catmint, Catnip. Europe,
Asia.
× *faassenii* fah-*sen*-ee-ee. *N. mussinii* × *N.
nepetella*. After J. H. Faassen, Dutch
nurseryman.
grandiflora grănd-i-*flō*-ra. large-flowered.
Caucasus.
mussinii see *N.* × *faassenii*
nepetella ne-pe-*tel*-la. Diminutive of
Nepeta. SW Europe.
nervosa ner-*vō*-sa. Conspicuously veined.
Kashmir.

Nephrolepis nef-rō-*lep*-is *Oleandraceae*. From
Gk. *nephros* (a kidney) and *lepis* (a scale)
referring to the shape of the indusium.
Tender ferns. Sword Fern.
cordifolia kor-di-*fo*-lee-a. With heart-
shaped leaves. Tropics and sub-tropics.
exaltata eks-al-*tah*-ta. Very tall. Tropics.
'Bostoniensis' bos-ton-ee-*en*-sis. Of
Boston. Boston fern.

Nerine nay-*ree*-nay *Amaryllidaceae*. After
Nerine, a sea nymph. Semi-hardy, Bulbous
herbs. S Africa.
bowdenii bow-den-ee-ee. After Mr
Athelston Bowden who introduced it.
filifolia fee-li-*fo*-lee-a. With thread-like
leaves.
flexuosa fleks-ew-*ō*-sa. Wavy (the perianth
lobes).
sarniensis sar-nee-*en*-sis. Of Guernsey
(Sarnia), where it has long been naturalised.
undulata un-dew-*lah*-ta. Wavy (the
perianth lobes).

Nerium nay-ree-um *Apocynaceae* The Gk.
name. Tender evergreen, poisonous shrub.
oleander o-lee-*ăn*-der. From *oleandra*, the
Italian name. Oleander. Mediterranean
region to E Asia.

Nertera ner-te-ra *Rubiaceae*. From Gk. *nerteros*
(lowly) referring to the dwarf habit. Tender
herb.
granadensis grăn-a-*den*-sis. (= *N. depressa*).
Of Granada, Colombia. Bead Plant. S
America, New Zealand and Tasmania.

Nettle Tree see *Celtis*
Never-never Plant see *Ctenanthe
oppenheimiana*
New Zealand Burr see *Acaena*
New Zealand Daisy see *Celmisia*
New Zealand Flax see *Phormium tenax*
New Zealand Lilac see *Hebe hulkeana*

Nicandra ni-*kăn*-dra *Solanaceae*. After
Nikander of Colophon, Greek physician
and poet c 137 B.C. Annual herb.
physalodes fi-sa-*lō*-deez. Like *Physalis*.
Apple of Peru. Peru.

Nicotiana nee-kō-tee-*ah*-na *Solanaceae*. After
Jean Nicot (1530–1600) who introduced
the tobacco plant to France. Annual herbs.
alata ah-*lah*-ta. (= *N. affinis*). Winged (the
petioles). S America.
glauca glow-ka. Glaucous (the shoots and
leaves). Bolivia, Argentina.
× *sanderae* sahn-da-rie. *N. alata* × *N.
forgetiana*. After Mrs Sander.
sylvestris sil-*ves*-tris. Of woods. Argentina.
tabacum ta-*băk*-um. Said to be the
Caribbean name for a pipe or from
Haitian *taina*, a roll of tobacco in a maize
leaf. Tobacco Plant. Cult.

Nidularium need-ew-*lah*-ree-um
Bromeliaceae. From L. *nidus* (a nest), the
flowers are borne in a nest-like depression in
the centre of a cluster of bracts. Tender,
evergreen herbs. Brazil.
fulgens ful-gens. Shining (the bracts).
innocentii in-o-*sent*-ee-ee. Said to be
named after Pope Innocenti.
striatum stree-*ah*-tum. (= *N. striatum*).
Striped (the leaves).
purpureum pur-*pewr*-ree-um. Purple (the
leaves).
rutilans ru-ti-lănz. Reddish (the flowers).

Nierembergia nee-e-ram-*berg*-ee-a *Solanaceae*.
After Juan Eusebio Nieremberg
(1595–1658), a Spanish Jesuit. Perennial
herbs. Cup Flower.
hippomanica hip-o-*măn*-i-ka. From Gk.
hippomanes, a plant that drives horses mad
or they love to eat. Argentina.
violacea vee-o-*lah*-kee-a. (= *N. caerulea*).
Violet (the flowers)

Nierembergia (continued)
repens ree-pens. (= *N. rivularis*). Creeping.
S America.

Nigella ni-*gel*-la *Ranunculaceae*. Diminutive of
L. *niger* (black) referring to the black seeds.
Annual herbs.
arvensis ar-*ven*-sis. Of fields. N Africa,
Europe, W Asia.
damascena dăm-a-*skay*-na. Of Damascus.
Love-in-a-mist. S Europe, N Africa.
hispanica his-*pah*-ni-ka. Spanish. Spain,
Portugal.
integrifolia in-teg-ri-*fo*-lee-a. With entire
(lower) leaves. Turkestan.
orientalis o-ree-en-*tah*-lis. Eastern. SW
Asia.
sativa sa-*teev*-a. Cultivated (the seeds are
used in seasoning). SE Europe, W Asia.

Ninebark se *Physocarpus*
Nirre see *Nothofagus antarctica*

Nolana nō-lah-na *Nolanaceae*. From L. *nola*
(a small bell) referring to the shape of the
corolla. Perennial herbs. Chilean Bellflower.
Chile.
acuminata see *N. paradoxa* ssp *atriplicifoliia*
paradoxa pă-ra-*doks*-a. Unusual.
atriplicifolia ă-tri-pli-ki-*fo*-lee-a (= *N.
acuminata*). *Atriplex*-leaved, Peru, Chile.

Nomocharis no-mō-*kă*-ris *Liliaceae*. From
Gk. *nomos* (a meadow) and *charis* (grace).
Bulbous perennials.
aperta a-*per*-ta. Closed. China.
farreri *fă*-ra-ree. After Farrer, see
Viburnum farreri. Burma.
mairei see *N. pardanthina*
pardanthina par-dan-*theen*-a. Like
Pardanthina (Belamcanda), the spotted
flowers. China.
saluenensis săl-ew-en-*en*-sis. From near the
Nu Jiang (Salween River), W Yunnan.
Burma, China, Tibet.

Nopalxochia nō-pal-*ho*-kee-a *Cactaceae*.
From a Mexican name. Mexico.
ackermannii ă-ker-*măn*-ee-ee.
(=*Epiphyllum ackermannii*). After Georg
Ackermann who introduced it. Orchid
Cactus.
phyllanthoides fil-lanth-*oi*-deez. Like
Phyllanthus.

Norfolk Island Pine see *Araucaria*
heterophylla

Nothofagus no-thō-*fah*-gus *Fagaceae*. From

Nothofagus (continued)
Gk. *nothos* (false) and *fagus* (beech) but
notofagus (southern beech) may have been
intended. Deciduous and evergreen trees.
Southern Beech.
antarctica ăn-*tark*-ti-ka. Of Antarctic
regions. Nirre. Southern S America.
betuloides bet-ew-*loi*-deez. Like *Betula*. S
Chile, S Argentina.
dombeyi dom-*bee*-ee. After Dombey, see
Dombeya. Chile, Argentina.
fusca *fus*-ka. Brown. New Zealand.
menziesii men-*zeez*-ee-ee. After Menzies
who collected the type specimen, see
Menziesia. New Zealand.
obliqua o-*blee*-kwa. Oblique (the leaf base).
Roblé. Chile, Argentina.
procera prō-*kay*-ra. Tall. Rauli. Chile,
Argentina.
solanderi so-*lăn*-de-ee. After Daniel Carl
Solander (1736–82), a botanist on Cook's
first voyage. New Zealand.

Notocactus no-tō-*kăk*-tus *Cactaceae*. From
Gk. *notos* (southern) and *Cactus* q.v.
referring to their southerly distribution.
An obsolete name, see *Parodia*

Notospartium no-tō-*spar*-tee-um
Leguminosae. From Gk. *notos* (southern) and
Spartium q.v. Semi-hardy shrub.
carmichaeliae kar-mie-*keel*-ee-ie. From the
resemblance to *Carmichaelia*. New
Zealand.

Nuphar new-far *Nymphaeaceae*. From the
Arabic name. Aquatic perennial herbs.
advena ad-*ven*-a. Adventive. Spatterdock. E
and C United States.
lutea loo-tee-a. Yellow. Brandy Bottle,
Yellow Water Lily. N hemisphere.
pumila *pew*-mi-la. Dwarf. Europe, Asia.

Nyctocereus nik-tō-*kay*-ree us *Cactaceae*.
From Gk. *nyktos* (night) and *Cereus* q.v.,
they flower at night.
chontalensis chon-ta-*len*-sis. From the
territory of the Chontal Indians, Oaxaca.
S Mexico.
serpentinus see *Peniocereus serpentinus*

Nymphaea nimf-*ie*-a *Nymphaeaceae*. The
classical name after Nymphe, a water
nymph. Aquatic perennial herbs.
alba *ăl*-ba. White. White Water Lily.
Europe, N Africa, Asia.
caerulea kie-*ru*-lee-a. Deep blue (the
flowers). Blue Egyptian Lotus. N and C
Africa.

Nymphaea (continued)
candida kǎn-di-da. White. N Europe, N
Asia.
capensis ka-pen-sis. Of the Cape of Good
Hope. Cape Blue Water Lily. S Africa.
lotus lō-tus. A Gk. name for several plants.
White Egyptian Lotus. Egypt.
× *marliacea* mar-lee-ǎ-kee-a. After Joseph
Latour Marliac (born 1830).
odorata o-dō-rah-ta. Scented. E United
States.
pygmaea see *N. tetragoma*.
stellata ste-*lah*-ta. Star-like. S and E Asia.
tetragona tet-ra-gōn-a. (= *N. pygmaea*). Four
angled. NE Asia, N America.
tuberosa tew-be-rō-sa. Tuberous. N
America.

Nymphoides nimf-*oi*-deez *Menyanthaceae* Like
Nymphaea. Aquatic perennial herb.
peltata pel-*tah*-ta. With the petiole attached
to the lower surface of the leaf blade,
literally, shield-like. Europe, Asia.

Nyssa ni-sa *Nyssaceae*. After Nyssa (Nysa) a
water nymph, the first-described species, *N.
aquatica* grows in swamps. Deciduous trees.
sinensis si-*nen*-sis. Of China.
sylvatica sil-*vǎ*-ti-ka. Of woods. Black
Gum, Tupelo. EN America.

O

Oak see *Quercus*
  Black see *Q. velutina*
  Black Jack see *Q. marilandica*
  Common see *Q. robur*
  Cork see *Q. suber*
  Daimio see *Q. dentata*
  Durmast see *Q. petraea*
  Golden, of Cyprus see *Q. alnifolia*
  Holm see *Q. ilex*
  Hungarian see *Q. frainetto*
  Lebanon see *Q. libani*
  Lucombe see *Q.* × *hispanica*
  'Lucombeana'
  Pedunculate see *Q. robur*
  Pin see *Q. palustris*
  Red see *Q. rubra*
  Scarlet see *Q. coccinea*
  Sessile see *Q. petraea*
  Shingle see *Q. imbricaria*
  Tanbark see *Lithocarpus densiflorus*
  Turkey see *Q. cerris*
Oat Grass see *Arrhenatherum elatius*
Obedient Plant see *Physostegia virginiana*

Ocean Spray see *Holodiscus discolor*

Ocimum ō-ki-mum *Labiatae*. From *okimom*
the Gk. name for an aromatic herb. Annual
herbs. SE Asia.
basilicum ba-*si*-li-kum. The classical name,
meaning royal or princely. Basil.
'Minimum' *mi*-ni-mum. Smaller. Bush
Basil.

Oconee Bells see *Shortia galacifolia*

× **Odontioda** o-don-tee-ō-da *Orchidaceae*.
Intergeneric hybrids, from the names of the
parents. *Cochlioda* × *Odontoglossum*.
Greenhouse orchids.

×**Odontocidium** o-don-to-*kid*-ee-um
Orchidaceae. Intergeneric hybrids, from the
names of the parents. *Odontoglossum* ×
Oncidium. Greenhouse orchids.

Odontoglossum o-don-to-*glos*-um
Orchidaceae. From Gk. *odontos* (a tooth) and
glossa (a tongue) referring to the toothed lip.
Greenhouse orchids.
bictoniense see *Lemboglossum bictoniense*
cervantesii see *Lemboglossum cervantesii*
citrosmum see *Cuitlauzina pendula*
cordatum see *Lemboglossum cordatum*
crispum kris-pum. Wavy-edged (the petals).
Lace orchid. Colombia.
cristatum kris-*tah*-tum. Crested (the disc).
Colombia, Ecuador.
grande see *Rossioglossum grande*
harryanum hǎ-ree-*ah*-num. After Sir Harry
Veitch. Colombia, Peru.
laeve lie-vee. Smooth. Mexico, Guatemala.
nobile nō-bi-lee. (= *O. pescatorei*) Notable.
Colombia.
pendulum see *Cuitlauzina pendula*
pescatorei see *O. nobile*
pulchellum see *Osmoglossum pulchellum*
rossi see *Lemboglossum rossii*
schlieperianum shlee-pa-ree-*ah*-num. After
Adolph Schlieper, an orchid collector.
Costa Rica, Panama.
triumphans tree-*um*-fanz. Splendid.
Colombia.
uroskinneri see *Lemboglossum uroskinneri*

× **Odontonia** o-don-*ton*-ee-a *Orchidaceae*.
Intergeneric hybrids, from the names of the
parents. *Miltonia* × *Odontolglossum*.
Greenhouse orchids.

Oemleria oom-*ler*-ee-a *Rosaceae*. For Herr
Oemler of Dresden. Deciduous shrub,
related to *Prunus*.

Oemleria (continued)
cerasiformis ke-ra-si-*form*-is (= *Osmaronia cerasiformis*). Cherry-shaped (the fruit). Oso–Berry. California.

Oenothera oy-nō-*the*-ra *Onagraceae*. From *oinotheras* the Gk. name of a plant. Biennial and perennial herbs.
acaulis a-*kaw*-lis (= *O. taraxacifolia*). Stemless. Chile.
berlandieri see *O. speciosa*
biennis bee-*en*-is. Biennial. Evening Primrose. E N America.
caespitosa kie-spi-*tō*-sa. Tufted. W N America.
    riparia see *O. tetragona riparia*
erythosepala see *O. glazioviana*
frutcosa froo-ti-*kō*-sa. Shrubby (which it isn't). Sundrops. E United States.
    glauca glow-ka (= *O. tetragona*). Glaucous (grey-green).
glazioviana gley-zee-o-vi-*ah*-na (= *O. lamarkiana, O. riparia*). After Auguste François Marie Glaziou (1828–1906), botanist. N W Europe.
grandiflora see *Clarkia amoena* ssp *lindleyi*
laciniata la-kin-ee-*ah*-ta. Deeply cut (the leaves). United States.
lamarckiana see *O. glazioviana*
macrocarpa mak-rō-*kar*-pa (= *O. missouriensis*). With large fruit. SC United States.
missouriensis see *O. macrocarpa*.
perennis pe-*ren*-is. Perennial. E N America.
speciosa spe-kee-*ō*-sa. Showy. S Central United States.
stricta strik-ta. Erect. S America.
taraxacifolia see *O. acaulis*
tetragona see *O. fruticosa*
ssp. *glauca*
    riparia see *O. fruticosa*

Okra see *Abelmoschus esculentus*
Old Maid see *Catharanthus roseus*
Old Man's Beard see *Clematis vitalba*
Old Woman see *Artemisia stelleriana*

Olea o-lee-a *Oleaceae*. The L. name. Semi-hardy, evergreen tree.
    europaea oy-rō-*pie*-a. European. Olive. SW Asia.

Oleander see *Nerium oleander*

Olearia o-lee-*ah*-ree-a *Compositae*. After Adam Öschläger (Olearius) (1603–71). Evergreen shrubs. Daisy Bush.
    avicenniifolia ă-vi-sen-ee-i-*fo*-lee-a. With leaves like *Avicennia*. New Zealand.

Olearia (continued)
chathamica cha-*tăm*-i-ka. Of the Chatham Islands.
frostii frost-ee-ee. After Charles Frost. Victoria (Australia).
gunniana see *O. phlogopappa*
× *haastii* hahst-ee-ee. *O. avicenniifolia* × *O. moschata*. After Sir Johann Franz Julius von Haast (1824–87) who collected the type specimen. New Zealand.
ilicifolia ee-li-ki-*fo*-lee-a. *Ilex*-leaved. New Zealand.
insignis in-*sig*-nis. (= *Pachystegia insignis*). Notable. New Zealand.
macrodonta măk-rō-*don*-ta. With large teeth (the leaves). New Zealand.
× *mollis* mol-lis. *O. ilicifolia* × *O. lacunosa* Softly hairy. The plant commonly grown under this name is of the parentage *O. ilicifolia* × *O. moschata*.
nummulariifolia num-ew-lah-ree-i-*fo*-lee-a. With coin-shaped leaves. New Zealand.
phlogopappa flog-ō-*pă*-pa. (= *O. gunniana*. *O. stellulata* hort.). With a *Phlox*-like pappus. SE Australia, Tasmania.
    Spendens *splen*-denz. (= *O. gunniana* 'Spendens'. *O. stellulata* 'Splendens'). Splendid.
× *scillonienis* si-lon-ee-*en*-sis. *O. lirata* × *O. phlogopappa*. Of the Scilly Isles, it was raised at Tresco.
semidentata se-mee-den-*tah*-ta. Half toothed, the leaves are toothed in the upper half. Chatham Islands.
stellulata 'Spendens' see *O. phlogopappa* Spendens
traversii tra-*vers*-ee-ee. After W. T. L. Travers (1819–1903). Chatham Islands.
virgata vir-*gah*-ta. Twiggy. New Zealand.
    'Wiakariensis' wie-kah-ree-*en*-sis. Of Waikari, New Zealand.
    'Zennorensis' zen-or-*ren*-sis. *O. ilicicfolia* × *O. lacunosa*. Of Zennor, Cornwall.

Olive see *Olea europaea*

Olsynium ol-*sin*-ee-um *Iridaceae*. From Gk. word meaning scarcely united (the stamens). Perennial herbs closely related to *Sisyrinchium*.
    douglasii dūg-*lăs*-ee-ee (= *Sisyrinchium douglasii*). After Douglas, see *Iris douglasiana* W N America.

Omphalodes omf-a-*lō*-deez *Boraginaceae*. From Gk. *omphalos* (a navel) referring to a navel-like depression in the seeds. Perennial herbs. Navelwort.

Omphalodes (continued)
cappadocica kăp-a-*do*-ki-ka. Of Cappadocia (Turkey). W Asia.
linifolia lee-ni-*fo*-lee-a. *Linum*-leaved. Venus's Navelwort. SW Europe.
luciliae loo-*sil*-ee-ie. After Lucile Boissier. Greece, W Asia.
verna *ver*-na. Of Spring (flowering). Blue-eyed Mary. SE Europe.

Oncidium ong-kid-ee-um *Orchidaceae*. From Gk. *onkos* (a tumour) referring to a swelling on the lip. Greenhouse orchids.
altissimum ăl-*tis*-i-mum. Tallest. W Indies.
cavendishianum kă-van-dish-ee-*ah*-num. After William George Spencer Cavendish, 6th Duke of Devonshire. Mexico, Guatemala.
cebolleta see *O. longifolium*
cheriophorum kay-*ro*-fo-rum. Hand-bearing. Colombia Buttercup. Costa Rica, Panama.
concolor kon-ko-lor. Similarly coloured (the petals and sepals). Brazil.
crispum kris-pum. Finely wavy (the petals). Brazil.
cucullatum ku-kew-*lah*-tum. Hood-like. Colombia, Ecuador.
flexuosum fleks-ew-*ō*-sum Tortuous. Dancing Doll Orchid. Brazil.
incurvum in-*kur*-vum. Incurved. Mexico.
longifolium long-gi-*fo*-lee-um. (= *O. cebolleta*). Long-leaved. Tropical America.
longipes long-gi-pays. Long-stalked. Brazil.
luridum loo-ri-dum. Pale yellow. Tropical America.
macranthum ma-krănth-um. Large-flowered. Ecuador.
marshallianum mar-shăl-ee-ah-num. After Mr W Marshall of Enfield who grew the type specimen. Brazil.
ornithorhyncum or-ni-thō-*ring*-kum. Like a bird's beak. Mexico to S America.
papilio see *Psychopsis papilio*
phalaenopsis fă-lie-*nop*-sis. Moth-like. Colombia, Ecuador.
pulchellum pul-*kel*-um. Pretty. Jamaica, Guyana.
pumilum *pew*-mi-lum. Dwarf. Brazil.
pusillum pu-*sil*-um. Dwarf. C and S America.
sarcodes sar-*kō*-deez. Flesh-like. Brazil.
splendidum *splen*-di-dum. Splendid. Guatemala.
triquetrum tri-*kwee*-trum. Three-angled (the pseudobulbs). Jamaica.
varicosum vah-ri-*ko*-sum. With dilated veins. Brazil.

Oncidium (continued)
wentworthianum went-wurth-ee-*ah*-num. After Lord Fitzwilliam. Guatemala.

Onion see *Allium cepa*
Tree see *A. cepa* Proliferum
Welsh see *A. fistulosm*

Onoclea o-nok-lee-a *Dryopteridaceae* (*Athyriaceae*) From *onokleia*, Gk. name for another plant, from *onos* (a vessel) and *kleio* (to close), the pinnules of the fertile fronds curl round the sori, enclosing them. Fern.
sensibilis sen-*si*-bi-lis. Sensitive (to early frosts). Sensitive Fern. Temperate N hemisphere.

Ononis o-*nō*-nis *Leguminosae (Papilionoideae)*. The Gk. name. Perennial herbs and sub-shrubs.
aragonensis ă-ra-gon-*en*-sis. Of Aragon, NE Spain. Pyrenees, Spain, N Africa.
fruticosa froo-ti-*kō*-sa. Shrubby. SW Europe.
rotundifolia ro-tun-di-*fo*-lee-a. With rounded leaves (leaflets). C and SW Europe.

Onopordum o-nō-*por*-dum *Compositae*. From *onopordon*, the Gk. name. Biennial herbs.
acanthium a-kănth-ee-um. Spiny. Scotch Thistle, Europe, W Asia.
nervosum ner-*vō*-sum (= *O. arabicum* hort.). Veined (the undersides of the leaves). Spain, Portugal.
tauricum tow-ri-kum. Of the Crimea. Balkan peninsula, Black Sea region.

Onosma o-*nos*-ma *Boraginaceae*. From Gk. *onos* (an ass) and *osme* (smell) referring to the roots. Perennial herbs.
alboroseum ăl-nō-*ros*-ee-um. White and rose-coloured, the flowers change from white to red. W Asia.
echioides e-kee-*oi*-deez. Like *Echium*. Italy, SE Europe.
stellulatum stel-ew-*lah*-tum. With small stars, referring to the star-shaped hairs. Yugoslavia.
tauricum tow-ri-kum. Of the Crimea. Golden Drop. SE Europe. W Asia.

Ophiopogon o-fee-ō-*pō*-gon *Liliaceae* (*Convallariaceae*). From Gk. *ophis* (a snake) and *pogon* (a beard). Perennial herbs with grass-like foliage.
japonicus ja-*pon*-i-kus. Of Japan. Japan, Korea.

Ophiopogon (continued)
planiscapus plahn-i-*skah*-pus. With a flat
scape. Japan.
'Nigrescens' ni-*gres*-enz. Blackish (the
leaves).

Ophrys of-ris *Orchidaceae*. Gk. name for an
orchid. Hardy, terrestrial orchids.
apifera a-*pi*-fe-ra. Bee-bearing, the flowers
resemble bees. Bee Orchid. Europe.
fusca fus-ka. Brown. Mediterranean region.
speculum see *O. vernixia*
vernixia ver-*nix*-ee-a (= *O. speculum*).
Varnished; the flowers are iridescent.
Mediterranean region.

Oplismenus op-*lis*-men-us *Gramineae*. From
Gk. *hoplismos* (a weapon) referring to the
awns. Tender grass.
hirtellus hir-*tel*-us. Rather hairy. Basket
Grass. Texas to S America.

Opuntia o-*pun*-tee-a *Cactaceae*. Gk. name of
a plant that grew near Opus (Opuntis) in
Ancient Greece. Prickly Pear.
articulata ar-tik-ew-*lah*-ta. Jointed (the
stem). Argentina.
'Diademata' dee-a-day-*mah*-ta. (= *O.
diademata*). Crowned.
basilaris bă-si-*lah*-ris. Basal, it branches from
the base. SW United States, N Mexico.
bergeriana ber-ga-ree-*ah*-na. After Alwyn
Berger (1871–1931). Cult.
bigelovii big-a-*lov*-ee-ee. After Jacob
Bigelow (1787–1879). SW United States,
N Mexico.
brasiliensis bra-zil-ee-*en*-sis. (=
Brasiliopuntia brasiliensis). Of Brazil.
compressa kom-*pres*-a (= *O. humifusa*)
Flattened sideways. E & C USA,
naturalized in Switzerland.
cylindrica si-*lin*-dri-ka. (=
Austrocylindropuntia cylindrica). Cylindrical
(the stem). Cane Cactus. Ecuador, Peru.
decumbens day-*kum*-benz. Prostrate.
Mexico, Guatemala.
diademata see *O. articulata* 'Daidemata'
ficus-indica fee-kus-in-di-ka. Fig of India.
Indian Fig Cactus. Cult.
humifusa see *O. compressa*.
humilis see *O. tuna*
imbricata im-bri-*kah*-ta. Densely
overlapping. Chain-link Cactus. SW
United States. Mexico.
kleiniae klien-ee-ie. From the resemblance
to *Kleinia*. SW United States.
leucotricha loo-*ko*-tri-ka. With white hairs.
Mexico.

Opuntia (continued)
macrorhiza măk-rō-*ree*-za. With large roots.
United States.
microdasys mik-rō-*dăs*-is. From Gk. *mikros*
(small) and *dasys* (shaggy) referring to the
small areoles. N Mexico.
'Albispina' ăl-bi-*speen*-a. White-spined.
rufido roo-fi-da. (= *O. rufida*). Reddish (the
spines). Cinnamon Cactus. Texas, N
Mexico.
monacantha mon-a-kan-tha (= *O. vulgaris*)
Single-spined. Brazil, Argentina.
ovata ō-*vah*-ta. Ovate (the fruit). Andes.
polyacantha po-lee-a-*kănth*-a. Many-
spined. W N America.
robusta rō-*bus*-ta. Robust. Mexico.
rufida roo-fi-da. Red-brown (the
glochids). SW US, N Mexico.
salmiana săl-mee-*ah*-na. After Prince
Joseph Salm-Reifferscheid-Dyck
(1773–1861). German authority on
succulents. Brazil, Argentina.
scheeri shear-ree. After Frederick Scheer (c.
1792–1868) an amateur botanist who
grew cacti. Mexico.
subulata soo-*bew*-lah-ta. Awl-shaped (the
leaves). Eve's Pin Cactus. Argentina.
sulphurea sul-*fa*-ree-a. Sulphur-yellow (the
flowers). S America.
tomentosa tō-men-*tō*-sa. Hairy. Mexico.
tuna too-na. (= *O. humilis*). Mexican name
for the *Opuntia* fruit. Jamaica.
tunicata tun-i-*kah*-ta. Coated, the white,
papery sheaths on the spines. Texas to
Chile.
verschaffeltii vair-sha-*felt*-ee-ee. After
Verschaffelt. N Bolivia.
vestita ves-*tee*-ta. Clothes with hairs (the
areoles). Cotton-pole Cactus. Bolivia.
vulgaris see *O. monacantha*

Orache see *Atriplex hortensis*
Orange, Bitter see *Citrus aurantium*
    Seville see *C. aurantium*
    Sweet see *C. sinensis*

Orbea or-*bee*-a *Asclepiadaceae* From Lat. *orbis*,
a disc; referring to the rim of the corolla.
Perennial herbs related to and sometimes
included in *Stapelia*.
variegata vă-ree-a-*gah*-ta (= *Stapelia
variegata*) Varigated (the corolla). Starfish
Plant, Toad Plant. Cape Province.

Orchid, Bee see *Ophrys apifera*
    Black see *Coelogyne pandurata*
    Butterfly see *Psychopsis papilio*
    Common Spotted see *Dactylorhiza fuchsii*
    Cradle see *Anguloa*

Orchid (continued)
  Dancing Doll see *Oncidium flexuosum*
  Early Purple see *Orchis mascula*
  Fox-tail see *Aerides*
  Lace see *Odontoglossum crispum*
  Lady's Slipper see *Cypripedium*
  Lily of the Valley see *Osmoglossum pulchellum*
  Meadow see *Dactylorhiza incarnata*
  Moth see *Phalaenopsis*
  Pansy see *Miltonia*
  Ram's Head Lady's Slipper see *Cypripedium arietinum*
  Slipper see *Paphiopedilum*
  Soldier see *Orchis militaris*
  Star of Bethlehem see *Angraecum sesquipedale*
  Tiger see *Rossioglossum grande*
Orchid Bush see *Bauhinia acuminata*
Orchid Tree, Purple see *Bauhinia variegata*

Orchis or-kis *Orchidaceae*. The Gk. name. hardy terrestrial orchids.
  elata see *Dactylorhiza elata*
  foliosa see *Dactylorhiza foliosa*
  fuchsii see *Dactylorhiza fuchsii*
  incarnata see *Dactylorhiza incarnata*
  maderensis see *Dactylorhiza foliosa*
  mascula mahs-kew-la. Male, compared to less robust 'female' species. Early Purple orchid. Europe, N Africa, N and W Asia.
  militaris mee-li-tah-ris. Like a soldier. Soldier orchid. Europe. W Asia, Siberia.
  praetermissa see *Dactylorhiza praetermissa*
  spectabilis spek-tah-bi-lis. Spectacular. E N America.

Oregon Grape see *Mahonia aquifolium*

Origanum o-ree-*gah*-num *Labiatae*. The Gk. name. Perennial herbs.
  amanum a-*mah*-num. Of the Amanus Mts., S Turkey.
  dictamnus dik-*tăm*-nus. The Gk. name. Cretan Dittany. Crete.
  × *hybridum* hib-ri-dum. *O. dictamnus* × *O. sipyleum*. Hybrid.
  libanoticum li-ba-*no*-ti-kum. Of Lebanon.
  majorana ma-jo-*rah*-na. An old name from the Gk. name *amarakus*. Sweet Marjoram. N Africa, SW Asia.
  onites o-nee-teez. Gk. name for a kind of marjoram. Pot Marjoram. Mediterranean region, W Asia.
  rotundifolium ro-tun-di-*fo*-lee-um. Round-leaved. Turkey.
  scabrum skăb-rum. Rough. S Greece.
  vulgare vul-*gah*-ree. Common. Wild Marjoram. Europe to C Asia

Ornithogalum or-ni-*tho*-ga-lum *Liliaceae (Hyacinthaceae)*. From Gk. *ornis* (a bird) and *gala* (milk). Bulbous, perennial herbs.
  arabicum a-*ră*-bi-kum. Arabian. Mediterranean region.
  balansae see *O. oligophyllum*
  montanum mon-*tah*-num. Of mountains. SE Europe, Turkey.
  nutans new-tănz. Nodding (the flowers). SE Europe, Turkey.
  oligophyllum ō-lee-go-*fil*-um (= *O. balansae*). Few-leaved. Balkans, Turkey, Georgia.
  thyrsioides thur-*soi*-deez. With flowers in a thyrse (a type of inflorescence). Chincherinchee. S Africa.
  umbellatum um-bel-*ah*-tum. The flowers appear to be in umbels. Star of Bethlehem. Europe, N Africa.

Orontium o-*ron*-tee-um *Araceae*. Classical name for a water plant growing in the Syrian river Orontes. Aquatic perennial herb.
  aquaticum a-*kwah*-ti-kum. Growing in water. Golden Club. E United States.

Oroya o-*roy*-a *Cactaceae*. After La Oroya in the Peruvian Andes where the following grows.
  peruviana pe-roo-vee-*ah*-na. Of Peru.

Orris see *Iris germanica florentina*
Osage Orange see *Maclura pomifera*
Osier, Common see *Salix viminalis*
  Purple see *S. purpurea*

Osmanthus os-*mănth*-us *Oleaceae*. From Gk. *osme* (fragrance) and *anthos* (a flower) referring to the fragrant flowers. Evergreen shrubs.
  armatus ar-*mah*-tus. Spiny (the leaves). W China.
  × *burkwoodii* burk-*wud*-ee-ee. *O. delavayi* × *O. decorus*. (= × *Osmarea burkwoodii*). After Burkwood and Skipwith, the raisers.
  decorus de-*kō*-rus. (= *Phillyrea decora*). Beautiful. Lazistan, on the SE coast of the Black Sea.
  delavayi del-a-*vay*-ee. After Delavay who introduced it to France in 1890, see *Abies delavayi*. China.
  × *fortunei* for-*tewn*-ee-ee. *O. fragrans* × *O. heterophyllus*. After Robert Fortune who introduced it in 1862, see *Fortunella*.
  fragrans frah-granz. Fragrant. China.
    aurantiacus ow-răn-tee-*ah*-kus. Orange (the flowers).
  heterophyllus he-te-rō-*fil*-lus. (= *O.*

Osmanthus (continued)
aquifolium. O. ilicifolius). With variable
leaves. Japan.
yunnanensis yoo-nan-*en*-sis. Of Yunnan,
China.

× **Osmarea burkwoodii** see *Osmanthus* ×
burkwoodii

Osmaronia see *Oemleria*

Osmoglossum os-mo-*glos*-um *Orchidaceae*.
From Gk. *Osme*, odour, and *glossa*, tongue,
referring to the sweet scent of some species.
Epiphytic orchids. formerly included in
Odontoglossum.
pulchellum pul-*kel*-um (= *Odontioglossum*
pulchellum). Pretty. Lily of the Valley orchid.
Guatemala. Mexico, El Salvador.

Osmunda os-*mun*-da *Osmundaceae*.
Derivation obscure. Ferns.
cinnamomea kin-a-*mō*-mee-a. Cinnamon-
coloured (the fronds). Cinnamon Fern.
Widely distributed.
claytoniana klay-ton-ee-*ah*-na. After John
Clayton (1686–1773), Virginian botanist.
Interrupted Fern. N America, Asia.
regalis ray-*gah*-lis. Royal. Royal Fern.
Europe, Asia.

Oso Berry see *Oemleria cerasiformis*

Osteospermum ost-ee-ō-*sperm*-um
Compositae. From Gk. *osteon* (a bone) and
sperma (a seed). Semi-hardy sub-shrubs. S
Africa.
ecklonis ek-*lon*-is. (= *Dimorphotheca*
ecklonis). After Christian Friedrich Ecklon
(1795–1868), German apothecary.
jucundum yoo-*kun*-dum. (= *Dimorphotheca*
barberiae). Pleasing.

Ostrowskia os-*trov*-skee-a *Campanulaceae*.
After Michael Nicholazewitsch von
Ostrowsky, a patron of botany. Perennial
herb.
magnifica mahg-*ni*-fi-ka. Magnificent.
Giant Bellflower. Turkestan.

Ostrya os-*tree*-a *Carpinaceae*. From *ostrys* the
Gk. name. Deciduous trees.
carpinifolia kar-peen-i-*fo*-lee-a. With leaves
like *Carpinus*. Hop Hornbeam. S Europe,
W Asia, Caucasus.
japonica ja-*pon*-i-ka. Of Japan. Japan,
China, Korea.
virginiana vir-jin-ee-*ah*-na. Of Virginia.
Iron Wood. E N America.

Oswego Tea see *Monarda didyma*

Othonna ō-*thon*-a *Compositae*. From Gk.
othone, linen, referring to the soft leaves.
Perennials and small shrubs.
cheirifolia kay-ri-*fō*-lee-a (= *Hertia*
cheirifolia, Othonnopsis cheirifolia). With
leaves like *Cheiranthus cheiri* (= *Erysimum*
cheiri) N Africa.

Othonnopsis cheirifolia see *Othonna cheirifolia*

Ourisia ow-*ris*-ee-a *Scrophulariaceae*. After
Ouris, a governor of the Falkland Islands
where the first species was found.
Herbaceous perennials.
alpina ăl-*peen*-a. Alpine. Andes.
coccinea kok-*kin*-ee-a. Scarlet. Andes.
elegans ay-le-gahnz. Elegant. Chile.
macrophylla măk-rō-*fil*-la. Large-leaved.
New Zealand.

Our Lady's Milk Thistle see *Silybum*
marianum

Our Lord's Candle see *Yucca whipplei*

Oxalis oks-*ah*-lis *Oxalidaceae*. The Gk. name
for sorrel, from *oxys* (acid). Hardy and
tender herbs.
acetosella a-kay-to-*se*-la. L. name for plants
with acid leaves. Wood Sorrel. Europe,
N Asia.
adenophylla a-den-o-*fil*-la. With glandular
leaves. Chile, Argentina.
articulata ar-tik-ew-*lah*-ta. Jointed.
Paraguay.
cernua see *O. pes-caprae*
chrysantha kris-*ănth*-a. With golden
flowers. Brazil.
depressa day-*pres*-a. (= *O. inops*). Low-
growing. S Africa.
dispar dis-par. Unusual. Guiana.
enneaphylla en-ee-a-*fil*-la. With nine
leaflets. Falkland Islands.
hirta hir-ta. Hairy. S Africa.
inops see *O. depressa*
laciniata la-kin-ee-*ah*-ta. Deeply cut (the
leaves). Patagonia.
magellanica mă-ge-*lăn*-i-ka. Of the
Magellan region. Patagonia, Australia,
New Zealand.
oregona o-ree-*gō*-na. Of Oregon. W
United States.
ortgiesii ort-*geez*-ee-ee. After Eduard
Ortigies (1829–1916). Peru.
pes-caprae pays-*kăp*-rie. (= *O. cernua*). Like
a goat's foot (the leaves). S Africa.

Oxlip see *Primula elatior*
Ox-tongue Lily see *Haemanthus coccineus*

Oxycoccus see *Vaccinium*

Oxydendrum oks-ee-*den*-drum *Ericaceae*.
From Gk. *oxys* (acid) and *dendron* (a tree)
referring to the acid-tasting leaves.
Deciduous tree.
    arboreum ar-*bo*-ree-um. Tree-like. Sorrel
    Tree. E N America.

Oxypetalum see *Tweedia*

Ozothamnus o-zō-*thăm*-nus *Compositae*.
From Gk. *ozo* (a smell) and *thamnos* (a shrub).
Evergreen shrubs.
    coralloides ko-ra-*loi*-deez. (= *Helichrysum
    coralloides*). Coral-like (the shoots). New
    Zealand.
    ledifolius lay-di-*fo*-lee-us. (= *Helichrysum
    ledifolium*). With leaves like *Ledum*.
    Tasmania.
    rosmarinifolius rōs-ma-reen-i-*fo*-lee-us. (=
    Helichrysum rosmarinifolium). With leaves
    like *Rosmarinus*. SE Australia, Tasmania.
    selago se-*lah*-go. (= *Helichrysum selago*).
    Like *Lycopodium selago*. New Zealand.

P

Pachistima see *Paxistima*

Pachycereus pă-kee-*kay*-ree-us *Cactaceae*.
From Gk. *pachys* (thick) and *Cereus* q.v.
referring to the thick shoots. W Mexico.
    marginatus mar-gi-*nah*-tus (=
    Lemaireocereus marginatus). Margined, the
    ribs are margined with white wool. C
    Mexico.
    pecten-aboriginum pek-ten-ă-bo-*ree*-gi-
    num. As the common name, Native's
    Comb.
    pringlei pring-*gal*-ee. After G. G. Pringle
    who collected in Mexico c 1887.

Pachyphragma pă-kee-*frăg*-ma *Cruciferae*.
*From Gk. *pachys* (thick) and *phragma* (a
partition) referring to the stout-ribbed
septum of the pod. Herbaceous perennial.
    macrophyllum măk-rō-*fil*-lum. Large-
    leaved. Caucasus.

Pachyphytum pă-kee-*fi*-tum *Crassulaceae*.
From Gk. *pachys* (thick) and *phyton* (a plant).
Tender succulents. Mexico.

Pachyphytum (continued)
    amethystinum see *Graptopetalum
    amethystinum*
    bracteosum brak-tee-ō-sum. With
    conspicuous bracts (on the flower stem).
    glutinicaule gloo-tin-i-*kaw*-lee. (= *P.
    brevifolium* hort.). With sticky stems.
    Sticky Moonstones.
    oviferum ō-*vi*-fe-rum. Egg-bearing,
    referring to the egg-shaped leaves.
    Moonstones.

Pachysandra pă-kis-*ăn*-dra *Buxaceae*. From
Gk. *pachys* (thick) and *andros* (male),
referring to the thick stamens. Evergreen
sub-shrubs.
    procumbens prō-*kum*-benz. Prostrate.
    Allegheny Spurge. SE United States.
    terminalis ter-mi-*nah*-lis. Terminal (the
    flower spikes). Japan.

Pachystachys pă-kee-*stă*-kis *Acanthaceae*.
From Gk. *pachys* (thick) and *stachys* (a spike)
referring to the dense inflorescences. Tender
shrubs.
    coccinea kok-*kin*-ee-a. (= *Jacobinia coccinea*).
    Scarlet. W Indies, S America.
    lutea *loo*-tee-a. Yellow (the bracts).
    Lollipop Plant. Peru.

Pachystegia insignis see *Olearia insignis*

Pachystima see *Paxistima*

× **Pachyveria** pă-kee-*ve*-ree-a *Crassulaceae*.
Intergeneric hybrid, from the names of the
parents. *Echeveria* × *Pachyphytum*. Tender
succulent.
    pachyphytoides pă-kee-fit-*oi*-deez. *Echeveria
    gibbiflora* × *Pachyphytum bracteosum*. Like
    Pachyphytum.

Paeonia pie-*on*-ee-a *Paeoniceae*. From the Gk.
name *paionia*, meaning of Paion, physician
to the gods. Herbaceous perennials and
shrubs. Paeony.
    anomala a-*nom*-a-la. Unusual. W and C
    Asia.
    arietina see *P. mascula arietina*
    clusii *clooz*-ee-ee. After Clusius, see
    Gentiana clusii. Crete.
    delavayi de-la-*vay*-ee. After Delavay, see
    Abies delavayi. W China.
    emodi e-*mō*-dee. Of *Emodi Montes* (the
    Himalaya).
    lactiflora lăk-ti-*flō*-ra. With milky flowers.
    NE Asia.
    × *lemoinei* la-*mwŭn*-ee-ee. *P. lutea* × *P.*

Paeonia (continued)
suffruticosa. After Messrs Lemoine who
raised some forms.
lutea loo-tea. Yellow. SW China.
ludlowii lūd-*lō*-ee-ee. After Frank
Ludlow (1885–1972) who, with
George Sherriff, introduced it in 1936.
SE Tibet.
mascula mahs-kew-la. Male, used for
vigorous species. Europe.
arietina a-ree-e-*tee*-na. (= *P. arietina*).
Like a ram's head. E Europe, W Asia.
mlokosewitschii mlo-ko-sa-*vich*-ee-ee. After
Ludwig Franzevich Mlokosewitsch
(1831–1909), who discovered it. Caucasus.
obovata ob-ō-*vah*-ta. Obovate (the terminal
leaflet). E Asia.
officinalis o-fi-ki-*nah*-lis. Sold as a herb.
Europe.
peregrina pe-re-*green*-a. Foreign. S and E
Europe.
potaninii po-tah-*nin*-ee-ee. After
Nicolaevich Potanin (1835–1920),
Russian explorer and plant collector.
China.
  trollioides trol-ee-*oi*-deez. Like *Trollius*.
× *smouthii smooth*-ee-ee *P. lactiflora* × *P.*
tenuifolia. After M Smouth.
suffruticosa suf-froo-ti-*kō*-sa. Sub-shrubby.
Moutan. N China.
tenuifolia ten-ew-i-*fo*-lee-a. With slender
leaves (leaf segments). SE Europe,
Caucasus.
wittmanniana vit-măn-ee-*ah*-na. After
Wittmann, who collected in the
Caucasian Taurus c 1840. Caucasus.

Painted Daisy see *Chrysanthemum carinatum*
Painted Drop-tongue see *Aglaonema crispum*
Painted Feather see *Vriesia carinata*
  Dwarf see *V. psittacina*
Painted Tongue see *Salpiglossis sinuata*
Painted Wood-lily see *Trillium undulatum*
Painter's Palette see *Anthurium andraeanum*

Paliurus pă-lee-*ew*-rus *Rhamnaceae*. The Gk.
name. Deciduous shrub or small tree.
spina-christi speen-a-*kris*-tee. Christ's
Thorn, it is believed to have been used
for the crown of thorns. S Europe, W
Asia.

Palm, Australian Fan see *Livistona australis*
  Bamboo see *Chamaedorea erumpens*
  Betel Nut see *Areca catechu*
  Burmese Fishtail see *Caryota mitis*
  Canary Island Date see *Phoenix
  canariensis*
  Chinese Fan see *Livistona chinensis*

Palm (continued)
  Chusan see *Trachycarpus fortunei*
  Curly Sentry see *Howea belmoreana*
  Date see *Phoenix dactylifera*
  Desert Fan see *Washingtonia filifera*
    Dwarf Fan see *Chamaerops humilis*
  Fan see *Trachycarpus fortunei*
  Fishtail see *Caryota*
  Lady see *Rhapis*
  Miniature Date see *Phoenix roebelinii*
  Paradise see *Howea forsteriana*
  Parlour see *Chamaedorea elegans*
  Sago see *Cycas revoluta*
  Sentry see *Howea*
  Thread see *Washingtonia robusta*
  Toddy see *Caryota urens*
  Weddell see *Microcoelum weddellianum*
  Wine see *Caryota urens*
  Yatay see *Butia yatay*
  Yellow see *Chrysalidocarpus lutescens*

Pamianthe păm-ee-*ănth*-ee *Amaryllidaceae*.
After Major Albert Pam (1875–1955) to
whom the following was sent in 1926.
Tender, bulbous herb.
peruviana pe-roo-vee-*ah*-na. Of Peru.

Pampas Grass see *Cortaderia selloana*

Pancratium păn-*krăt*-ee-um *Amaryllidaceae*.
Gk. name for a bulbous plant. Bulbous
perennials.
canariense ka-nah-ree-*en*-see. Of the
Canary Islands.
illyricum i-*li*-ri-kum. Of Illyria (W
Yugoslavia). Corsica, Sardinia, Capri etc.
maritimum ma-*ri*-ti-mum. Growing near
the sea. Sea Daffodil, Sea Lily.
Mediterranean region.

Pandanus păn-da-nus *Pandanaceae*. From
pandan the Malayan name. Tender,
evergreen shrubs and trees. Screw Pine.
baptistii băp-*tist*-ee-ee. After Baptist. New
Britain Islands.
candelabrum kăn-day-*lah*-brum. Like a
candelabra. W Africa.
pygmaeus pig-*mie*-us. Dwarf. Madagascar.
sanderi sahn-da-ree. After the Sander
nursery. Timor.
veitchii veech-ee-ee. After the Veitch
nursery. Polynesia.

Panda Plant see *Kalanchoe tomentosa,
Philodendron bipennifolium*

Pandorea păn-do-ree-a *Bignoniaceae*. After
Pandora of Gk. mythology. Tender,
evergreen shrubs or trees.

Pandorea (continued)
jasminoides yăs-min-*oi*-deez. Jasmine-like.
Bower Plant. Australia.
pandorana păn-do-*rah*-na. After Pandora.
Wonga-wonga Vine. SE Asia to Australia.

Panicum pah-ni-kum *Gramineae*. The L.
name for millet. Annual grasses.
capillare kă-pi-*lah*-ree. Hair-like. Witch
Grass. E N America.
miliaceum mi-lee-*ah*-kee-um. Like *Milium*.
Asia.
virgatum vir-*gah*-tum. Wand-like. N and C
America.

Pansy see *Viola*
   Garden see *V.* × *wittrockiana*
Panther see *Lilium pardalinum*

Papaver pa-*pah*-ver *Papaveraceae*. The L.
name. Annual, biennial and perennial herbs.
Poppy.
alpinum ăl-*peen*-um. Alpine. Alpine
Poppy. Origin and identity of cultivated
plants uncertain.
bracteatum brak-tay-*ah*-tum. With bracts.
Caucasus, Asia minor.
commutatum kom-ew-*tah*-tum.
Changeable. Caucasus. W Asia.
glaucum glow-kum. Glaucous (the leaves).
Tulip Poppy. W Asia.
heldreichii hel-*driek*-ee-ee (= *P. spicatum*)
After Heldreich. C Mediterranean.
miyabeanum see *P. nudicaule*
nudicaule new-di-*kaw*-lee. (= *P.
miyabeanum*). Bare-stemmed. Arctic
Poppy, Iceland Poppy. N America,
Europe, Asia.
orientale o-ree-en-*tah*-lee. (= *P. bracteatum*).
Eastern. W Asia.
pilosum pi-*lō*-sum. Hairy. W Asia.
rupifragum roo-*pi*-fra-gum. Rock-
breaking, i.e. growing in rock crevices.
Spanish Poppy. S Spain.
somniferum som-*ni*-fe-rum. Sleep-bearing.
Opium Poppy. SE Europe. W Asia.
spicatum see *P. heldreichii*

Paphiopedilum pă-fee-ō-*pe*-di-lum
Orchidaceae. Gk. Paphos, site of a temple on
Cyprus where Aphrodite was worshipped
and *pedilon* (a slipper). Greenhouse orchids,
sometimes listed under *Cypripedium*. Slipper
Orchid.
   acmodontum ăk-mō-*don*-tum. With a sharp
   tooth (on the front margin of the lip).
   Philippines.
   barbatum bar-*bah*-tum. Bearded (the warts
   on the petals). Malaya.

Paphiopedilum (continued)
bellatulum be-*lah*-tew-lum. Pretty. SE Asia.
callosum ka-*lō*-sum. Calloused (the petals).
Thailand.
concolor *kon*-ko-lor. Similarly coloured (the
petals and sepals). SE Asia.
fairrieanum fair-ree-*ah*-num. After Mr
Fairrie of Liverpool who bought it in a
sale of Assam plants.
hirsutissimum hir-soo-*tis*-i-mum. Very
hairy (the scape). Assam to S China.
insigne in-*sig*-nee. Remarkable. Assam.
niveum niv-ee-um. Snow-white (the lip).
Thailand, Malaya.
philippinense fi-li-peen-*en*-see. Of the
Philippines.
purpuratum pur-pew-*rah*-tum. Purplish
(the flowers). Hong Kong, S China.
rothschildianum roths-chield-ee-*ah*-num.
After Rothschild. Borneo.
spicerianum spie-sa-ree-*ah*-num. After Mr
Spicer, a tea planter who introduced many
orchids. Assam, N Burma.
sukhakulii soo-ka-*koo*-lee-ee. After P.
Sukhakuli. a Thai nurseryman. NE
Thailand.
tonsum *tōn*-sum. Smooth. Sumatra.
venustum ve-*nus*-tum. Handsome.
Himalaya.
villosum vi-*lō*-sum. Softly hairy (the scape).
Assam, China, Thailand.

Papilionanthe pă-pi-li-ō-*năn*-the.
Orchidaceae. From Lat. *papilio*, meaning
moth, and Gk. *anthos*, flower. Epiphytic and
terrestrial orchids.
teres te-reez (= *Vanda teres*). Cylindrical.
Himalaya, N India, Burma.
vandarum văn-*dah*-rum Like *Vanda*.
Himalaya.

Papyrus see *Cyperus papyrus*

Paradisea pă-ra-*dees*-ee-a *Liliaceae*
(Asphodelaceae). After Count Giovanni
Paradisi (1760–1826). Perennial herb.
liliastrum lee-le-*ăs*-trum. Like *Lilium*. St
Bruno's Lily. S Europe.

Parahebe pă-ra-*hay*-bay *Scrophulariaceae*. From
Gk. *para* (close to) and *Hebe* q.v. Dwarf,
evergreen shrubs related to *Hebe*. New
Zealand.
catarractae ka-ta-*răk*-tie. (= *Hebe cararractae*).
Of waterfalls.
decora de-*kō*-ra. Beautiful
lyallii lie-*ăl*-ee-ee. After David Lyall
(1817–95), naval surgeon and naturalist
who collected the type specimen.

Paris Daisy see *Argyranthemum frutescens*

Parkinsonia park-kin-son-ee-a *Leguminosae (Caesalpinoideae)* After John Parkinson (1567–1650), London apothecary and botanical author. Tender, evergreen tree.
aculeata a-kew-lee-*ah*-ta. Prickly. Jerusalem Thorn. Tropical America.

Parnassia par-*năs*-ee-a *Saxifragaceae*. From the 16th century name *Gramen Parnassi*, referring to Mt Parnassus, Greece. Bog garden perennial herb.
palustris pa-*lus*-tris. Of marshes. Grass of Parnassus. N temperate regions.

Parochetus pa-*ro*-ke-tur *Leguminosae (Papilionoideae)*. From Gk. *para* (near) and *ochetus* (a brook) referring to the habitat. Perennial herb.
communis kom-*ew*-nis. Common. Shamrock Pea. Himalaya, E Africa.

Parodia pa-*rō*-dee-a *Cactaceae*. After Lorenzi Raimondo Parodi (1895–1966), Argentine botanist. N Argentina.
apricus see *P. concinna*
aureispina ow-ree-i-*speen*-a. With golden spines. Golden Tom Thumb.
chrysacanthion kris-a-*kănth*-ee-on. With golden spines.
concinna kon-*kin*-a. Elegant. S Brazil, Uruguay.
haselbergii hah-zal-*berg*-ee-ee. After Dr von Haselberg of Stralsind, a cactus grower. Scarlet Ball Cactus. S Brazil.
leninghausii len-ing-*howz*-ee-ee. After Leninghaus. Golden Ball Cactus. Brazil.
maassii mahs-ee-ee. After W. Maass. Also Bolivia.
mammulosus măm-ew-*lō*-sus. Bearing nipples. S America.
microsperma mik-rō-*sperm*-a. With small seeds.
muricatus see *P. concinna*
mutabilis mew-*tah*-bi-lis. Changeable.
nivosa ni-*vō*-sa. Snow-white (the spines).
ottonis o-*tō*-nis. After Friedrich Otto (1782–1856), curator of Berlin Botanic Garden. SE South America.
sanguiniflora See *P. microsperma*
scopa skō-pa. Broom-like. Silver Ball Cactus. S Brazil, Uruguay.
scopaoides see *P. microsperma*
submammulosus sub-măm-ew-*lō*-sus. With small nipples. N Argentina.
tabularis see *P. concinna*

Parrotia pa-*rŏt*-ee-a *Hamamelidaceae*. After F.

Parrotia (continued)
W. Parrot (1792–1841), a German naturalist. Deciduous tree.
persica per-si-ka. Of Iran (Persia). Persian Ironwood. SW Caspian Sea region.

Parrotiopsis pa-rŏt-ee-*op*-sis *Hamamelidaceae*. From *Parrotia* q.v. and Gk. *-opsis* indicating resemblance. Deciduous tree.
jacquemontiana zhahk-a-mont-ee-*ah*-na. After Victor Jacquemont (1801–32), French naturalist. W. Himalaya.

Parsley see *Petroselinum crispum*
Parsley Vine see *Vitis vinifera* 'Apiifolia'
Parsnip see *Pastinaca sativa*

Parthenocissus par-then-ō-*kis*-us. *Vitaceae*. From Gk. *parthenos* (a virgin) and *kissos* (ivy) referring to the common name, Virginia Creeper. Deciduous climbers.
henryana hen-ree-*ah*-na. After Augustine Henry who discovered it, see *Illicium henryi*. C China.
himalayana him-a-lay-*ah*-na. Of the Himalaya.
inserta in-*ser*-ta. Inserted, it needs to be attached to walls. N America.
quinquefolia kwing-kwee-*fo*-lee-a. With five leaves (leaflets). Virginia Creeper. N America.
thomsonii see *Cayratia thomsonii*
tricuspidata tri-kus-pi-*dah*-ta. Three-pointed (the leaves). Boston Ivy. Japan, China.

Partridge Brry see *Mitchella repens*
Pasque Flower see *Pulsatilla*

Passiflora pă-si-*flō*-ra *Passifloraceae*. From L. *passio* (passion) and *flos* (a flower). The parts of the flower have been compared with various aspects of the crucifixion of Christ. Evergreen, mainly tender, climbers. Passion Flower.
× *allardii* a-*lard*-ee-ee. *P. caerulea* 'Constance Elliott' × *P. quadrangularis*. After Edgar John Allard (c 1877–1918), gardener and hybridist.
antioquiensis ăn-tee-ō-kee-*en*-sis. Of Antioquia, Colombia.
caerulea kie-*ru*-lee-a. Blue (the flowers). Blue Passion Flower. S Brazil.
× *caponii* ka-pon-ee-ee. *P. quadrangularis* × *P. racemosa*. After W. J. Capon who raised it in 1953.
coccinea kok-*kin*-ee-a. Scarlet (the flowers). Red Granadilla. N A America

Passiflora (continued)
edulis e-*dew*-lis. Edible (the fruit).
Granadilla. Brazil.
x *exoniensis* eks-ō-nee-*en*-sis. *P.
antioquiensis* x *P. mollissima*. Of Exeter.
incarnata in-kar-*nah*-ta. Flesh-coloured
(the flowers). May Apple, Maypop. S
United States.
laurifolia low-ri-*fo*-lee-a. *Laurus*-leaved.
Yellow Granadilla. W Indies, S America.
manicata măn-i-*kah*-ta. Long sleeved. N S
America.
mixta *miks*-ta. Mixed. N S America.
mollissima mol-*lis*-i-ma. Very softly hairy
(the shoots and undersides of the leaves).
N S America.
quadrangularis kwod-rang-gew-*lah*-ris.
Four angled (the shoots). Giant
Granadilla. S America.
racemosa ră-kay-*mō*-sa. With flowers in
racemes. Brazil.
umbilicata um-bi-lee-*kah*-ta. With a navel.
S America.

Passion Flower see *Passiflora*
  Blue see *P. caerulea*

Pastinaca păs-ti-*nah*-ka *Umbelliferae*. The L.
name for parsnip and carrot, from *pastus*
(foot). Biennial herb.
sativa se-*tee*-va. Cultivated. Parsnip.
Europe, Asia.

Paulownia pow-*lō*-nee-a *Scrophulariacea*.
After Anna Paulowna (1795–1865),
daughter of Czar Paul I of Russia. Deciduous
trees. China.
tomentosa tō-men-*tō*-sa. Hairy (the leaves).
'Lilacina' li-la-*keen*-a. (= *P. fargesii*
hort.). Lilac (the flowers).

Pavonia pa-*von*-ee-a *Malvaceae*. After Jose
Antonia Pavon (1754–1840), Spanish
botanist. Tender, evergreen shrub.
multiflora mul-tee-*flō*-ra. Many flowered.
Brazil.

Pawpaw see *Asimina triloba*

Paxistima păks-*i*-sti-ma *Celastraceae*. (=
Pachistima. *Pachystima*). From Gk. *pachys*
(thick) and *stigma*. Evergreen shrubs.
canbyi *kăn*-bee-ee. After its discoverer
William Marriott Canby (1831–1904). E
United States.
myrtifolia mur-ti-*fo*-lee-a. *Myrtus*-leaved.
NW North America.

Pea, Garden see *Pisum sativum*

Pea Tree see *Caragana arborescens*
Peach see *Prunus persica*
Peacock Plant see *Calathea makoyana*
Peacock Tiger Flower see *Tigridia pavonia*
Pear see *Pyrus*.
  Common see *P. communis*
  Willow-leaved see *P. salicifolia*
Pearl Fruit see *Margyricarpus pinnatus*
Pearl Grass see *Briza maxima*
Pearl Plant see *Haworthia margaritifera*
Pearlwort see *Sagina*
Pearly Everlasting see *Anaphalis*

Pedilanthus pe-di-*lănth*-us *Euphorbiaceae*.
From Gk. *pedilon* (a slipper) and *anthos* (a
flower) referring to the shape of the
involucre. Tender succulents.
tithymaloides ti-thee-mah-*loi*-deez. Like
Tithymalus (= *Euphorbia*). Ribbon
Cactus, Slipper Flower. C and N South
America.
  smallii *smawl*-ee-ee. Jacob's Ladder.
  After John Kunkel Small (1869–1938),
  American botanist. Florida, Cuba.

Pelargonium pe-lar-*gon*-ee-um *Geraniaceae*.
From Gk. *pelargos* (a stork) referring to the
storksbill-like fruit. Tender perennial herbs
and shrubs. Geranium. S Africa.
crispum *kris*-pum. Finely wavy (the leaves).
Lemon-scented Pelargonium.
denticulatum den-tik-ew-*lah*-tum. Toothed
(the leaves).
x *domesticum* do-*mes*-ti-kum. Cultivated.
Regal Pelargonium.
Fragrans *frah*-granz. *P. exstiplatum* x *P.
odoratissimum*. Fragrant (the leaves).
Nutmeg Pelargonium.
fulgidum *ful*-gi-dum. Shining (the scarlet
flowers).
graveolens gra-*vee*-o-lenz. Aromatic. Rose
Pelargonium.
x *hortorum* ho-*to*-rum. Of gardens. Zonal
Pelargonium.
odoratissimum o-dō-ra-*tis*-i-mum. Highly
scented (the leaves). Apple Pelargonium.
peltatum pel-*tah*-tum. Peltate (the leaves).
Ivy-leaved Pelargonium.
quercifolium kwer-ki-*fo*-lee-um. *Quercus*-
leaved. Oak-leaved Pelargonium.
tetragonum tet-ra-*gō*-num. Four-angled
(the stems).
tomentosum tō-men-*tō*-sum. Hairy.
Peppermint-scented Pelargonium.
triste *tris*-tee. Sad, the sombre-coloured
flowers.
zonale zō-*nah*-lee. Zoned (the leaves).
Mainly grown as hybrids.

Pellaea pe-*lie*-a *Pteridaceae (Adiantaceae)*.
From Gk. *pellaios* (dark) referring to the
often dark stalks. Ferns.
   atropurpurea aht-rō-pur-*pewr*-ree-a. Deep
   purple (the stalks). Purple Cliff Brake. N
   America.
   rotundifolia ro-tund-i-*fo*-lee-a. With round
   leaves (pinnae). Button Fern. New
   Zealand.
   viridis *vi*-ri-dis. Green. Green Cliff Brake.
   Africa.

Pellionia pe-lee-*on*-ee-a *Urticaceae*. After
Alphonse Odet Pellion (1796–1868), a French
Admiral. Tender, evergreen herbs.
   daveauana see *P. repens*
   pulchra *pul*-kra. Pretty. Rainbow Vine.
   Vietnam.
   repens ree-*pens* (= *P. daveauana*). Creeping.
   SE Asia.

Peltiphyllum see *Darmera*

Peniocereus pee-ni-ō-*kay*-ree-us. *Cactaceae*.
From Gk. *penios*, a thread, and *Cereus* q.v.
Shrub-like-cacti. C America to SW US.
   serpentinus ser-pen-*teen*-us (= *Nyctocereus
   serpentinus*). Snake-like. Mexico.

Pennisetum pen-i-*say*-tum *Gramineae*. From
L. *penna* (a feather) and *seta* (a bristle)
referring to the feathery bristles around the
spikelets. Perennial grasses.
   alopecuroides ă-lō-pek-ew-*roi*-deez. Like
   Alopecurus. Asia.
   orientale o-ree-en-*tah*-lee. Eastern.
   Abyssinia.
   setaceum say-*tah*-kee-um. Bristly. Fountain
   Grass. Africa.
   villosum vi-*lō*-sum. Softly hairy. Abyssinian
   Feathertop. Africa.

Pennyroyal see *Mentha pulegium*

Penstemon pen-*stay*-mon *Scrophulariaceae*.
From Gk. *pente* (five) and *stemon* (a stamen)
referring to the five stamens. Perennial herbs
and shrubs.
   alpinus ăl-*peen*-us. Alpine. W United
   States.
   barbatus bar-*bah*-tus (= *Chelone barbata*).
   Bearded (the lower lip of the corolla).
   SW United States, Mexico.
   barrettiae ba-*ret*-ee-ie. After Mrs Barret
   who discovered it. Oregon.
   cordifolius kor-di-*fo*-lee-us. With heart-
   shaped leaves. S California.
   davidsonii day-vid-*son*-ee-ee. (= *P.*

Penstemon (continued)
   menziesii). After Davidson. W N
   America.
   × *edithae* ee-dith-ie. *P. barretiae* × *P. rupicola*.
   After Edith Hardin English.
   fruticosus froo-ti-kō-sus. Shrubby. W N
   America.
   hartwegii hart-*weg*-ee-ee. After Carl
   Theodore Hartweg (1812–71). Mexico.
   heterophyllos he-te-rō-*fil*-lus. With variable
   leaves. California.
   menziesii men-*zeez*-ee-ee. After Menzies,
   see *Menziesia*
   newberryi new-*be*-ree-ee. After J. S.
   Newberry who discovered it. W United
   States.
      humilior hu-*mil*-ee-or. (= *P. roezlii*
      hort.). Low-growing.
   ovatus ō-*vah*-tus. Ovate (the leaves). W N
   America.
   pinifolius peen-i-*fo*-lee-us. With leaves like
   Pinus. SW United States, Mexico.
   roezlii hort. see *P. newberryi humilior*
   rupicola roo-*pi*-ko-la. Growing on rocks.
   W N America.
   scouleri skool-a-ree. After Dr John Scouler
   (1804–71), who collected with David
   Douglas.
   virens *vi*-renz. Green. W United States.

Pentapterygium serpens see *Agapetes serpens*

Pentas *pen*-tas *Rubiaceae*. From Gk. *pentas* (a
series of five), it differs from related genera
in having the floral parts in fives. Tender sub-
shrub.
   lanceolata lăn-kee-o-*lah*-ta. Lance-shaped
   (the leaves). Egyptian Star Cluster. E
   Africa, Arabia.

Pen Wiper see *Kalanchoe marmorata*

Peperomia pe-pe-*rom*-ee-a *Piperaceae*. From
Gk. *peperi* (pepper) and *homoios*
(resembling), it is closely related to the
pepper plant. Tender, evergreen, succulent
herbs.
   argyreia ar-gi-*ree*-a. Silvery (the leaves).
   Watermelon Peperomia, Rugby Football
   Plant. Tropical S America.
   caperata kă-pe-*rah*-ta. Wrinkled (the
   leaves). Brazil.
   fraseri *fray*-za-ree. After Fraser who
   collected the type specimen. Mignonette
   Peperomia. Ecuador.
   glabella gla-*bel*-la. Rather glabrous. Wax
   Privet. Tropical America.
   griseoargentea gri-see-ō-ar-*gen*-tee-a. Grey-

Peperomia (continued)
silver (the leaves). Ivy-leaf Peperomia.
Brazil.
magnoliifolia măg-nol-ee-i-*fo*-lee-a.
Magnolia-leaved. Desert Privet. W Indies,
S America.
obtusifolia ob-tew-si-*fo*-lee-a. Blunt-
leaved. Baby Rubber Plant. Tropical
America.
orba *or*-ba. (= *P.* 'Princess Astrid'). L. For
orphan, it is of unknown origin.
scandens skăn-denz. Climbing. Identity and
origin of cultivated plants uncertain.
verticillata ver-ti-ki-*lah*-ta. Whorled (the
leaves). W Indies.

Pepper, Chilli see *Capsicum frutescens*
 Christmas see *C. annuum*
 Sweet see *C. annuum*
Peppermint, Black see *Mentha* × *piperita*
 White see *M.* × *piperita officinalis*

Pereskia pe-*res*-kee-a *Cactaceae.* After
Nicholas Claude Fabre de Pieresc
(1580–1637), French naturalist. Leafy Cacti.
aculeata a-kew-lee-*ah*-ta. Prickly. Barbados
Goosebery. Tropical America.
grandifolia grănd-i-*fo*-lee-a. (= *P.*
grandiflora). Large-leaved. Brazil.

Pericallis per-i-*kal*-is *Compositae.* From Gk.
peri, around, and *kallos*, meaning beauty.
Perennial herbs and shrubs.
cruenta kroo-en-tă (= *Senecio cruentus,*
Cineraria cruentus). Blood-red (the
flowers). Canary Islands.
× *hybrida* hib-rid-ă (= *Senecio* × *hybrida,*
Cineraria × *hybrida*). Hybrid. Florists'
cineraria. Hybrids probably between *P.*
lanata, P. cruenta, and possibly other
species.

Perilla pe-*ril*-la *Labiatae* Derivation obscure.
Annual herb.
frutescens froo-*tes*-enz. Shrubby. Himalaya,
E Asia.

Peristrophe pe-*ri*-stro-fay *Acanthaceae.* From
Gk. *peri* (around) and *strophe* (turning)
referring to the twisted corolla tube. Tender
perennial herbs.
hyssopifolia hi-sō-pi-*fo*-lee-a. (= *P.*
angustifolia. P. salicifolia). *Hyssopus*-leaved.
Java.
speciosa spe-kee-ō-sa. Showy. India.

Periwinkle see *Vinca*
 Greater see *V. major*
 Lesser see *V. minor*

Periwinkle (continued)
 Madagascar see *Catharanthus roseus*

Pernettya see *Gaultheria*

Perovskia pe-*rof*-skee-a *Labiatae.* After V A.
Perovsky (1794–*c* 1857). Sub-shrubs.
abrotanoides a-bro-ta-*noi*-deez. Like *Artemisia*
abrotanum. Russia, W Asia.
atriplicifolia ă-tri-pli-ki-*fo*-lee-a. *Atriplex*-
leaved. W Himalaya, Afghanistan.

Persea per-see-a *Lauraceae.* Gk. name of a
tree. Tender, evergreen tree.
americana a-me-ri-*kah*-na. (= *P. gratissima*).
American. Avocado Pear. C America.

Persian Ironwood see *Parrotia persica*
Persian Shield see *Strobilanthes dyerianus*
Persian Violet see *Exacum affine*
Persimmon see *Diospyros virginiana*
 Chinese see *D. kaki*

Petasites pe-ta-*see*-teez *Compositae.* From
Gk. *petasos* (a hat) referring to the large
leaves. Perennial herbs.
fragrans frah-granz. Fragrant. Winter
Heliotrope. W Mediterranean region.
japonicus ja-*pon*-i-kus. Of Japan. Korea,
China, Japan.

Petrea pet-ree-a *Verbenaceae.* After Lord
Robert James Petre (1713–43), botanical
and horticultural patron. Tender climber.
volubilis vol-*ew*-bi-lis. Twining. Purple
Wreath, Queen's Wreath, Mexico, C
America.

Petrocoptis pet-ro-*kop*-tis *Caryophyllaceae.*
From Gk. *petros* (a rock) and *kopto* (to
break), from the habit of growing in rock
crevices. Perennial herb.
glaucifolia glow-ki-*fo*-lee-a. (= *P. lagascae.*
Lychnis lagascae). With glaucous leaves.
Pyrenees, N Spain.

Petrorhagia pet-ro-*rah*-gee-a *Caryophyllaceae.*
From Gk. *petros* (a rock) and *rhagas* (a chink),
the following grows in cracks in rocks.
Perennial herb.
saxifraga săks-*if*-ra-ga. (= *Tunica saxifraga*).
The L. version of the Gk. name above.
C and S Europe to C Asia.

Petroselinum pet-ro-se-*leen*-um *Umbelliferae.*
From Gk. *petros* (a rock) and *selinon*
(parsley). Biennial herb.
crispum kris-pum. Finely wavy (the leaves).
Parsley. Europe, W Asia.

Petunia pe-*tewn*-ee-a *Solanaceae*. From *petun*, the Brazilian name for tobacco. Annual herbs.
  × *hybrida* hib-ri-da. Hybrid. Common Petunia.

Peyote see *Lophophora williamsii*

Pfeiffera see *Lepismium ianothele*

Phacelia fa-*kel*-ee-a *Hydrophyllaceae*. From Gk. *phakelos* (a bundle) referring to the clustered flowers. Annual herbs.
  campanularia kăm-păn-ew-*lah*-ree-a. Like *Campanula*. California Bluebell. W United States.
  tanacetifolia tăn-a-set-i-*fo*-lee-a. *Tanacetum*-leaved. SW United States, Mexico.
  viscida *vis*-ki-da. Sticky. California.

Phaius *fie*-us *Orchidaceae*. From Gk. *phaius* (dusky) referring to the flowers of one species. Greenhouse orchid.
  tankervilliae tang-ka-*vil*-ee-ie. (= *P. wallichii*). After Emma Lady Tankerville (c 1750–1836). Himalaya.

Phalaenopsis fă-lie-*nop*-sis *Orchidaceae*. From Gk. *phalaina* (a moth) and -*opsis* indicating resemblance. Greenhouse orchids. Moth Orchid.
  amabilis a-*mah*-bi-lis. Beautiful. SE Asia to N Australia.
  esmeralda see *Doritis pulcherrima*
  lueddemanniana loo-da-măn-ee-*ah*-na. After Lueddemann, a Paris orchid grower. Philippines.
  parishii pă-*rish*-ee-ee. After the Rev. Charles Samuel Parish, an army surgeon who collected orchids in Burma. Burma.
  sanderiana sahn-da-ree-*ah*-na. After the Sander orchid nursery. Philippines.
  schilleriana shi-la-ree-*ah*-na. After Herr Schiller of Hamburg, an orchid grower. Philippines.
  stuartiana stew-art-ee-*ah*-na. After Stuart Low, see *Cymbidium lowianum*. Philippines.

Phalaris fa-*lah*-ris *Gramineae*. Gk. name for a grass. Annual and perennial grasses.
  arundinacea a-run-di-*nah*-kee-a. Reed-like. Reed Canary Grass. Europe, Asia, N America.
    'Picta' *pik*-ta. Painted (the leaves). Gardener's Garters.
  canariensis ka-nah-ree-*en*-sis. Of the Canary Islands. Canary Grass. Canary Islands, N Africa.

Phaseolus fa-*see*-o-lus *Leguminosae* (*Papilionoideae*). From *phaselos* the Gk. name for another bean. Annual herbs. Tropical America.
  coccineus kok-*kin*-ee-us. Scarlet. Scarlet Runner Bean.
  vulgaris vul-*gah*-ris. Common. French Bean, Runner Bean.

Pheasant's Eye see *Adonis autumnalis*

Phellodendron fe-lo-*den*-dron *Rutaceae*. From Gk. *phellos* (cork) and *dendron* (a tree) referring to the corky bark. Deciduous trees.
  amurense ăm-ew-*ren*-see. Of the Amur region. NE Asia.

Philadelphus fil-a-*del*-fus *Philadelphaceae*. The Gk. name. Deciduous shrubs. Sometimes referred to as Syringa q.v.
  'Burfordensis' bur-fad-*en*-sis. Of Burford Court where it was raised.
  coronarius ko-rō-*nah*-ree-us. Used in garlands. Mock Orange. SE Europe, W Asia.
  delavayi de-la-*vay*-ee. After Delavay who discovered and introduced it in 1887, see *Abies delavayi*. W China, SE Tibet, N Burma.
  microphyllus mik-rō-*fil*-lus. Small-leaved. SW United States.

Philesia fi-*leez*-ee-a *Liliaceae*. From Gk. *phileo* (to love) referring to the attractive flowers. Evergreen shrub.
  magellanica mă-ge-*lăn*-i-ka. Of the region of the Magellan Straits. S Chile.

Phillyrea fi-*li*-ree-a *Oleaceae*. The Gk. name. Evergreen shrubs and trees.
  angustifolia ang-gus-ti-*fo*-lee-a. Narrow-leaved. S Europe, N Africa.
    'Rosmarinifolia' rŏs-ma-reen-i-*fo*-lee-a. With leaves like *Rosmarinus*.
  decora see *Osmanthus decorus*
  latifolia lah-tee-*fo*-lee-a. Broad-leaved. Mediterranean region.

Philodendron fi-lo-*den*-dron *Araceae*. From Gk. *phileo* (to love) and *dendron* (a tree) referring to their tree-climbing habit. Tender, evergreen climbers.
  angustisectum ang-gus-ti-*sek*-tum. With narrow divisions (the leaves). Colombia.
  bipennifolium bi-pen-i-*fo*-lee-um. With bipinnate leaves. Panda Plant. Brazil.
  bipinnatifidum bi-pi-nah-ti-*fi*-dum. (= *P. panduriforme* hort.). Bipinnately divided (the leaves). Brazil.

Philodendron (continued)
domesticum do-*mes*-ti-kum. (= *P. hastatum*
hort.). Cultivated. cult.
erubescens e-roo-*bes*-enz. Blushing, the rosy
stipules and young leaves. Colombia.
hastatum hort. see *P. domesticum*
imbe im-bee. The native name. Brazil.
laciniatum see *P. pedatum*
melanochrysum me-la-*no*-kris-um. Black-
gold, from the colour of the leaves.
Colombia.
panduriforme hort. see *P. bipinnatifidum*
pedatum pe-*dah*-tum. (= *P. laciniatum*).
Like a bird's foot (the leaves). S America.
pinnatifidum pin-ah-ti-*fi*-dum. Pinnately
divided (the leaves). Venezuela, Trinidad.
sagittifolium să-gi-ti-*fo*-lee-um. With
arrow-shaped leaves. S Mexico.
scandens skăn-denz. Climbing. Tropical
America.
selloum see *P. bipinnatifidum*

Phlebodium *fleb*-ō-dee-um *Polypodiaceae*.
From Gk. *phlebion*, a vein; the fronds are
conspicuously veined. Tropical Ferns.
Formerly included in *Polypodium*.
aureum ow-ree-um. Golden (the sori). Hare's-
foot Fern. Tropical America.

Phlomis *flo*-mis *Labiatae*. Gk. name for a
plant. Perennial herbs and evergreen shrubs.
chrysophylla kris-o-*fil*-la. Golden-leaved
(referring to dried specimens). W Asia.
fruticosa froo-ti-*kō*-sa. Shrubby. Jerusalem
Sage. Mediterranean region.
italica ee-*tăl*-i-ka. Italian. Balearic Islands.
russelliana rū-sel-ee-*ah*-na. After Russell, a
physican at Aleppo. W Asia.
samia sah-mee-a. Of the island of Samos
in the Aegean. Greece.

Phlox floks *Polemoniaceae*. From Gk. *phlox* (a
flame). Annual and perennial herbs.
adsurgens ăd-*sur*-genz. Erect, the flowering
stems from postrate shoots. Oregon,
California.
amoena a-*moy*-na. Pleasant. SE United
States.
bifida bi-fi-da. Divided into two (the
petals). E United States.
divaricata di-vah-ri-*kah*-ta. Spreading. E N
America.
douglasii dŭg-*lăs*-ee-ee. After Douglas, see
Iris douglasii. W United States.
drummondii drū-*mond*-ee-ee. After
Thomas Drummond (c 1790–1835),
who collected in N America. Pride of
Texas. Texas.

Phlox (continued)
maculata măk-ew-*lah*-ta. Spotted (the
stems). E United States.
nana nah-na. Dwarf. SW United States, N
Mexico.
paniculata pa-nik-ew-*lah*-ta. With flowers
in panicles. E United States.
× *procumbens* prō-*kum*-benz. *P. stolonifera* ×
P. subulata. Prostrate.
stolonifera sto-lō-*ni*-fe-ra. Bearing stolons.
E United States.
subulata soob-ew-*lah*-ta. Awl-shaped (the
leaves).

Phoenix *fee*-niks *Palmae*. The Gk. name.
Tender palms.
canariensis ka-nah-ree-*en*-sis. Of the
Canary Islands. Canary Island Date Palm.
dactylifera dăk-ti-*li*-fe-ra. Finger-bearing.
Date Palm. W Asia, N Africa.
roebelinii rur-be-*lin*-ee-ee. After M.
Robelin who collected for Sander's in SE
Asia. Miniature Date Palm. Laos.

Phormium *for*-mee-um *Agavaceae*. From Gk.
phormion (a mat), fibre is produced from the
leaves of *P. tenax*. Evergreen perennial herbs.
New Zealand.
colensoi kō-len-so-ee (= *P. cookianum*).
After William Colenso (1811–99), botanist.
Mountain Flax.
cookianum see *P. colensoi*.
tenax ten-ahks. Tough. New Zealand Flax.

Photinia fō-*tin*-ee-a *Rosaceae*. From Gk.
photos (light) referring to the shining leaves
of some species. Deciduous and evergreen
shrubs and trees.
beauverdiana bō-ver-dee-*ah*-na. After
Gustave Beauverd (1867–1942), botanist
and artist. W China.
davidiana dă-vid-ee-*ah*-na. (= *Stranvaesia
davidiana*). After David who discovered it
in 1869, see *Davidia*. China.
× *fraseri* fray-zee-ree. *P. glabra* × *P. serrulata*.
After the Fraser nurseries, Alabama, who
raised it.
glabra glăb-ra. Glabrous (the leaves). Japan,
China.
serrulata se-ru-*lah*-ta. With small teeth (the
leaves). China.
villosa vi-*lō*-sa. Softly hairy (the young
leaves and shoots). Japan, China, Korea.
laevis lie-vis. Smooth.

Phuopsis foo-*op*-sis *Rubiaceae*. From Gk. *phou*
(a kind of valerian) and *-opsis* indicating
resemblance. Perennial herb.

Phuopsis (continued)
 stylosa sti-*lō*-sa. With a prominent style.
 Caucasus.

Phygelius foo-*gay*-lee-us *Scrophulariaceae*.
Probably from Gk. *phyge* (flight) and *helios*
(the sun). Semi-hardy sub-shrubs. S Africa.
 aequalis ie-*kwah*-lis. Equal, the corolla
 lobes, which are not in *P. capensis*. *capensis*
 ka-*pen*-sis. Of the Cape of Good Hope.
 'Coccineus' kok-*kin*-ee-us. Scarlet.

Phyla *fi*-la *Verbenaceae*. From Gk. *phyla* (a
tribe) referring to the compound flower
heads. Perennial herb.
 nodiflora nō-di-*flō*-ra. (= *Lippia nodiflora*).
 With flowers born from the nodes. Widely
 distributed.

× **Phylliopsis** fil-lee-*op*-sis *Ericaceae*.
Intergeneric hybrid, from the names of the
parents. *Kalmiopsis* × *Phyllodoce*. Evergreen
shrub.
 hillieri hi-lee-a-ree. *Kalmiopsis leachina* ×
 Phyllodoce breweri. After Hillier's in whose
 nursery it was found.

Phyllitis see *Asplenium*

Phyllocladus fi-*lo*-kla-dus *Phyllocladaceae*.
From Gk. *phyllon* (a leaf) and *klados* (a
branch) referring to the flattened, leaf-like
shoots. Semi-hardy conifer.
 alpinus ǎl-*peen*-us. Alpine. Celery Pine.
 New Zealand.

Phyllodoce fi-*lo*-do-kee *Ericaceae*. After
Phyllodoce, a sea nymph. Dwarf, evergreen
shrubs.
 aleutica a-*loo*-ti-ka. Of the Aleutian Islands.
 Japan to Alaska.
 breweri broo-a-ree. After Brewer, who
 discovered it in 1862. California.
 caerulea kie-*ru*-lee-a. Deep blue (the
 corolla). Arctic and alpine N hemisphere.
 empetriformis em-pet-ri-*form*-is. Like
 Empetrum. W N. America.
 × *intermedia* in-ter-*med*-ee-a. *P.*
 empetriformis × *P. glanduliflora*.
 Intermediate (between the parents). W N
 America.
 nipponica ni-*pon*-i-ka. Of Japan.

Phyllostachys fi-*lo*-sta-kis *Gramineae*. From
Gk. *phyllon* (a leaf) and *stachys* (a spike)
referring to the leafy inflorescences.
Bamboos. China.
 aurea ow-ree-a. Golden (the canes).

Phyllostachys (continued)
 bambusoides băm-bew-*soi*-deez. Like
 Bambusa, a related genus.
 flexuosa fleks-ew-*ō*-sa. Zig-zag (the stems).
 nigra *nig*-ra. Black (the old stems). Black
 Bamboo.

Physalis *fi*-sa-lis *Solanaceae*. From Gk. *physa*
(a bladder) referring to the bladder-like fruits.
Perennial herbs.
 alkekengi ǎl-ke-*ken*-jee. (= *P. franchetii*).
 From *al kakendi* the Arabic name.
 Chinese Lantern. SE Europe to Japan.
 peruviana pe-roo-vee-*ah*-na. Of Peru.
 Cape Gooseberry. S America.

Physocarpus fi-sō-*kar*-pus *Rosaceae*. From
Gk. *physa* (a bladder) and *karpon* (a fruit)
referring to the inflated fruits. Deciduous
shrubs. Ninebark.
 opulifolius op-ew-li-*fo*-lee-us. With leaves like
 Viburnum opulus. E N America.

Physoplexis see *Phyteuma*

Physostegia fi-sō-*stee*-gee-a *Labiatae*. From
Gk. *physa* (a bladder) and *stege* (covering)
referring to the inflated calyx which covers
the fruit. Herbaceous perennial.
 virginiana vir-jin-ee-*ah*-na. Of Virginia.
 Obedient Plant. E N America.

Phyteuma fi-*tew*-ma *Campanulaceae*. The Gk.
name for *Reseda phyteuma*. Perennial herbs.
 comosum ko-*mō*-sum. (= *Physoplexis*
 comosa). Tufted. Devil's Claw. S Alps.
 hemisphaericum he-mi-*sfie*-ri-kum.
 Hemispherical (the flower heads). S
 Europe.
 orbiculare or-bik-ew-*lah*-ree. Orbicular
 (the flower heads). Rampion. Europe.
 spicatum spee-*kah*-tum. With flowers in
 spikes. Spiked Rampion. Europe.

Phytolacca fi-tō-*lă*-ka *Phytolaccaceae*. From
Gk. *phyton* (a plant) and L. *lacca* referring to
the lac insect *Laccifer lacca* from which a dye
is obtained. A red dye is obtained from the
berries. Perennial herb.
 americana a-me-ri-*kah*-na. American.
 Pokeweed. United States, Mexico.

Picea pi-kee-a *Pinaceae*. L name for a pitch-
producing pine, from *pix* (pitch). Evergreen
conifers. Spruce.
 abies *ă*-bee-ayz. L. name for fir (*Abies*).
 Norway Spruce. Europe.
  'Acrocona' ăk-rō-*kō*-na. With terminal
  cones.

Picea (continued)
'Clanbrassiliana' klăn-bra-sil-ee-*ah*-na.
After Lord Clanbrassil who grew the
original plant.
'Nidiformis' nee-di-*form*-is. Nest-
shaped.
asperata ă-spe-*rah*-ta. Rough (the foliage).
Dragon Spruce. W China.
brachytyla bră-kee-*tee*-la. With short
swellings. China.
breweriana broo-a-ree-*ah*-na. After William
Henry Brewer (1828–1910), American
botanist who discovered it. Brewer Spruce.
California, Oregon.
glauca glow-ka. Glaucous (the leaves).
White Spruce. N America.
jezoensis ye-zō-*en*-sis. Of Hokkaido
(Yezo), Japan. Yezo Spruce. NE Asia.
    hondoensis hon-dō-*en*-sis. Of Honshu
(Hondo), Japan. Hondo Spruce.
koyamae koy-*ah*-mie. After Mitsuo
Koyama (1885–1935) who discovered it
in 1911. Japan, Korea.
likiangensis li-kee-ang-*gen*-sis. Of Likiang,
Yunnan. W China, Tibet.
    purpurea see *P. purpurea*
mariana mă-ree-*ah*-na. Of Maryland.
Black Spruce. N America.
omorika o-*mo*-ri-ka. the native name.
Serbian Spruce. Yugoslavia.
orientalis o-ree-en-*tah*-lis. Eastern. Turkey,
Caucasus.
pungens pung-genz. Sharp-pointed (the
leaves). Colorado Spruce. W United States.
    Glauca *glow*-ka. Glaucous (the leaves).
purpurea pur-*pewr*-ree-a. (= *P. likiangensis
purpurea*). Purple (the young cones). China.
sitchensis sit-*ken*-sis. Of Sitka, Alaska. Sitka
Spruce. N California to Alaska.
smithiana smith-ee-*ah*-na. After Sir James
Edward Smith (1759–1828). W
Himalaya.

Pickerel Weed see *Pontederia cordata*

Picrasma pi-*krăs*-ma *Simaroubaceae*. From Gk.
pikra (a bitter taste) referring to the bitter
leaves and wood. Deciduous tree.
ailanthoides ie-lanth-*oy*-deez (= *P.
quassioides*). Like *Ailanthus* (the leaves).
Japan, N. China, Korea.
quassioides see *P. ailanthoides*

Pieris *pee*-e-ris *Ericaceae*. From *Pierides* a
name of the Muses (goddesses of the arts).
Evergreen shrubs.
floribunda flō-ri-*bun*-da. Profusely
flowering. SE United States.

Pieris (continued)
formosa for-*mō*-sa. Beautiful. Himalaya,
China.
    forrestii fo-*rest*-ee-ee. After Forrest who
introduced it, see *Abies delavayi forrestii*.
W China.
japonica ja-*pon*-i-ka. Of Japan.
taiwanensis see *P. japonica*.

Piggy-back Plant see *Tolmiea menziesii*

Pilea *pee*-lee-a *Urticaceae*. From L. *pileus* (a
cap), referring to the female flowers. Tender,
evergreen herbs.
cadierei kă-dee-e-ree-ee. After R. P.
Cadiere who collected it in 1938.
Aluminium Plant. Vietnam.
involucrata in-vo-loo-*krah*-ta. With an
involucre, the leaves are clustered beneath
the flowers. Friendship Plant. C and S
America.
microphylla mik-rō-*fil*-la. Small-leaved.
Artillery Plant. Tropical America.
nummulariifolia num-ew-lah-ree-i-*fo*-lee-a.
With coin-shaped leaves. Creeping
Charlie. C and S America.
repens ree-penz. Creeping. W Indies.
spruceana see *P. involucrata*

Pileostegia pee-lee-ō-*stee*-gee-a
Hydrangeaceae. From Gk. *pilos* (a cap) and
stege (a covering) referring to the corolla.
Evergreen climber.
viburnoides vee-burn-*oi*-deez. Like
Viburnum (the flower heads). NE India,
China, Taiwan.

Pilocereus pi-lo-*kay*-ree-us *Cactaceae*. From
Lat. *pilosus*, hairy and *Cereus*; formerly
included in *Cephalocereus*. Shrubby or tree-
like Cacti. SE US, Mexico and tropical S
America.
palmeri *pahl*-ma-ree (= *Cephalocereus
palmeri*). After E. Palmer (1831–1911),
who collected in the S US and Mexico.
E. Mexico.

Pilosella pi-lo-*sell*-a *Compositae*. Diminutive
of Lat. *pilosus* meaning hairy. Perennials.
Eurasia and NW Africa.
aurantiaca ow-răn-tee-*ah*-ka (= *Hieracium
aurantiacum*). Orange (the flowers). Devil's
Paintbrush. Europe.

Pimelea pi-*me*-lay-a *Thymelaeaceae*. From Gk.
pimele (fat) referring to the high oil content
of the seeds. Tender, evergreen shrubs.
ferruginea fe-roo-*gin*-ee-a. Rusty. W
Australia.

Pimelea (continued)
longiflora long-gi-*flo*-ra. Long-flowered. W Australia.
prostrata pros-*trah*-ta. Prostrate. New Zealand.
rosea ro-see-a. Rose-coloured (the flowers). W Australia.
spectabilis spek-*tah*-bi-lis. Spectacular. W Australia.

Pimpernel, Bog see *Anagallis tenella*
   Scarlet see *A. arvensis*
Pincushion Flower see *Hakea laurina*
Pine see *Pinus*
   Aleppo see *P. halepensis*
   Arolla see *P. cembra*
   Austrian see *P. nigra*
   Balkan see *P. peuce*
   Beach see *P. contorta*
   Bhutan see *P. wallichiana*
   Big-cone see *P. coulteri*
   Bishop see *P. muricata*
   Bosnian see *P. leucodermis*
   Bristle-cone see *P. aristata*
   Chinese see *P. tabuliformis*
   Chinese White see *P. armandii*
   Corsican see *P. nigra maritima*
   Digger see *P. sabiniana*
   Dwarf Mountain see *P. mugo*
   Eastern White see *P. strobus*
   Himalayan see *P. wallichiana*
   Japanese Black see *P. thunbergii*
   Japanese Red see *P. densiflora*
   Japanese White see *P. parviflora*
   Knobcone see *P. attenuata*
   Lacebark see *P. bungeana*
   Limber see *P. flexilis*
   Lodgepole see *P. contorta latifolia*
   Macedonian see *P. peuce*
   Maritime see *P. pinaster*
   Mexican White *see P. ayacahuite*
   Monterey see *P. radiata*
   Northern Pitch see *P. rigida*
   Scots see *P. sylvestris*
   Stone see *P. pinea*
   Sugar see *P. lambertiana*
   Umbrella wee *P. pinea*
   Western White see *P. monticola*
   Western Yellow see *P. ponderosa*
   Weymouth see *P. strobus*
   Whitebark see *P. albicaulis*
   Yunnan see *P. yunnanensis*
Pineapple see *Ananas comosus*
   Wild see *A. bracteatus*

Pinguicula ping-*gwi*-kew-la *Lentibulariaceae*. From L. *pinguis* (fat) referring to the greasy appearance of the leaves. Carnivorous herbs. Butterwort.

Pinguicula (continued)
alpina ăl-*peen*-a. Alpine. Europe.
caudata see *P. moranensis*
grandiflora grănd-i-*flō*-ra. Large-flowered. W Europe.
gypsicola gip-*si*-ko-la. Lime-loving. Mexico.
moranensis mor-an-*en*-sis (= *P. caudata*). From Moran, Texas. Mexico.
vulgaris vul-*gah*-ris. Common. N America, Europe, Asia.

Pink see *Dianthus*
   Cheddar see *D. gratianopolitanus*
   Clove see *D. caryophyllus*
   Fringed see *D. superbus*
   Glacier see *D. glacialis*
   Indian see *D. chinensis*
   Maiden see *D. deltoides*
Pink Sand Verbena see *Abronia umbellata*
Pink Siris see *Albizia julibrissin*

Pinus pee-nus *Pinaceae*. The L. name. Evergreen conifers. Pine.
albicaulis ăl-bi-*kaw*-lis. White-stemmed. Whitebark Pine. W N America.
aristata ă-ris-*tah*-ta. Awned (the slender cone bristles). Bristlecone Pine. SW United States.
armandii ar-*mond*-ee-ee. After Armand David who discovered it, see *Davidia*, Chinese White Pine. W and C China.
attenuata a-ten-ew-*ah*-ta. Drawn out (the cones). Knobcone Pine. W United States, N Mexico.
ayacahuite ie-a-ka-*hee*-tee. The Mexican name, Mexican White Pine. Mexico, Guatemala.
bungeana bung-gee-*ah*-na. After Alexander von Bunge (1803–90), Russian botanist who collected the type specimen. Lacebark Pine. China.
cembra kem-bra. The Italian name. Arolla Pine. Alps to the Carpathians.
contorta kon-*tor*-ta. Twisted (the young shoots). Beach Pine. W N America (coastal).
   latifolia lah-tee-*fo*-lee-a. Broad-leaved. Lodgepole Pine. Rocky Mountains.
coulteri kool-ta-ree. After its discoverer. Thomas Coulter (1793–1843), an Irish physician. Big-cone Pine. California.
densiflora den-si-*flō*-ra. Densely flowered. Japanese Red Pine. Japan, Korea, China.
   'Oculus-draconis' *ok*-ew-lis-dra-*kō*-nis. Dragon's eye.
   'Umbraculifera' um-brahk-ew-*li*-fe-ra. Umbrella-bearing.

Pinus (continued)
flexilis fleks-i-lis. Flexible (the young shoots). Limber Pine. W N America.
halepensis hă-le-*pen*-sis. Of Aleppo (Syria). Aleppo Pine. S Europe, W Asia.
heldreichii hel-*driek*-ee-ee. After Theodor von Heldreich (1822–1902). SE Europe.
jeffreyi jef-ree-ee. After John Jeffrey (1826–54), a gardener at the Edinburgh Botanic Garden, who discovered it. W United States, N Mexico.
lambertiana lăm-bert-ee-*ah*-na. After Aylmer Bourke Lambert (1761–1842). Sugar Pine. W United States.
leucodermis see *P. heldreichii*
montezumae mon-tee-*zoo*-mie. After Montezuma, an Aztec Emperor. S and C Mexico.
monticola mon-*ti*-ko-la. Growing on mountains. Western White Pine. W N America.
mugo mew-gō. An old Tyrolese name. Dwarf Mountain Pine. C and SE Europe.
muricata mew-ri-*kah*-ta. Rough with spines (the cone). Bishop Pine. California.
nigra nig-ra. Black (the bark). Austrian Pine. SC and SE Europe.
　maritima ma-*ri*-ti-ma. Growing near the sea. Corsican Pine. S Italy, Sicily, Corsica, N Africa.
　'Hornibrookiana' horn-i-bruk-ee-*ah*-na. After Murray Hornibrook, an authority on dwarf conifers.
parviflora par-vi-*flō*-ra. Small-flowered. Japanese White Pine. Japan.
peuce poy-kay. Classical name of a pine. Macedonian Pine, Balkan Pine. Yugoslavia.
pinaster peen-*ăs*-ter. L. name meaning inferior to the cultivated pine, *P. pinea*. Maritime Pine, SW Europe, N Africa.
pinea pee-nee-a. L. name for pine-nuts, which in this tree are edible. Stone Pine, Umbrella Pine. S Europe, W Asia.
ponderosa pon-de-*rō*-sa. Heavy (the wood). Western Yellow Pine. W N America.
radiata ră-dee-*ah*-ta. Radiating (lines on the cone scales). Monterey Pine. Monterey, California.
rigida ri-gi-da. Rigid (the leaves). Northern Pitch Pine. E N America.
sabiniana sa-been-ee-*ah*-na. After Sabine. Digger Pine. California.
strobus stro-bus. L. name of a gum-yielding tree. Eastern White Pine, Weymouth Pine. E N America.
sylvestris sil-*ves*-tris. Of woods. Scots Pine. N Europe, N Asia.

Pinus (continued)
　'Beuvronensis' burv-ron-*en*-sis. Of Beuvron, near Orléans, France.
tabuliformis tăb-ew-li-*form*-is. Flat-topped. Chinese Pine. China.
yunnanensis see *P. yunnanensis*
thunbergii thun-*berg*-ee-ee. After Thunberg, see *Thunbergia*. Japanese Black Pine. Japan.
wallichiana wo-lik-ee-*ah*-na. After Nathaniel Wallich (1785–1854). Danish surgeon and botanist with the E India Co. Bhutan Pine, Himalayan Pine. Himalaya.
yunnanensis yoo-nan-*en*-sis. (= *P. tabuliformis yunnanensis*). Of Yunnan. Yunnan Pine. W China.

Piptanthus pip-*tănth*-us *Leguminosae*. From Gk. *pipto* (to fall) and *anthos* (a flower), the flowers fall intact. Evergreen shrub.
nepalensis ne-pa-*len*-sis. (= *P. laburnifolius*). Of Nepal. Himalaya.

Pistacia pis-*tah*-kee-a *Anacardiaceae*. From Gk. *pistake*, pistachio nut, obtained from another species. Deciduous shrub or tree.
chinensis chi-*nen*-sis. Of China.

Pistia pis-tee-a *Araceae*. From Gk. *pistos* (water) referring to its habitat. Floating, tender perennial herb.
stratiotes stră-tee-ō-teez. The Gk. name. Water Lettuce. Tropics and subtrops.

Pisum pee-sum *Leguminosae* (*Papilionoideae*). The L. name. Annual herb.
sativum sa-*teev*-um. Cultivated Garden Pea. Europe, Asia.

Pitcher Plant see *Nepenthes, Sarracenia*
　California see *Darlingtonia californica*
　Northern see *Sarracenia purpurea*
　Yellow see *S. flava*

Pittosporum pi-*tos*-po-rum *Pittosporaceae*. From Gk. *pitta* (pitch) and *sporum* (a seed) referring to the sticky seeds. Tender and semi-hardy, evergreen trees and shrubs.
adaphniphylloides a-daf-ni-fil-*oi*-deez. Not *P. daphniphylloides* which it was originally thought to be. China.
crassifolium krăs-i-*fo*-lee-um. Thick-leaved. New Zealand.
dallii dăl-ee-ee. After J Dall (died 1912), who discovered it in 1905. New Zealand.
eugenioides ew-jeen-ee-*oi*-deez. Like *Eugenia*. New Zealand.
'Garnettii' gar-*net*-ee-ee. *P. ralphii* × *P. tenuifolium*. After Garnett.

Pittosporum (continued)
ralphii rălf-ee-ee. After Dr Ralph who collected the type specimen. New Zealand.
tenuifolium ten-ew-i-*fo*-lee-um. With thin leaves. New Zealand.
tobira to-*bi*-ra. The native name. E Asia.
undulatum un-dw-*lah*-tum. Wavy-edged (the leaves). E Australia.

Plagianthus plă-gee-*ánth*-us *Malvaceae.* From Gk. *plagios* (oblique) and *anthos* (a flower) referring to the asymmetrical flowers. Deciduous shrubs and trees. New Zealand.
betulinus see *P. regius*
divaricatus di-vah-ri-*kah*-tus. Spreading.
regius ree-jee-us (= *P. betulinus*). Royal.

Plane see *Platanus*
  London see *P.* × *acerifolia*
  Oriental see *P. orientalis*
Plantain Lily see *Hosta*

Platanus plă-ta-nus *Platanaceae.* From *platanos* the Gk. name for *P. orientalis.* Deciduous trees. Plane.
× *acerifolia* a-ke-ri-*fo*-lee-a. (= *P. hispanica*). Possibly *P. occidentalis* × *P. orientalis.* With leaves like *Acer.* London Plane. Cult.
orientalis o-ree-en-*tah*-lis. Eastern. Oriental Plane. SE Europe.

Platycarya plă-tee-*kă*-ree-a *Juglandaceae.* From Gk. *platys* (broad) and *karyon* (a nut) referring to the fruit. Deciduous tree.
strobilacea stro-bi-*lah*-kee-a. Cone-like (the fruit). E Asia.

Platycerium plă-tee-*ke*-ree-um *Polypodiaceae.* From Gk. *platys* (broad) and *keras* (a horn) referring to the flattened, horn-like. Tender Ferns.
bifurcatum bi-fur-*kah*-rum (= *P. alcicorne* hort.). Forked into two. Elk's-horn Fern, Stag's-horn Fern. Australia, Polynesia.
grande grǎnd-ee. Large. Australia.

Platycladus plat-i-klă-dus *Cupressaceae.* From Gk. *platys*, broad, and *klados*, a young branch, a reference to the spreading branches. A single species of conifer formerly included in *Thuja.*
orientalis ŏ-ree-en-*tah*-lis (= *Thuja orientalis*). Eastern. Oriental Thuja. W China, N Korea.

Platycodon plă-tee-*kŏ*-don *Campanulaceae.* From Gk. *platys* (broad) and *kodon* (a bell)

Platycodon (continued)
referring to the shape of the corolla. Perennial herbs.
grandiflorus grănd-i-*flŏ*-rus. Large flowered. Balloon Flower. E Asia.
mariesii ma-*reez*-ee-ee. After Maries, see *Davallia mariesii.*

Platystemon plă-tee-*stay*-mon *Papaveraceae.* From Gk. *platys* (broad) and *stemon* (a stamen) referring to the broad stamens. Annual herb.
californicus kăl-*forn*-i-kus. Of California. Cream Cups. S California, N Mexico.

Plectostachys ple-ko-sta-kis *Compositae.* From Gk. *plectos* (to plait) and *stachys* (a spike) referring to the inflorescence. Annual herb or semi-hardy sub-shrub.
serpyllifolia ser-pi-li-*fo*-lee-a. (= *Helichrysum microphyllum* hort.). Thyme-leaved. S Africa.

Plectranthus plek-*tránth*-us. *Labiatae.* From Gk. *plectron* (a spur) and *anthos* (a flower) referring to the spurred flowers of the type species. Tender perennial herbs.
australis ow-*strah*-lis. Southern. Swedish Ivy. SE Australia.
coleoides see *P. forsteri*
forsteri for-*ste*-ree. (= *P. coleoides*). After Forster. New Caledonia, Fiji, E Australia.
nummularius see *P. verticillatus*
oertendahlii ur-tan-*dahl*-ee-ee. After Oertendahl. Candle Plant. Cult.
verticillatus ver-ti-ki-*lah*-ta (= *P. nummularius*). Arranged in whorls (the flowers). Eastern S Africa to Mozambique.

Pleioblastus ply-ŏ-*blǎst*-us *Graminae.* From Gk. *pleios* many, and *blastos*, buds, referring to the several branches at each node. Bamboo. China, Japan.
auricoma aw-rik-ŏ-mǎ (= *Arundinaria viridistrata*). With golden hairs. Japan.
humilis hew-mil-is. Low growing.
pumilus pew-mil-us (= *Arundinaria pumila*). Dwarf; an even smaller variety of the species. Japan.
variegatus vǎ-ree-a-*gah*-tus (= *Arundinaria variegata*). Variegated (the leaves). Japan.

Pleione play-o-nay *Orchidaceae.* After Pleione, wife of Atlas and mother of the Pleiades. Semi-hardy or cool greenhouse orchids.
bulbocodioides bul-bŏ-kŏ-dee-*oi*-deez.

Pleione (continued)
Like *Bulbocodium*. Tibet. China,
Taiwan.
formosana for-mō-*sah*-na (= *P. pricei*).
From Taiwan (Formosa).
hookeriana huk-a-ree-*ah*-na. After Sir
Joseph Hooker. Himalaya, Assam,
Tibet.
humilis hu-mi-lis. Low-growing.
Himalaya, Burma, Tibet.
praecox prie-koks. Early (flowering).
Himalaya to W China and Thailand.
pricei see *P. formosana*

Pleiospilos play-*os*-pi-los *Aizoaceae*. From
Gk. *pleios* (many) and *spilos* (a spot) referring
to the spotted leaves. Tender succulent.
bolusii bō-*lus*-ee-ee. After Harry Bolus
(1834–1911). S Africa.

Pleomele reflexa see *Dracaena reflexa*
Plover Eggs see *Adromischus cooperi*
Plum see *Prunus domestica*
Plum Yew see *Cephalotaxus*

Plumbago plum-*bah*-gō *Plumbaginaceae*. The
L. name from *plumbum* (lead). Tender
climbers. Leadwort.
auriculata ow-rik-ew-*lah*-ta. (= *P.
capensis*). With auricles (the leaves).
Cape Leadwort. S Africa.
indica in-di-ka. Indian. Scarlet
Leadwort. SE Asia.

Plume Bush see *Calomeria amaranthoides*

Plumeria ploo-*me*-ree-a *Apocynaceae*. After
Charles Plumier (1646–1704), French
monk and botanist. Tender tree.
rubra rub-ra. Red (the flowers).
Frangipani. Mexico, C America.

Poached Egg Flower see *Limmanthes douglasii*

Podocarpus pod-o-*kar*-pus *Podocarpaceae*.
From Gk. *podos* (a foot) and *karpos* (a fruit)
referring to the stalked fruit. Evergreen
conifers.
alpinus ăl-*peen*-us. Alpine. Tasmania,
New South Wales.
andinus see *Prumnopitys andina*
macrophyllus măk-rō-*fil*-lus. Large-
leaved. Japan.
nivalis ni-*vah*-lis. Growing near snow.
New Zealand.
nubigenus new-bi-*gen*-us. From among
the clouds. Chile, Argentina.
salignus sa-*lig*-nus. Willow-like (the
leaves). Chile.

Podocarpus (continued)
totara tō-*tah*-ra. The native name. New
Zealand.

Podophyllum pod-o-*fil*-lum *Berberidaceae*.
From *anapodophyllum*, Gk. *anas* (a duck),
podos (a foot) and *phyllon* (a leaf), referring
to the leaves of *P. peltatum*. Perennial herbs.
hexandrum heks-*ăn*-drum. (= *P. emodi*).
With six stamens. Himalaya, W China.
peltatum pel-*tah*-tum. Peltate (the
leaves). E N America.

Poinsettia see *Euphorbia pulcherrima*
Annual see *E. cyathophora*
Poke Weed see *Phytolacca americana*

Polemonium po-lee-*mō*-nee-um
Polemoniaceae. From *polemonion* the Gk.
name of a plant. Perennial herbs.
caeruleum kie-*ru*-lee-um. Deep blue (the
flowers). Jacob's Ladder. Europe, Asia.
carneum kar-nee-um. Flesh-coloured
(the flowers). Oregon, California.
foliosissimum fo-lee-ō-*sis*-i-mum. Very
leafy. W United States.
pauciflorum paw-si-*flō*-rum. Few-
flowered. SW United States, Mexico.
reptans rep-tănz. Creeping. E United
States.

Polianthes po-lee-*ănth*-eez *Agavaceae*. From
Gk, *polios* (grey) and *anthos* (a flower).
Tender perennial herb.
tuberosa tew-be-*rō*-sa. Tuberous.
Tuberose. Cult.

Policeman's Helmet see *Impatiens
glandulifera*
Polka-dot Plant see *Hypoestes phyllostachya*
Polyanthus see *Primula* Polyantha

Polygala po-*li*-ga-la *Polygalaceae*. The classical
name from Gk. *polys* (much) and *gala* (milk),
they were thought to enhance milk secretion.
Herbs and shrubs.
calcarea kăl-*kah*-ree-a. Growing on
chalk. Europe.
chamaebuxus kā-mie-*buks*-us. Dwarf
Buxus. Europe.
vayredae vay-*ree*-die. After Vayreda who
rediscovered it in 1877. Pyrenees.

Polygonatum po-li-go-*nah*-tum *Liliaceae*.
From *polygonaton*, the Gk. name, from *polys*
(many) and *gony* (a knee), referring to the
jointed rhizome. Perennial herbs. Solomon's
Seal.

Polygonatum (continued)
  biflorum bi-*flō*-rum. Two-flowered. E United States.
  commutatum see *P. biflorum*
  hookeri hu-ka-ree. After Sir Joseph Hooker. Himalaya, W China.
  × *hybridum* hib-ri-dum. *P. multiflorum* × *P. odoratum*. Hybrid.
  multiflorum mul-tee-*flō*-rum. Many-flowered. Europe, Asia.
  odoratum o-dō-*rah*-tum. Scented. Europe, Asia.
  verticillatum ver-ti-ki-*lah*-tum. Whorled (the leaves). Europe to the Himalaya.

Polygonum po-*li*-go-num *Polygonaceae*. From Gk. *polys* (many) and *gony* (a knee) referring to the jointed stems. Perennial herbs and climbers.
  affine a-*fee*-nee. Related to. Afghanistan, Himalaya, Tibet.
  amphibium ăm-*fi*-bee-um. Growing in water or on land. Widely distributed.
  amplexicaule ăm-pleks-i-*kaw*-lee. With stem-clasping leaves. Afghanistan, Himalaya, China.
  aubertii ō-*bair*-tee-ee. After the French missionary Aubert who introduced it. China.
  baldschuanicum băld-shoo-*ăn*-i-kum. Of Balzhuan, C Asia. S Tadzhikstan.
  bistorta bis-*tor*-ta. A medieval name, from L. *bis* (twice) and *tortus* (twisted) referring to the twisted roots. Bistort. Europe to C Asia.
  campanulatum kăm-păn-ew-*lah*-tum. Bell-shaped (the corolla). Himalaya.
  capitatum kă-pi-*tah*-tum. In a dense head (the flowers). Himalaya to W China.
  macrophyllum măk-rō-*fil*-lum. Large-leaved. Himalaya, China.
  multiflorum mul-tee-*flō*-rum. Many-flowered. Japan.
  sphaerostachium see *P. macrophyllum*
  tenuicaule ten-ew-i-*kaw*-lee. Slender-stemmed. japan.
  vacciniifolium va-keen-ee-i-*fo*-lee-um. *Vaccinium*-leaved. Himalaya, Tibet.

Polypodium po-lee-*pod*-ee-um *Polypodiaceae*. From Gk. *polys* (many) and *podos* (a foot) referring to the branched rhizome. Ferns.
  aureum see *Phlebodium aureum*
  vulgare vul-*gah*-re. Common. Common Polypody. Widespread.

Polypody, Common see *Polypodium vulgare*
  Limestone se *Gymnocarpium robertianum*

Polyscias po-*lis*-kee-as *Araliaceae*. From Gk. *polys* (many) and *skias* (an umbel) referring to the numerous umbels in the inflorescence. Tender, evergreen shrubs.
  scutellaria sku-tel-*ah*-ree-a. Resembling *Scutellaria* q.v.
  'Balfourii' băl-*for*-ree-ee. After Sir Isaac Bayley Balfour (1853–1922), who collected in Socotra. New Caledonia.
  guifoylei gil-*foyl*-ee-ee. After W. R. Guifoyle (1840–1912), who collected in the S Pacific Islands. Polynesia.

Polystichum po-*li*-sti-kum *Dryopteridaceae* From Gk. *polys* (many) and *stichos* (a row) referring to the arrangement of the sori. Ferns.
  acrostichoides a-kro-sti-*koi*-deez. Like *Acrostichum*, a related genus. Christmas Fern, Holly Fern. E N America.
  aculeatum a-kew-lee-*ah*-tum. Prickly. Hard Shield Fern. Widely distributed.
  falcatum see *Cyrtomium falcatum*
  munitum mew-*nee*-tum. Armed (with teeth). American Sword Fern. W N America.
  setiferum say-*ti*-fe-rum. Bristly. Soft Shield Fern. Europe.
  tsu-simense tsoo-see-*men*-see. Of Tsu-shima, Japan. E Asia.

Pomegranate see *Punica granatum*

Poncirus pon-*si*-rus *Rutaceae*. From the French name of a kind of citron. Deciduous shrub.
  trifoliata tri-fo-lee-*ah*-ta. With three-leaves (leaflets). China.

Pond Cypress see *Taxodium ascendens*

Pontederia pon-te-*de*-ree-a *Pontederiaceae*. After Guilo Pontedera (1688–1757), a professor of botany at Padua. Aquatic perennial herb.
  cordata kor-*dah*-ta. Heart-shaped (the leaves. Pickerel Weed. E N America.

Poor Man's Orchid see *Schizanthus*
Poplar see *Populus*
  Balsam see *P. balsamifera*
  Black see *P. nigra*
  Grey see *P.* × *canescens*
  Lombardy see *P. nigra* 'Italica'
  White see *P. alba*
Poppy see *Papaver*
  Alpine see *Papaver alpinum*
  Arctic see *P. nudicaulis*
  Californian see *Eschscholzia californica*

Poppy (continued)
Celandine see *Stylophorum diphyllum*
Crested see *Argemone platyceras*
Harebell see *Meconopsis quintuplinervis*
Himalayan Blue see *Meconopsis betonicifolia*
Iceland see *Papaver nudicaule*
Mexican Tulip see *Hunnemannia fumariifolia*
Opium see *Papaver somniferum*
Plume see *Macleaya microcarpa*
Prickly see *Argemone mexicana*
Red Horned see *Glaucium corniculatum*
Spanish see *Papaver rupifragum*
Tree see *Romneya*
Tulip see *Papaver glaucum*
Welsh see *Meconopsis cambrica*
Yellow Horned see *Glaucium flavum*

Populus pō-pu-lus *Salicaceae*. The L. name. Deciduous trees. Poplar.
alba ăl-ba. White (the undersides of the leaves). Abele, White Poplar. Europe, N Africa to C Asia.
balsamifera băl-sa-mi-fe-ra. Balsam-bearing Poplar. N America.
× *berolinensis* be-ro-leen-*en*-sis. *P. laurifolia* × *P. nigra* 'Italica'. Of Berlin, where it originated.
× *canadensis* kan-a-*den*-sis. *P. deltoides* × *P. nigra*. Of Canada.
'Robusta' rō-*bus*-ta. Robust.
'Serotina' se-*ro*-ti-na. Late (into leaf).
candicans see *P.* × *jackii* 'Gileadensis'
× *canescens* kah-*nes*-enz. Grey-hairy (the leaves). Grey Poplar. Europe.
× *jackii* jak-ee-ee (= *P. candicans*) *P. balsamifera* × *P. deltoides*. After Jack, derivation not found. C & EN America.
'Gileadensis'. *gil*-ee-a-*den*-sis. Of Gilead. Balm of Gilead.
lasiocarpa lă-see-ō-*kar*-pa. With woolly fruits. China.
nigra nig-ra. Black (the bark). Black Poplar. Europe, W Asia.
betulifolia bet-ew-li-*fo*-lee-a. With leaves like *Betula*. W Europe.
'Italica' ee-*tăl*-i-ka. Italian. Lombardy Poplar.
tremula trem-ew-la. Trembling (the leaves). Aspen. Europe, N Africa, N Asia.
trichocarpa tri-kō-*kar*-pa. With hairy fruits. W N America.
wilsonii wil-*son*-ee-ee. After Wilson who introduced it in 1907, see *Magnolia wilsonii*. China.

Porroglossum-por-o-*gloss*-um *Orchidaceae*. From Gk. *porro*, forward, and *glossa*, tongue, referring to the shape of the lip. Epiphytic and terrestrial orchids. C & S America.
muscosa moo-*sko*-sa. (= *Masdevallia muscosa*). Mossy. Perhaps *muscipula* was intended as the lip of the flower traps small insects. Colombia.

Portugal Laurel see *Prunus lusitanica*

Portulaca por-tew-*lah*-ka *Portulacaceae*. The L. name for *P. oleraceae*. Succulent annuals.
grandiflora grănd-i-*flō*-ra. Large-flowered. Sun Plant. S America.
oleraceae o-le-*rah*-kee-a. Vegetable-like. Purslane, India.

Potamogeton po-ta-mo-*gay*-ton *Potamogetonaceae*. From the classical name, from Gk. *potamos* (a river) and *geiton* (a neighbour). Aquatic herbs. Widely distributed.
crispus kris-pus. Wavy (the leaves).
pectinatus pek-ti-*nah*-tus. Comb-like (the arrangement of the leaves).
perfoliatus per-fo-lee-*ah*-tus. With the leaf base surrounding the stem.

Potato see *Solanum tuberosum*

Potentilla po-ten-*til*-la *Rosaceae*. From L. *potens* (powerful) referring to medicinal properties. Perennial herbs and shrubs.
alba al-ba. White (the flowers). Europe, Caucasus.
alchemilloides ăl-ke-mil-*oi*-des. Like *Alchemilla*. Pyrenees.
atrosanguinea aht-rō-sang-*gwin*-ee-a. Deep red (the flowers).
argyrophylla ar-gi-rō-*fil*-la. With silvery leaves. Himalaya.
aurea ow-ree-a. Golden (the flowers). Europe.
chrysocraspeda kris-ō-*krăs*-pe-da. (= *P. ternata*). Golden-fringed. SE Europe.
calabra ka-*lăb*-ra. Of Calabria. Italy, Sicily.
cinerea ki-*ne*-ree-a. (= *P. tommasiniana*). Grey-hairy. Alps.
crantzii krănts-ee-ee. After H. J. N. von Crantz (1722–99) a botanical writer. Europe, W Asia.
eriocarpa e-ree-ō-*kar*-pa. With woolly fruits. Himalaya.
fragiformis see *P. megalantha*
fruticosa froo-ti-*kō*-sa. Shrubby. N temperate regions.

Potentilla (continued)
arbuscula ar–*bus*–kew–la. Like a small tree. Himalaya, China.
'Beesii' *beez*–ee–ee. After Messrs Bees who distributed it.
megalantha meg–al–*anth*–a (= *P. fragiformis*). Large-flowered. NE Asia.
nepalensis ne–pa–*len*–sis. Of Nepal. Himalaya.
neumanniana new–man–ee–ah–na (= *P. tabernaemontana*). After Alfred Neumann (1916–73), German botanist. N, W & C Europe.
nitida ni–ti–da. Shining (the flowers). Alps.
recta rek–ta. Erect. Europe, N Africa.
ruperstris roo–*pes*–tris. Growing on rocks. Europe, Asia, W N America.
tabernaemontani see *P. neumanniana*
ternata see *P. aurea chrysocraspeda*
tommasiniana see *P. cinera*

× **Potinara** po–ti–*nah*–ra Orchidaceae. Intergeneric hybrid, after M. Potin, president of the French orchid society in 1922.
Brassavola × *Cattleya* × *Laelia* × *Sophronitis*. Greenhouse orchids.

Powder Puff see *Mammillaria bocasana*

Pratia prah–tee–a Lobeliaceae. After Ch. L. Prat-Bernon, a French naval officer. Perennial herb.
angulata ang–gew–*lah*–ta. (= *P. treadwellii*). Angled. New Zealand.

Prayer Plant see *Maranta leuconerura*
Prickly Moses see *Acacia verticillata*
Prickly Pear see *Opuntia*
Prickly Thrift see *Acantholimon*
Pride of Texas see *Phlox drummondii*
Primrose see *Primula vulgaris*
  Fairy see *P. malacoides.*

Primula preem–ew–la Primulaceae. From L. *primus* (first), referring to the early flowers. Perennial herbs.
acaulis see *P. vulgaris*
allionii ă–lee–ō–nee–ee. After Carlo Allioni (1705–1804), Italian botanist. Maritime Alps.
alpicola ăl–*pi*–ko–la. Growing on mountains. Tibet.
amoena a–*moy*–na. Pleasant. Caucasus.
aurantiaca ow–răn–tee–*ah*–ka. Orange (the flowers). China.
auricula ow–*rik*–ew–la. An old name, from L. *auricula* (an ear), the leaves have been likened to bear's ears. Auricula. Europe.

Primula (continued)
beesiana beez–ee–*ah*–na. After Bees Nursery. China.
bracteosa brăk–tee–*o*–sa. With conspicuous bracts. W China.
bulleyana bu–lee–*ah*–na. After A. K. Bulley, see *Iris bulleyana*. China.
burmanica bur–*măn*–i–ka. Of Burma. China, Burma.
capitata kă–pi–*tah*–ta. In a dense head (the flowers). Himalaya.
chionantha kee–on–*ănth*–a. With snow-white flowers. China.
clarkei klark–ee–ee. After C. B. Clarke who discovered it in 1876. Kashmir.
cockburniana kō–burn–ee–*ah*–na. After H. Cockburn of the Consular Service, Chungking and the Rev. G. Cockburn who assisted the collector, A. E Pratt. China.
cortusiodes kor–tew–*soi*–deez. Like *Cortusa*. W Siberia.
denticulata den–tik–ew–*lah*–ta. Slightly toothed. Afghanistan, Himalaya, Burma.
edgeworthii ej–*werth*–ee–ee. After Edgeworth, see *Edgeworthia*. Himalaya.
elatoir ay–*lah*–tee–or. Taller. Oxlip. Europe, W Asia.
florindae flo–*rin*–die. Discovered by Kingdon-Ward and named after his first wife Florinda. SE Tibet.
frondosa fron–*dō*–sa. Leafy. Bulgaria.
helodoxa see *P. prolifera*
hirsuta hir–*soo*–ta. (= *P. rubra*). Hairy. Alps. Pyrenees.
japonica ja–*pon*–i–ka. Of Japan.
'Kewensis' kew–*en*–sis. *P. floribunda* × *P. verticillata*. Of Kew where it was raised in 1898.
littoniana see *P. vialii*
malacoides mă–la–*koi*–deez. Mallow-like. Fairy Primrose. China.
marginata mar–gi–*nah*–ta. Margined (the leaves). SW Alps.
minima mi–ni–ma. Smaller. E Europe.
nutans new–tanz. Nodding (the flowers). China.
obconica ob–*kō*–ni–ka. Like an inverted cone (the calyx). China.
Polyantha po–lee–*ănth*–a. Many flowered. Polyanthus.
polyneura po–lee–*new*–ra. With many veins. China.
prolifera prō–*li*–fe–ra. Proliferous. Assam.
pulverulenta pul–ve–ru–*len*–ta. Mealy. China.
rosea ro–see–a. Rose-coloured. Himalaya.
rubra see *P. hirsuta.*

Primula (continued)
secundiflora se-kun-di-*flō*-ra. With flowers on one side of the stalk. Himalaya.
sieboldi see-*bōld*-ee-ee. After Siebold, see *Acanthopanax sieboldii*. Japan.
sikkimensis si-kim-*en*-sis. Of Sikkim. Himalaya, W China.
sinensis si-*nen*-sis. Of China.
spectabilis spek-*tah*-bi-lis. Spectacular. S Alps.
veris *ve*-ris. Of spring. Cowslip. Europe.
vialii vee-*ahl*-ee-ee. (= *P. littoniana*). After Pere Vial. China.
vulgaris vul-*gah*-ris. (= *P. acaulis*). Common. Primrose. Europe.
yargongensis yar-gong-*en*-sis. Of the Yargong gorge, Tibet. W China, Burma, Tibet.

Prince Albert's Yew see *Saxegothaea conspicua*
Prince's Feather see *Amaranthus hybridus*
Princess of the Night see *Selenicereus pteranthus*
Princess Vine see *Cissus sicyoides*

Pritzelago prit-sel-*ay*-gō *Cruciferae*. For George August Pritzel (1815–74), German botanist. One species of perennial herb. C & S Europe (mountains).
alpina *ăl*-peen-a (= *Hutchinsia alpina*) Alpine.
  auerswladii ow-erz-*văld*-ee-ee. (= *H. auerswaldii*). After Auerswald.

Privet see *Ligustrum*
Common see *L. vulgare*

Proboscidea pro-bos-*ki*-dee-a *Pedaliaceae*. From Gk. *proboskis* (an elephant's trunk) referring to the long, curved beak of the fruit. Annual and perennial herbs.
fragans frah-granz. Fragrant. Mexico.
louisianica loo-eez-ee-ee-*ah*-ni-ka. (= *Martynia louisianica*). Of Louisiana. S United States.

Promenaea pro-men-*ie*-a *Orchidaceae*. After Promeneia, a priestess of the temple of Dodona. Greenhouse Orchids.
stapelioides sta-pel-ee-*oi*-deez. Like *Stapelia*. Brazil.

Propeller Plant see *Crassula falcata*
Prophet Flower see *Arnebia pulchra*

Prostanthera pros-*tănth*-*e*-ra *Labiatae*. From Gk. *prosthema* (an appendage) and *anthera*

Prostanthera (continued)
(an anther) referring to the spurred anthers. Semi-hardy, evergreen shrubs.
ovalifolia ō-vah-li-*fo*-lee-a. With oval leaves. New South Wales.
rotundifolia ro-tun-di-*fo*-lee-a. Round-leaved. SE Australia.

Protea *prō*-tee-a *Proteaceae*. After Proteus a Gk. sea god with the power of prophecy. Tender, evergreen shrubs. S Africa.
barbigera bar-*bi*-ge-ra. Bearded.
cynaroides si-nah-*roi*-deez. Like *Cynara*. King Protea.
eximea eks-i-*mee*-a. Distinguished.
grandiceps *grănd*-i-keps. Large-headed.
nerifolia nay-ree-i-*fo*-lee-a. With leaves like *Nerium*.
speciosa spe-kee-*ō*-sa. Showy.

Prumnopitys proo-*nō*-*pit*-is *Podocarpaceae*. From Gk. *prumne*, stern, and *pitys*, pine; meaning unclear. Evergreen conifer formerly included in *Podocarpus*.
andina an-*deen*-a (= *Podocarpus andinus*). Of the Andes. Chile, Argentina.

Prunella proo-*nel*-la *Labiatae*. Possibly from L. *prunum* (purple) referring to the flowers or from German *Braüne* (quinsy) which they were said to cure. Perennial herbs. Self Heal.
grandiflora grănd-i-*flō*-ra. Large-flowered. Europe.
× *webbiana* web-ee-*ah*-na. After Webb.

Prunus proo-nus *Rosaceae*. The L. name for the plum tree. Deciduous trees, deciduous and evergreen shrubs.
× *amygdalo-persica* a-mig-da-lō-*per*-si-ka. *P. dulcis* (= *P. amygdalus*) × *P. persica*. From the names of the parents.
  'Pollardii' po-*lard*-ee-ee. After Mr Pollard, the raiser.
armeniaca ar-men-ee-*ah*-ka. Of Armenia. Apricot. N China.
avium *ă*-vee-um. Of birds. Gean, Mazzard. Europe.
× *blireana* bli-ree-*ah*-na. Of Bléré, France.
cerasifera ke-ra-*si*-fe-ra. Cherry-bearing. Cherry Plum, Myrobalan. Cult.
  'Pissardii' pee-*sard*-ee-ee. After M. Pissard, gardener to the Shah of Persia, who sent it to France on 1886.
cerasus ke-ra-sus. L. name for cherry. Sour Cherry. Cult.
× *cistena* sis-*teen*-a. *P. cerasifera* 'Pissardii' × *P. pumila*. The Sioux name for baby, from the dwarf habit.

Prunus (continued)
conradinae kon-ra-*deen*-ie. Name by
Koehne after his wife Conradine. China.
davidiana dă-vid-ee-*ah*-na. After David
who introduced it to France in 1865, see
Davidia. China.
x *domestica* do-*mes*-ti-ka. Cultivated. Plum.
Cult.
dulcis *dul*-kis. Sweet. Almond. SE Europe,
N Africa, W Asia.
glandulosa glăn-dew-*lō*-sa. Glandular. NE
Asia.
x *hillieri* hi-lee-a-ree. After Hillier's who
raised it.
incisa in-*kee*-sa. Deeply cut (the leaves).
Fuji Cherry. Japan.
    'Praecox' *prie*-koks. (Early flowering).
laurocerasus low-rō-*ke*-ra-sus. As the
common name, Cherry Laurel. SE Europe,
W Asia.
    'Magnoliifolia' măg-nol-ee-i-*fo*-lee-a.
    Magnolia-leaved.
    'Schipkaensis' ship-ka-*en*-sis. Of the
    Schipka Pass, Bulgaria.
    'Zabeliana' za-bel-ee-*ah*-na. After
    Zabel.
lusitanica loo-si-*tah*-ni-ka. Of Portugal.
Portugal Laurel. Spain, Portugal.
mume mew-mee. From *ume* the Japanese
name. Japanese Apricot. SW China.
padus pă-us. Gk. name of a wild cherry.
Bird Cherry. N Europe, N Asia.
persica per-si-ka. Of Persia. Peach. China.
pumila pew-mi-la. Dwarf, NE United
States.
sargentii sar-*jent*-ee-ee. After Charles
Sprague Sargent (1841–1927), first
director of the Arnold Arboretum. N
Japan.
serrula se-ru-la. Saw-toothed (the leaves).
W China.
serrulata se-ru-lah-ta. With small teeth (the
leaves). Cult.
spinosa spee-*nō*-sa. Spiny. Sloe, Blackthorn.
Europe, N Asia.
subhirtella sub-hir-*tel*-la. Somewhat hairy.
Cult.
    'Autumnalis' ow-tum-*nah*-lis. Of
    autumn (flowering).
tenella te-*nel*-la. Dainty. Dwarf Russian
Almond. E Europe, SW Russia.
triloba tri-*lō*-ba. Three-lobed (the leaves).
China.
    'Multiplex' *mul*-ti-pleks. Much-folded
    (the double flowers).
x *yedoensis* ye-dō-*en*-sis. Of Tokyo (Yedo).
Yoshino Cherry.
    'Ivensii' ie-*venz*-ee-ee. After Arthur J.
    Ivens of Hillier's, who raised it.

Pseuderanthemum soo-de-*rănth*-e-mum
Acanthaceae. From Gk. *pseudo* (false) and
Eranthemum, a related genus. Tender shrubs.
Polynesia.
    atropurpureum aht-rō-pur-*pewr*-ree-um.
    Deep purple (the leaves).
    reticulatum ray-tik-ew-*lah*-tum. Net-
    veined (the leaves).

Pseudolarix soo-dō-*lă*-riks *Pinaceae*. From
Gk. *pseudo* (false) and *Larix* q.v. Deciduous,
larch-like conifer.
    amabilis a-*mah*-bi-lis. Beautiful. Golden
    Larch. E China.

Pseudopanax soo-dō-*păn*-ăks. *Araliaceae*.
From Gk. *pseudo* (false) and *Panax* a related
genus, see *Acanthopanax*. Semi-hardy,
evergreen shrubs or trees.
    crassifolius krăs-i-*fo*-lee-us. Thick-leaved.
    New Zealand.
    davidii see *Metapanax davidii*
    ferox fe-rōks. Spiny (the leaves). New
    Zealand.

Pseudosasa soo-dō-*say*-sa. *Graminae* From
Gk. *pseudo* and *Sasa*. Bamboo. Japan and
Korea.
    japonica ja-*pon*-i-ka. Of Japan. Arrow
    bamboo.

Pseudotsuga soo-dō-tsoo-ga *Pinaceae*. From
Gk. *pseudo* (false) and *Tsuga* q.v. Evergreen
conifer.
    menziesii men-*zeez*-ee-ee. After Menzies,
    see *Menziesia*. Douglas Fir. W N
    America.
        'Fletcheri' *flech*-a-ree. After Fletcher's
        nursery who distributed it.

Pseudowintera soo-dō-win-*te*-ra *Winteraceae*.
From Gk. *pseudo* (false) and *Wintera*.
Originally named *Wintera*, a name which had
already been used for *Drimys*, see *D. winteri*
Evergreen shrub.
    colorata ko-lo-*rah*-ta. Coloured (the leaves).
    New Zealand.

Psidium si-dee-um *Myrtaceae*. From *psidion*
the Gk. name for a pomegranate. Tender,
evergreen shrub.
    guajava gwah-*hah*-va. The native name.
    Guava. Tropical America.

Psychopsis si-*kop*-sis *Orchidaceae*. From Gk.
psyche, butterfly, and *opsis*, appearance.
Epiphytic orchids formerly included in
Oncidium. C & S America.

Psychopsis (continued)
  papilio pah-*pi*-lee-o (= *Oncidium papilio*).
Butterfly. Butterfly Orchid. S America.

Psylliostachys si-lee-*o*-sta-kis *Plumbaginaceae*.
From Gk. *psyllion* (a type of plantain) and
stachys (a spike). Annual herbs. W and C
Asia.
  spicata spee-*kah*-ta. (= *Limonium spicatum*).
With flowers in spikes.
  suworowii soo-vo-*rov*-ee-ee. (= *Limonium
suworowii*). After Ivan Petrowitch
Suworow, a medical inspector in Turkestan
c 1886.

Ptelea tel-ee-a *Rutaceae*. Gk name for an elm,
the winged fruits are similar. Deciduous
tree.
  trifoliata tri-fo-lee-*ah*-ta. With three leaves
(leaflets). Hop Tree. S Canada. E United
States.

Pteris te-ris *Pteridaceae*. From Gk. *pteris* (a
fern). Tender ferns.
  cretica kray-ti-ka. Of Crete. Cretan Brake.
Tropics.
    'Albo-lineata' ăl-bō-li-nee-*ah*-ta.
White-lined (the fronds).
  ensiformis ayn-si-*form*-is. Sword-shaped
(the pinnae). Sword Brake. E Asia to
Australia.
  multifida mul-ti-*fi*-da. Divided many times.
E Asia.
  quadriaurita kwod-ree-ow-*ree*-ta. Four-
eared (the base of the fronds). Tropics.
    'Argyraea' ar-gi-*ree*-a. Silvery.
  tremula trem-ew-la. Trembling. Australian
Brake. Australia. New Zealand.

Pterocactus te-rō-kăk-tus *Cactaceae*. From
Gk. *pteron* (a wing) and *Cactus* q.v. referring
to the winged seeds.
  kuntzei kŭnt-see-ee (= *P. tuberosus*) After
Kuntze, Carl Ernste (1843–1907),
botanist. Argentina.

Pterocarya te-rō-*kă*-ree-a *Juglandaceae*. From
Gk. *pteron* (a wing) and *karyon* (a nut)
referring to the winged fruit. Deciduous
trees. Wing-nut.
  fraxinifolia fraks-i-ni-*fo*-lee-a. *Fraxinus*-
leaved. Caucasian Wing-nut. Caucasus,
N Iran.
  × *rehderiana* ray-da-ree-*ah*-na. *P.
fraxinifolia* × *P. stenoptera*. After Rehder,
see *Rehderodendron*.

Pterocephalus te-rō-*kef*-a-lus *Dipsacaceae*.
From Gk. *pteron* (a wing) and *kephale* (a

Pterocephalus (continued)
head) referring to the feathery fruiting heads.
Perennial herb.
  perennis pe-*ren*-is. (= *P. parnassi*).
Perennial. S and E Greece.

Pteropogon ter-ō-*po*-gon *Compositae*. From
Gk. *pteron*, wing or feather, and *pogon*,
beard; probably referring to the feathery
pappus. Annual herbs. Everlasting.
  humboldtianum hum-bolt-ee-*ah*-num (=
Helipterum humboldtianum). After W. H.
Alexander von Humboldt (1769–1859). S
Australia.

Pterostyrax te-rō-*sti*-răks *Styracaceae*. From G.
pteron (a wing) and *Styrax* q.v. Deciduous
shrub.
  hispida his-pi-da. Bristly (the fruit). Japan,
China.

Ptilotrichum see *Alyssum*

Pulmonaria pul-mon-*ah*-ree-a *Boraginaceae*.
From L. *pulmo* (the lung), the leaves were
used to treat bronchial infections.
Herbaceous perennials. Lungwort.
  angustifolia ang-gus-ti-*fo*-lee-a. Narrow-
leaved. C and E Europe.
  longifolia long-gi-*fo*-lee-a. Long-leaved. W
Europe.
  montana mon-*tah*-na (= *P. rubra*). Of
mountains. SE Europe.
  officinalis o-fi-ki-*nah*-lis. Sold as a herb.
Europe.
  saccharata să-ka-*rah*-ta. Appearing
sprinkled with sugar (the leaves). SE
France, N Italy.

Pulsatilla pul-sa-*til*-la *Ranunculaceae*. From L.
pulso (to strike). Perennial herbs. Pasque
Flower.
  alpina ăl-*peen*-a. (= *Anemone alpina*).
Alpine. C and S Europe.
  halleri hăl-a-ree. (= *Anemone halleri*). After
Albrecht von Haller (1708–77) a Swiss
botanist. Alps.
  rubra rub-ra. Red (the flowers). France,
Spain.
  vernalis ver-*nah*-lis. (= *Anemone vernalis*).
Of spring. Europe.
  vulgaris vul-*gah*-ris. (= *Anemone pulsatilla*).
Common. Europe.

Pummelo see *Citrus maxima*
Pumpkin see *Cucurbita maxima*

Punica pew-ni-ka *Punicaceae*. The L. name.
Semi-hardy evergreen shrub or tree.

Punica (continued)
granatum grah-*nah*-tum. Many-seeded.
Iran, Afghanistan. Pomegranate.

Purple Bell Vine see *Rhodochiton atrosanguineum*
Purple Heart see *Tradescantia pallida*
Purple Moor Grass see *Molinia caerulea*
Purple Passion Vine see *Gynura aurantiaca*
Purple Top see *Verbena bonariensis*
Purple Wreath see *Petrea volubilis*
Purslane see *Portulaca oleracea*
  Rock see *Calandrinia*
  Tree see *Atriplex halimus*
Pussy Ears see *Cyanotis somaliensis*, *Kalanchoe tomentosa*

Puschkinia push-*kin*-ee-a *Liliaceae*. After
Count Apollos Mussin-Puschkin (died
1815), who collected in the Caucasus in
1802. Bulbous perennial.
  scilloides skil-*loi*-deez. (= *P. libanotica*). Like
  Scilla. Caucasus, W Asia.

Pyracantha pi-ra-*kănth*-a *Rosaceae*. From Gk.
pyr (fire) and *akantha* (a thorn) referring to
the spiny shoots and red berries.
  angustifolia ang-gus-ti-*fo*-lee-a. Narrow-
  leaved. W China.
  atalantioides ă-ta-lăn-tee-*oi*-deez. Like
  Atalantia (Rutaceae). C China.
  coccinea kok-*kin*-ee-a. Scarlet (the fruits). S
  Europe, W Asia.
    'Lalandei' la-*lond*-ee-ee. After M.
    Lalande who raised it in France c 1874.
  rogersiana ro-jerz-ee-*ah*-na. After G. L.
  Coltman-Rogers of Stanage Park who
  first exhibited it. W China.

Pyrethropsis pi-reeth-*rop*-sis *Compositae*.
From *Pyrethrum*, and Gk. *opsis*, meaning
appearance. Perennial herbs and subshrubs,
usually rhizomatous. NW Africa.
  hosmariensis hos-mah-ree-*en*-sis (=
  Chrysanthemum hosmariense). Of Beni
  Hosmar, near Tetuán, Morocco.

Pyrethrum see *Tanacetum coccineum*

Pyrola *pi*-ro-la *Pyrolaceae*. Diminutive of
Pyrus, from the similar leaves. Perennial
herbs. Wintergreen.
  americana a-me-rik-*ah*-na (= *P. rotundifolia*)
  Of America. Eastern N America.
  asarifolia a-sah-ri-*fo*-lee-a. *Asarum*-leaved.
  N America.
  elliptica e-*lip*-ti-ka. Elliptic (the leaves). N
  America, Japan.

Pyrola (continued)
  media *me*-dee-a. Intermediate. Europe, W
  Asia.
  rotundifolia see *P. americana*

Pyrus *pi*-rus *Rosaceae*. The L. name.
Deciduous trees. Pear.
  calleryana ka-le-ree-*ah*-na. After J. Callery, a
  French missionary who collected the type
  specimen. China.
  communis kon-*ew*-nis. Common. Common
  Pear. Europe.
  salicifolia să-li-ki-*fo*-lee-a. *Salix*-leaved.
  Willow-leaved Pear. Caucasus, Turkey,
  NW Iran.

Q

Quaking Grass see *Briza media*
Quamoclit coccinea see *Ipomoea coccinea*
  lobata see *Ipomoea lobata*
  pennata see *Ipomoea quamoclit*
Queen of the Night see *Selenicereus grandiflorus*
Queen of the Prairie see *Filipenula rubra*
Queen's Wreath see *Petrea volubilis*
Queensland Umbrella Tree see *Schefflera actinophylla*

Quercus *kwer*-kus *Fagaceae*. The L. name.
Deciduous and evergreen trees. Oak.
  acutissima ă-kew-*tis*-i-ma. Very sharply
  pointed (the leaves). E Asia.
  agrifolia ăg-ri-*fo*-lee-a. With spiny leaves.
  California, N Mexico.
  alnifolia ă-ni-*fo*-lee-a. *Alnus*-leaved.
  Golden Oak of Cyprus. Cyprus.
  canariensis ka-nah-ree-*en*-sis. Of the
  Canary Islands. N Africa, S Portugal,
  Spain.
  castaneifolia kas-tăn-ee-i-*fo*-lee-a. *Castanea*-
  leaved. N Iran, SW Russia.
  cerris ke-ris. The L. name. Turkey Oak.
  Europe, Turkey.
  coccifera kok-*kif*-e-ra. Berry-bearing,
  referring to the kermes insects which
  breed on the tree. Kermes Oak. W
  Mediterranean region, N Africa.
  coccinea kok- *kin*-ee-a. Scarlet (autumn
  colour). Scarlet Oak. E N America.
  dentata den-*tah*-ta. Toothed (the leaves).
  Daimio Oak. Japan, NE Asia.
  frainetto fray-*net*-ō. In error for *farnetto*, the
  Italian name. Hungarian Oak. SE Europe.
  × *hispanica* his-*pah*-ni-ka. *Q. cerris* × *Q.*

Quercus (continued)
suber. Spanish, it was originally thought
to come from Spain.
'Ambrozyana' ăm-brō-zee-*ah*-na. After
Count Ambrozy of the Arboretum
Mlynany, who raised it.
'Lucombeana' lū-kom-bee-*ah*-na. After
Lucombe, an Exeter nurseryman, who
raised it. Lucombe Oak.
ilex ee-leks. The L. name. Holm Oak.
Mediterranean region.
imbricaria im-bri-*kah*-ree-a. From L. *imbrex*
(a tile) the wood was used for roof tiles.
Shingle Oak. E United States.
ithaburensis i-tha-bew-*ren*-sis. Of Ithabura.
Vallonea Oak. Syria, Palestine.
    macrolepis măk-rō-*lep*-is. With large
    scales (on the acorn). SE Europe, Turkey.
× *kewensis* kew-*en*-sis. *Q. cerris* × *Q.
wislizenii.* Of Kew, where it was raised.
libani li-ba-nee. Of Lebanon. Lebanon
Oak. W Asia.
macranthera ma-*krănth*-e-ra. With large
anthers. Caucasus, N Iran.
marilandica mă-ri-*lănd*-i-ka. Of Maryland.
Black Jack Oak. E United States.
myrsinifolia mur-si-ni-*fo*-lee-a. Myrsine-
leaved. E Asia.
palustris pa-*lus*-tris. Of swamps. Pin Oak.
E United States.
petraea pe-*trie*-a. Of rocky places. Durmast
Oak, Sessile Oak, Europe.
phillyreoides fi-li-ree-*oi*-deez. Like *Phillyrea.*
China, Japan.
pyrenaica pi-ray-*nah*-i-ka. Of the Pyrenees.
SW Europe, Morocco.
robur rō-bur. The L. name for the oak and
its wood. Common Oak, Pedunculate
Oak. Europe, Caucasus.
rubra nub-ra. Red (autumn colour). Red
Oak. E N America.
suber soo-ber. The L. name. Cork Oak. W
Mediterranean region.
× *turneri turn*-a-ree. *Q. ilex* × *Q. robur.*
After Mr Spencer Turner (c 1728–76) in
whose Essex nursery it was raised.
velutina vel-ew-*teen*-a. Velvety (the buds).
Black Oak. E N America.
wislizenii wiz-li-*zen*-ee-ee. After A.
Wislizenius. California.

Quince see *Cydonia oblonga*
  Ornamental see *Chaenomeles*

R

Rabbit's Foot see *Maranta leuconeura
kerchoveana*
Radish see *Raphanus sativus*
Rainbow Star see *Cryptanthus bromelioides*
Rainbow Vine see *Pellionia pulchra*

Ramonda ra-*mon*-da *Gesneriaceae.* After Louis
Francis Ramond (1753–1827), French
botanist. Perennial herbs.
    myconi mi-*kō*-nee. After Francisco Mico
    (born 1528), Spanish physician and
    botanist. Pyrenees, N Spain.
    nathaliae na-*tah*-lee-ie. After Queen
    Nathalia, wife of King Milan. S
    Yugoslavia, N Greece.
    serbica ser-bi-ka. Of Serbia. SE Europe.

Rampion see *Phyteuma orbiculare*
  Spiked see *P. spicatum*

Ranunculus rah-*nun*-kew-lus *Ranunculaceae.*
The L. name from *rana* (a frog) as many
grow in wet places. Annual and perennial
herbs. Buttercup.
    aconitifolius ă-kon-ee-ti-*fo*-lee-us.
    Aconitum-leaved. White Bachelors
    Buttons. C and S Europe.
        'Flore Pleno' *flō*-ree-*play*-nō. With
        double flowers. Fair Maids of France.
    acris ah-kris. Sharp-tasting. Europe, Asia.
        'Flore Pleno' *flō*-ree-*play*-nō. With
        double flowers. Yellow Bachelors
        Buttons.
    alpestris ăl-*pes*-tris. Of the lower
    mountains. Alps, Pyrenees.
    amplexicaulus ăm-pleks-i-*kaw*-lis. With
    leaves clasping the stems. SW Europe.
    aquatilis a-*kwah*-ti-l Growing in water.
    Water Crowfoot. Europe.
    asiaticus ah-see-*ah*-ti-kus. Asian. Persian
    Buttercup. Crete, SW Asia.
    bulbosus bul-*bō*-sus. Bulbous. Europe.
    calandrinioides kă-kan-dree-nee-*oi*-deez.
    Like *Calandrinia.* Morocco.
    ficaria fee-*kah*-ree-a. Like *Ficus,* the fig-like
    tubers. Lesser Celandine. Europe, W.
    Asia.
    glacialis glă-kee-*ah*-lis. Of icy regions.
    Europe, Greenland.
    gramineus grah-*min*-ee-us. Grass-like (the
    leaves).
    lingua ling-gwa. Tongue-like (the leaves).
    Greater Spearwort. Europe.
    montanus mon-*tah*-nus. Of mountains.
    Europe, Caucasus.

Ranunculus (continued)
parnassifolius par-năs-i-*fo*-lee-us. With
leaves like *Parnassia*. Pyrenees, Alps.
pyrenaeus pi-ray-*nie*-us. Of the Pyrenees.
Alps, Pyrenees, Corsica.

Raoulia rowl-ee-a *Compositae*. After Edward
Raoul (1815–52), French surgeon who
studied New Zealand plants. Perennial herbs
and sub-shrubs. New Zealand.
australis ow-*strah*-lis. (= *R. lutescens*).
Southern.
hookeri huk-a-ree. After Sir Joseph Hooker.
tenuicaulis ten-ew-i-*kaw*-lis. Slender-
stemmed.

Rape see *Brassica napus*

Raphanus ră-fa-nus *Cruciferae*. The L. name.
Biennial herb.
sativus sa-*teev*-us. Cultivated. Radish. Cult.

Raphiolepis see *Rhaphiolepis*
Raspberry see *Rubus idaeus*
Rattlesnake Plant see *Calathea lancifolia*
Rauli see *Nothofagus procera*

Rebutia ra-*bew*-tee-a *Cactaceae*. After P.
Rebut, a French cactus dealer. Crown Cactus.
Unless otherwise stated, Argentina.
aureiflora ow-ree-i-*flo*-ra. Golden flowered.
deminuta day-mi-*new*-ta. (= *Aylostera
deminuta*). Small.
famatimensis fă-ma-tee-*men*-sis (= *Lobivia
famatimensis*). Of Famatima.
fiebrigii fee-*brig*-ee-ee. After Dr C. Fiebrig
of the Museum and Garden, Asuncion,
Paraguay. Bolivia.
kupperiana ku-pa-ree-*ah*-na. After
Professor W. Kupper of Munich.
minuscula min-*nus*-kew-la. Rather small.
Red Crown Cactus.
pseudodeminuta soo-dō-day-mi-*new*-ta.
False *R. deminuta*.
pygmaea pig-*mie*-a (= *Lobivia pygmaea*).
Dwarf. Bolivia, N Argentina.
senilis se-*nee*-lis. An old man, referring to
the white, bristle-like spines. Fire Crown
Cactus.
spegazziniana spe-ga-zeen-ee-*ah*-na. After
Professor Carlos Spegazzini (1858–1926),
an Argentine botanist.
spinosissima speen-ō-*sis*-i-ma. Very spiny.
violaciflora see *R. minuscula*
xanthocarpa zănth-ō-*kar*-pa. With yellow
fruits.

Rechsteineria cardinalis see *Sinningia cardinalis*
cyclophylla see *S. macropoda*

Rechsteineria cardinalis (continued)
leucotricha see *S. canescens*
Red Cape Tulip see *Haemanthus*
Red Ivy see *Hemigraphis alternata*
Red-hot Cat's Tail see *Acalypha hispida*
Red Hot Poker see *Kniphofia*
Red Nodding Bells see *Streptocarpus dunnii*
Red Ribbons see *Clarkia concinna*
Redbud see *Cercis canadensis*
Redwood, Coast see *Sequoia sempervirens*
   Giant see *Sequoiadendron giganteum*
Reed Canary Grass see *Phalaris arundinacea*
Reed Grass see *Glyceria maxima*
Reed-mace see *Typha*

Rehderodendron ray-da-rō-*den*-dron
Styracaceae. After Alfred Rehder
(1863–1949), American botanist, of the
Arnold Arboretum, and Gk. *dendron* (a tree).
Deciduous tree.
macrocarpum măk-rō-*kar*-pum. Large-
fruited. Omei Shan, China.

Rehmannia ray-*mahn*-ee-a *Gesneriaceae*. After
Josehp Rehmann (1753–1831), German
physician. Perennial herb.
elata ay-*lah*-ta. (= R. *angulata* hort.). Tall.
Chinese Foxglove. China.

Reineckia rie-*nek*-ee-a *Liliaceae*. After Johann
Heinrich Julius Reinecke (1799–1871).
Perennial herb.
carnea kar-nee-a. Flesh-pink (the flowers).
China, Japan.

Reinwardtia rein-*vart*-ee-a *Linaceae*. After
Caspar Reinwardt (1773–1854), founder of
Bogar Botanic Garden, Java. Tender sub-
shrub.
indica in-di-ka. (= R. *tetragyna*. R. *trigyna*).
Indian. Yellow Flax. Himalaya, E and SE
Asia.

Renanthera ray-nan-*the*-ra *Orchidaceae*. From
L. *renes* (kidneys) and *anthera* (an anther)
referring to the kidney-shaped pollinia.
Greenhouse orchids. SE Asia.
coccinea kok-*kin*-ee-a. Scarlet (the flowers).
imschootiana im-shoot-ee-*ah*-na. After A.
van Imschoot of Ghent.

Reseda re-*say*-da *Resedaceae*. The L. name
from *resedo* (to heal) referring to medicinal
properties. Annual herb.
odorata o-dō-*rah*-ta. Fragrant (the flowers).
Mignonette. N Africa.

Resurrection Lily see *Lycoris squamigera*

Resurrection Plant see *Selaginella lepidophylla*
Rex-begonia Vine see *Cissus discolor*

Rhamnus răm-nus *Rhamnaceae*. The Gk. name of a shrub. Deciduous and evergreen, trees and shrubs.
alaternus ă-la-*tern*-us. The L. name. S Europe.
catharticus ka-*thar*-ti-kus. Purging. Common Buckthorn. Europe, Asia.
frangula frang-gew-la. The L. name. Alder Buckthorn. Europe, N Africa, Asia.

Rhaphiolepis răf-ee-ō-*lep*-is *Rosaceae*. (*Raphiolepis*). From Gk. *rhaphis* (a needle) and *lepis* (a scale) referring to the narrow bracteoles of the inflorescence. Semi-hardy, evergreen shrubs.
× *delacourii* de-la-*kour*-ree-ee. R. *indica* × R. *umbellata*. After M. Delacour, the raiser.
indica in-di-ka. Indian. Indian Hawthorn. S China.
umbellata um-bel-*ah*-ta. The flowers appear to be in umbels. Japan, Korea.

Rhapis ră-pis *Palmae*. From Gk. *rhapis* (a needle) referring to the slender leaf segments. Tender, suckering palms. Lady Palm. S China.
excelsa eks-*kel*-sa. Tall.
humilis hu-mi-lis. Low-growing.

Rhazya see *Amsonia*

Rheum ray-um *Polygonaceae*. From *rheon*, the Gk. name for rhubarb. Perennial herbs.
alexandrae ă-leks-*ahn*-drie. After Queen Alexandra, wife of Edward VII. Himalaya.
× *cultorum* cŭl-*tor*-um. Cultivated. Rhubarb.
palmatum pahl-*mah*-tum. Lobed like a hand (the leaves). NE Asia.
rhabarbarum see R. × *cultorum*

Rhipsalidopsis see *Hatiora*

Rhipsalis rip-sa-lis *Cactaceae*. From Gk. *rhips* (wicker-work) referring to the supple shoots.
baccifera bah-*ki*-fe-ra (= R. *cassutha*). Berry-bearing. Mistletoe Cactus. S America, Ceylon, Africa.
capilliformis ka-pi-li-*form*-is. Thread-like (the stems). Brazil.
cereuscula kay-ree-*us*-kew-la. Like a small *Cereus*. Brazil, Uruguay.

Rhipsalis (continued)
crispata kris-*pah*-ta. Wavy-edged (the stems). Brazil.
houlletiana see *Lepismium houlletianum*
micrantha my-*kran*-tha (= R. *tonduzii*). Small-flowered. Costa Rica.
paradoxa pă-ra-*doks*-a. Unusual. Chain Cactus. Brazil.
pilocarpa pi-lō-*kar*-pa (= *Erythrorhipsalis pilocarpa*). With a hairy fruit. Brazil.
tonduzii see R. *micrantha*
warmingiana see *Lepismium warmingianum*

Rhodanthe ro-*dan*-thee *Compositae*. From Gk. *rhodon*, rose and *anthos*, flower. Annuals, perennials and shrubs.
manglesii man-*galz*-ee-ee (= *Helipterum manglesii*). After Mangles, see *Anizoganthus manglesii*. Swan River Everlasting. W Australia.

Rhodiola ro-*dee*-o-la *Crassulaceae*. From Gk. *rhodon* (a rose) referring to the rose-scented root of R. *rosea*. Perennial herbs.
fastigiata fa-stig-ee-*ah*-ta. (= *Sedum fastigiatum*). Upright. Himalaya, Tibet, China.
heterodonta he-te-rō-*don*-ta. (= *Sedum heterodontum*). Variably toothed (the leaves). Himalaya, Tibet.
rosea ro-see-a. (= *Sedum rhodiola*. *Sedum rosea*). Rose-like, see above. Roseroot. Arctic and alpine N hemisphere.

Rhodochiton ro-do-ki-ton *Scrophulariaceae*. From Gk. *rhodo–* (red) and *chiton* (a cloak) referring to the enveloping calyx. Semi-hardy climber.
atrosanguineum aht-ro-san-*gwin*-ee-um (= R. *volubile*). Dark red. (the corolla). Purple Bell Vine. S Mexico.

Rhododendron ro-do-*den*-dron *Ericaceae*. The Gk. name for *Nerium oleander* from *rhodo–* (red) and *dendron* a tree. Evergreen and deciduous shrubs. Azaleas are marked (Az).
aberconwayi ă-ba-*kon*-way-ee. After Lord Aberconway. E. Yunnan.
albrechtii ăl-*brekt*-ee-ee (Az). After Dr Albrecht of the Russian Consul at Hakodate, who collected the type specimen. C and N Japan.
arborescens ar-bo-*res*-enz (Az). Becoming tree-like. E N America.
arboreum ar-bo-ree-um. Tree-like. Himalaya.
atlanticum ăt-*lăn*-ti-kum (Az). Of the Atlantic coast (of N America). SE United States.

Rhododendron (continued)

augustinii aw-gus-*tin*-ee-ee. After
Augustine Henry who discovered it in
1886, see *Illicium henryi*. C and W China.
 chasmanthum kǎs-*mǎnth*-um. With
 gaping flowers.
auriculatum ow-rik-ew-*lah*-tum. Auricled
(the leaves). China.
barbatum bar-*bah*-tum. Bearded (the shoots
of some forms). Himalaya.
bullatum see *R. edgeworthii*
bureavii bew-*reev*-ee-ee. After Edouard
Bureau (1831–1918), a French botanist.
NW Yunnan.
calendulaceum ka-len-dew-*lah*-kee-um
(Az). Like *Calendula* (the flower colour).
E N America.
callimorphum kǎl-i-*morf*-um. Beautifully
shaped. W Yunnan, Upper Burma.
calophytum kǎl-o-*fi*-tum. Beautiful plant.
W Sichuan.
calostrotum kǎl-o-strō-tum. With a
beautiful covering (the leaves). Upper
Burma.
 keleticum kay-*lay*-ti-cum. (= *R.
 keleticum*). Charming. Tibet.
 radicans rah-di-kǎnz. (= *R. radicans*).
 With rooting stems.
caloxanthum see *R. campylocarpum
caloxanthum*
campylocarpum kǎm-pi-lō-*kar*-pum. With a
curved fruit. Himalaya.
 caloxanthum kǎ-loks-*ǎnth*-um. (= *R.
 caloxanthum*). A beautiful yellow. Upper
 Burma.
campylogynum kǎm-pi-*lo*-gi-num. With a
curved ovary. Himalaya, Upper Burma,
W China.
catawbiense ka-taw-bee-*en*-see. From near
the Catawba River. SE United States.
charitopes see *R. saluense* ssp. *saluense*
ciliatum ki-lee-*ah*-tum. Fringed with hairs
(the leaves). Himalaya.
cinnabarinum ki-na-ba-*reen*-um. Cinnabar
red. Himalaya.
 xanthocodon zǎnth-o-*kō*-don. (= *R.
 concatenans*, *R. xanthocodon*). A yellow
 bell.
concatenans see *R. cinnabarinum xanthocodon*
concinnum kon-*kin*-um. Elegant. W
Sichuan.
 pseudoyanthinum see *R. concinnum*
crassum see *R. maddenii crassum*
decorum de-*kō*-rum. Beautiful. China.
discolor see *R. fortunei discolor*
edgeworthii ei-*werth*-ee-ee. (= *R. bullatum*).
After Edgeworth, see *Edgeworthia*.
Himalaya, W China.
falconeri fawl-*kon*-a-ree. After Hugh

Rhododendron (continued)

Falconer (1808–65), Scottish doctor and
botanist with the E India Co. Himalaya.
fargesii see *R. oreodoxa fargesii*
fastigiatum fa-stig-ee-*ah*-tum. With
upright branches. Yunnan.
ferrugineum fe-roo-*gi*-nee-um. Rusty (the
shoots and undersides of the leaves). S
Europe (mountains).
fictolacteum see *R. rex fictolacteum*
forrestii fo-*rest*-ee-ee. After Forrest, see
Abies delavayi forrestii, W China, Upper
Burma.
 repens ree-pens. Creeping.
fortunei for-*tewn*-ee-ee. After Fortune who
discovered it. See *Fortunella*. E China.
 discolor dis-ko-lor. (= *R. discolor*). Two-
 coloured (the leaves). C China.
fulvum *ful*-vum. Tawny (the lower leaf
surface). E Himalaya, W China.
glaucophyllum glow-kō-*fil*-um. With
glaucous leaves (the undersides).
Himalaya.
haematodes hie-ma-*tō*-deez. Blood-red.
Yunnan.
hanceanum hǎns-ee-*ah*-num. After Henry
Fletcher Hance (1827–56), British
Consul in China. SW Sichuan.
hippophaeoides hi-po-fa-ee-*oi*-deez. Like
Hippophae. W China.
hirsutum hir-*soo*-tum. Hairy. Alps,
Yugoslavia.
impeditum im-pe-*dee*-tum. Tangled.
Yunnan.
imperator see *R. uniflorum imperator*
indicum *in*-di-kum (Az). Indian. S Japan.
insigne in-*sig*-nee. Distinguished. W
Sichuan.
kaempferi kempf-a-ree (Az). After Engelbert
Kaempfer (1651–1716), German
physician. Japan.
keleticum see *R. calostrotum keleticum*
kiusianum kee-oo-see-*ah*-num (Az). Of
Kyushu, Japan.
lepidostylum le-pi-dō-*sti*-lum. With a scaly
style. W China.
leucapsis loo-*kǎs*-pis. A white shield,
referring to the white, saucer-shaped
flowers. E Himalaya.
× *loderi*-lō-da-ree. *R. fortunei* × *R.
griffithianum*. After Edmund Loder who
raised it.
lutescens loo-*tes*-enz. Yellowish (the
flowers). W China.
luteum loo-tee-um. (Az). (= *Azalea
pontica*). Yellow (the flowers). W
Caucasus, Turkey.
macabeanum ma-kay-bee-*ah*-num. After

Rhododendron (continued)
Mr M'Cabe, Deputy Commissioner for the Naga Hills. Assam, Manipur.
maddenii ma-*den*-ee-ee. After Major Madden of the Bengal Civil Service. Himalaya.
crassum krăs-um. (= *R. crassum*). Thick (the leaves). NW Yunnan, Upper Burma.
moupinense moo-pin-*en*-see. Of Moupin. W Sichuan.
mucronulatum mew-kron-ew-*lah*-tum. With a short point (the leaves). NE Asia.
nakaharae nă-ka-*hah*-rie. After G. Nakahara, a Japanese botanist. Taiwan.
neriiflorum nay-ree-i-*flō*-rum. Nerium-flowered. E Himalaya, W China.
x *obtusum* ob-*tew*-sum (Az). Blunt (the leaves). Cult.
'Amoenum' a-*moy*-num. Pleasant.
occidentale ok-ki-den-*tah*-lee (Az). Western. W N America.
orbiculare or-bik-ew-*lah*-ree. Rounded (the leaves). W Sichuan.
oreodoxa o-ree-ō-*doks*-a. Glory of the mountain. W Sichuan.
fargesii far-*jee*-zee-ee. (= *R. fargesii*). After Farges who discovered it, see *Decaisnea fargesii*. Hubei, Sichuan.
oreotrephes o-ree-*o*-tre-feez. Growing on mountains. W China, Himalaya.
pemakoense pe-ma-kō-*en*-see. Of Pemako, Tibet.
polycladum po-lee-*klăd*-um. With many branches. China.
Scintillans *skin*-ti-lănz. (= *R. scintillans*). Gleaming.
ponticum pon-ti-kum. Of Pontus (NE Turkey). W Asia, Balkans.
poukhanense see *R. yedoense poukhanense*
pruniflorum proon-i-*flō*-rum. (= *R. tsangpoense pruniflorum*). Prunus-flowered. Upper Burma. Assam.
pseudochrysanthum soo-dō-kris-*ănth*-um. False *R. chrysanthum*. Taiwan.
quinquefolium kwing-kwee-*fo*-lee-um (Az). Five-leaved, the leaves are in terminal clusters of five. Japan.
racemosum ră-kay-*mō*-sum. With flowers in racemes. W China.
radicans see *R. calostrotum radicans*
reticulatum ray-tik-ew-*lah*-tum (Az). Net-veined (the lower surface of the leaves). Japan.
rex reks. King. W China.
fictolacteum fik-tō-*lak*-tee-um. (= *R. fictolacteum*). False *R. lacteum*. W China, SE Tibet.

Rhododendron (continued)
rubiginosum roo-bi-gi-*nō*-sum. Rusty (the lower leaf surface). W China, SE Tibet.
russatum rus-*ah*-tum. Russet, the lower leaf surface. W China.
saluenense săl-ew-en-*en*-see. Of the Nu Jiang (Salween River) W Yunnan. W China. SE Tibet.
sargentianum sar-jent-ee-*ah*-num. After Sargent, see *Prunus sargentii*. W Sichuan.
schlippenbachii shlip-an-*bahk*-ee-ee (Az). After Baron Schlippenbach who discovered it in 1854. Korea, E Russia, E China.
scintillans see *R. polycadum* Scintillans
simsii simz-ee-ee. After John Sims (1749–1831). S and C China, Taiwan, Burma.
sinograndе si-nō-*grăn*-dee. The Chinese *R. grande*. Burma. Assam, Tibet. W China.
souliei soo-lee-ee. After the French missionary Jean André Soulie (1858–1903) who discovered it. W Sichuan.
sutchuenense sūch-wen-*en*-see. Of Sichuan (Szechwan). W Hubei, E Sichuan.
tephropeplum tef-rō-*pep*-lum. Grey-cloaked. E Himalaya, Upper Burma.
thomsonii tom-*son*-ee-ee. After Thomson, see *Aster thomsonii*. Himalaya.
uniflorum ew-ni-*flō*-rum. One flowered. SE Tibet.
imperator im-*pe*-ra-tor. (= *R. imperator*). Emperor, it was originally named *R.* 'Purple Emperor'.
valentinianum vă-len-tin-ee-*ah*-num. After Père S. P. Valentin, missionary in China c 1919. Yunnan.
vaseyi vay-zee-ee (Az). After George R. Vasey (1822–93) who discovered it in 1878. N Carolina.
viscosum vis-*kō*-sum (Az). Sticky (the flowers). E N America.
wardii word-ee-ee. After Kingdon Ward who collected the type specimen, see *Cassiope wardii*. W China, Tibet.
williamsianum wil-yămz-ee-*ah*-num. After J. C. Williams of Caerhays. W Sichuan.
xanthocodon see *R. cinnabarinum xanthocodon*
Yakushimanum ya-koo-shee-*mah*-num. Of Yakushima, an island of S Japan.
yedoense ye-dō-*en*-see (Az). Of Tokyo (Yedo). Cult.
poukhanense poo-ka-*nen*-see. (= *R. poukhanense*). Of Poukhan-san, Korea. Korea.
yunnanense yoo-nan-*en*-see. Of Yunnan. W China, Tibet, Upper Burma.

Rhodohypoxis ro-do-hi-*poks*-is *Hypoxidaceae*.
From Gk. *rhodo*- (red) and *Hypoxis* q.v.
Semi-hardy perennial herbs.
   baurii bow-ree-ee. After the Rev. R. Baur
   who collected *Rhodohypoxis* in S Africa.
   S Africa.

Rhodotypos ro-do-*ti*-pos *Rosaceae*. From Gk.
rhodon (a rose) and *typos* (type) referring to
the rose-like flowers. Deciduous shrub.
   scandens skǎn-denz. Climbing. China,
   Japan.

Rhoeo see *Tradescantia*

Rhoicissus rō-i-*kis*-us *Vitaceae*. From L.
rhoicus (of *Rhus*) and *Cissus* q.v. Tender,
evergreen climber.
   capensis ka-*pen*-sis. (= *Cissus capensis*). Of
   the Cape of Good Hope. S Africa.
   rhomboidea hort. see *Cissus rhombifolia*

Rhombophyllum rom-bō-*fil*-lum *Aizoaceae*.
From Gk. *rhombos* (a rhombus) and a *phyllon*
(a leaf) referring to the shape of the leaves.
Tender succulents. S Africa.
   nelii nel-ee-ee. After G. C. Nel
   (1885–1960). Elk's Horns.
   rhomboideum rom-*boi*-dee-um. Diamond-
   shaped (the leaves).

Rhubarb see *Rheum* × *cultorum*

Rhus rus *Anacardiaceae*. The L. name of *R.
coriaria*. Deciduous shrubs and trees.
Sumach.
   cotinoides see *Cotinus obovatus*
   cotinus see *Cotinus coggygria*
   glabra glǎb-ra. Glabrous. Smooth Sumach.
   N America.
      'Laciniata' la-kin-ee-*ah*-ta. Deeply cut
      (the leaflets).
   typhina tee-*fee*-na. Like *Typha*. Stag's Horn
   Sumach. E N America.
      'Dissecta' di-*sek*-ta. (= 'Laciniata').
      Deeply cut (the leaflets).

Rhyncolaelia rin-kō-*lie*-lee-a *Orchidaceae*.
From Gk. *rhynchos*, beak, and *Laelia* q.v.; the
flowers have very prominent lips. Epiphytic
and lithophytic orchids, formerly included
in *Brassovola*.
   digbyana dig-bee-*ah*-na. After Edward St
   Vincent Digby, the first to flower it, in
   1846. C America.
   glauca glow-ka (= *Brassavola glauca*).
   Glaucous (the leaves). C America.

Ribbon Gum see *Eucalyptus viminalis*

Ribes rie-beez *Grossulariaceae*. From Arabic or
Persian *ribas* (acid-tasting), referring to the
fruit. Deciduous and evergreen shrubs.
   alpinum ǎl-*peen*-um. Alpine. Mountain
   Currant. Europe.
   gayanum gay-*ah*-num. After C Gay, a
   French botanist. Chile.
   laurifolium low-ri-*fo*-lee-um. *Laurus*-
   leaved. W China.
   nigrum nig-rum. Black (the fruit). Black
   Currant. Europe.
   odoratum o-dō-*rah*-tum. Fragrant (the
   flowers). Buffalo Currant. C United
   States.
   rubrum see *R. silvestre*
   sanguineum sang-*gwin*-ee-um. Blood-red.
   Flowering Currant. W N America.
   speciosum spe-kee-ō-sum. Showy.
   California.
   silvestre sil-*ves*-tree (= *R. rubrum*). Of
   woodlands. Red currant. W Europe.
   uva-crispa oo-va-*kris*-pa. A medieval name
   meaning crisp grape. Gooseberry.
   Europe, N Africa.

Richea ree-shee-a *Epacridaceae*. After C. A. G.
Riche, a French naturalist (died 1791).
Evergreen shrub.
   scoparia skō-*pah*-ree-a. Broom-like.
   Tasmania.

Ricinus ri-ki-nus *Euphorbiaceae*. The L. name
from *ricinus* (a tick), the seeds resemble ticks.
Tender tree or shrub grown as an annual.
   communis kom-*ew*-nis. Common. Castor-
   oil Plant. Tropical Africa.

Robinia ro-bin-ee-a *Leguminosae*
(*Papilionoideae*). After Jean Robin
(1550–1629), herbalist to Henry IV of
France. Deciduous trees and shrubs.
   hispida his-pi-da. Bristly (the shoots). Rose
   Acacia. SE United States.
   kelseyi kel-see-ee. After Mr Harlan P.
   Kelsey who introduced it to cultivation.
   E United States.
   pseudacacia sood-a-kay-see-a. False *Acacias*.
   Locust, False Acacia. E United States.
      'Frisia' *free*-see-a. Of Friesland.
      'Umbraculifera' um-brah-kew-*li*-fe-ra.
      (= 'Inermis'). Umbrella-bearing (the
      shape of the head). Mop-headed Acacia.
   × *slavinii* sla-*vin*-ee-ee. *R. kelseyi* × *R.
   pseudoacacia*. After B. H. Slavin, who
   collected the seed from which it was
   raised.
      'Hillieri' *hi*-lee-a-ree. (*R.* × *hillieri*).
      After Hillier's who raised it.

Roblé see *Nothofagus obliqua*

Rochea see *Crassula*

Rock Jasmine see *Androsace*
Rock Lily see *Arthropodium*
Rock Rose see *Cistus*

Rodgersia ro-*jerz*-ee-a *Saxifragaceae*. After
Rear Admiral John Rodgers (1812–82).
Perennial herbs.
   aesculifolia ie-skew-li-*fo*-lee-a. With leaves
   like *Aesculus*. China.
   pinnata pi-*nah*-ta. Pinnate (the leaves).
   China.
   podophylla po-dō-*fil*-la. With stoutly
   stalked leaves. China, Japan.
   sambucifolia săm-bew-ki-*fo*-lee-a.
   Sambucus-leaved. China.
   tabularis see *Astilboides tabularis*

Rohdea rō-dee-a *Liliaceae (Convallariaceae)*.
After Michael Rohde (1782–1812). Tender
perennial herb.
   japonica ja-*pon*-i-ka. Of Japan. Lily of
   China. China. Japan.

Romneya rom-*nee*-a *Papaveraceae*. After Dr
Thomas Romney Robinson (1792–1882),
Irish astronomer. Sub-shrubs. Tree Poppy.
   coulteri *kool*-ta-ree. After Dr Thomas
   Coulter who discovered it in 1833. S
   California.
   trichocalyx tri-kō-*kă*-liks. With a hairy
   calyx. S California, N Mexico.

Romulea rom-*ew*-lee-a *Iridaceae*. After
Romulus, founder and first King of Rome.
Cormous herbs.
   bulbocodium bul-bō-*kŏ*-dee-um. Like
   Bulbocodium. Mediterranean region, N
   Africa, W Asia.
   clusiana see *R. bulbocodium*
   requienii rek-wee-*en*-ee-ee. After Requien,
   see *Mentha requienii*. Mediterranean region.
   rosea ro-see-a. Rose-coloured (the
   flowers). S Africa.

Rooksbya euphorbioides see *Carnegiea
euphorbioides*

Rosa ro-sa *Rosaceae*. The L. name Shrubs and
climbers. Rose.
   × *alba* *ăl*-ba. White (the flowers). White
   Rose of York.
   banksiae *banks*-ee-ie. After Lady Dorothea
   Banks, wife of Sir Joseph Banks. Banksian
   Rose. China.
   brunonii broo-*non*-ee-ee. After Robert

Rosa (continued)
Brown (1773–1858), Scottish botanist.
Himalayan Musk Rose. Himalaya, W
China.
   canina ka-*neen*-a. Of dogs. Dog Rose,
   Europe, SW Asia.
   × *centifolia* kent-i-*fo*-lee-a. With 100 leaves
   (petals). Holland Rose, Provence Rose.
   Cult.
   chinensis chin-*en*-sis. Of China. Cult.
      'Mutabilis' see *R.* × *odorata* 'Mutabilis'
      'Viridiflora' vi-ri-di-*flō*-ra. With green
      flowers.
   × *damascena* dă-ma-*skay*-na. Of Damascus.
   Damask Rose. Cult.
      'Trigintipetala' tree-gin-ti-*pe*-ta-la.
      With thirty petals. Kazanlik Rose.
      'Versicolor' ver-*si*-ko-lor. Variously
      coloured. York and Lancaster Rose.
   ecae ee-kie. After Mrs E. C. Aitchinson
   (E.C.A.) whose husband inintroduced it to
   cultivation. C Asia.
   eglanteria see *R. rubiginosa*.
   elegantula ay-le-*gănt*-ew-la. (= *R. farreri*).
   Elegant. China.
      'Persetosa' per-say-*tō*-sa. Very bristly.
      Threepenny-bit Rose.
   farreri see *R. elegantula*
   filipes *fee*-li-pays. Slender-stalked (the
   flowers). W China.
   foetida *foy*-ti-da. Foetid (the flowers). C
   Asia.
      'Bicolor' *bi*-ko-lor. Two-coloured.
      Austrian Copper Briar.
      'Persiana' per-see-*ah*-na. Of Persia.
   gallica *gă*-li-ka. Of France. Red Rose.
   Europe, W Asia.
   glauca *glow*-ka. (= *R. rubrifolia*). Glaucous
   (the bloom on the stem and young
   leaves). S Europe.
   × *harisonii* hă-ri-*son*-ee-ee. *R. foetida* × *R.
   pimpinellifolia*. After George Folliot
   Harison (died 1846), of New York, who
   raised it.
   helenae he-len-ie. After Ernest Wilson's
   wife, Ellen. China.
   moyessi *moyz*-ee-ee. After the Rev. J.
   Moyes, a missionary in W China. W
   Sichuan.
      'Fargesii' far-*jeez*-ee-ee. Afer Farges, see
      Decaisnea fargesii
   nitida ni-ti-da. Shining (the leaves). E N
   America.
   × *odorata* o-dō-*rah*-ta. *R. chinensis* × *R.
   gigantea*. Scented.
      'Mutabilis' mew-*tah*-bi-lis. (= *R.
      chinensis* 'Mutabilis'). Changeable (the
      flower colour).
   omeiensis see *R. sericea omeiensis*

Rosa (continued)
   pteracantha see *R. sericea pteracantha*
   pimpinellifolia pim-pi-nel-i-*fo*-lee-a. (= *R.
   spinosissima*). *Pimpinella*-leaved. Burnet
   Rose. Europe to C Asia.
      'Grandiflora' grănd-i-*flō*-ra. (= *altaica*
      hort.). Large-flowered.
   rubiginosa roo-bij-in-*ō*-sa (= *R. eglanteria*).
   Rusty-red. Eglantine, Sweet Brier.
   rubrifolia see *R. glauca*
   rugosa roo-*gō*-sa. Wrinkled (the leaves).
   NE Asia.
   sericea say-*ri*-kee-a. Silky-hairy. Himalaya,
   China.
      omeiensis ō-may-*en*-sis. (= *R. omeiensis*).
      Of the Omei Shan, China.
      pteracantha te-ra-*kănth*-a. (= *R. omeiensis
      pteracantha*). With winged spines.
   spinosissima see *R. pimpinellifolia wichuraiana*
   vi-kewr-ra-ee-*ah*-na. After Max
   Wichura, a Prussian diplomat who
   collected the type specimen. Japan,
   Korea.
   xanthina zăn-*theen*-a. Yellow (the flowers).
   China, Korea.
      hugonis hew-*gō*-nis. After its introducer
      Father Hugh Scallan (Pater Hugo).
      NW China.

Rosary Vine see *Ceropegia woodii*

Roscoea ros-*kō*-ee-a *Zingiberaceae*. After
William Roscoe of Liverpool (1753–1831).
Semi-hardly perennial herbs.
   alpina ăl-*peen*-a. Alpine. Himalaya, Tibet.
   cautleoides kawt-lee-*oi*-deez. Like *Cautleya*.
   W China.
   humeana hewm-ee-*ah*-na. After David
   Hume of Edinburgh Botanic Garden. W
   China.
   purpurea pur-*pewr*-ree.a. Purple (the
   flowers). Himalaya.

Rose see *Rosa*
   Banksian see *R. banksiae*
   Burnet see *R. pimpinellifolia*
   Damask see *R. × damascena*
   Dog see *R. canina*
   Himalayan Musk see *R. brunonii*
   Holland see *R. × centifolia*
   Kazanlik see *R. × damascena*
   'Trigintipetala'
   Provence see *R. × centifolia*
   Threepenny-bit see *R. elegantula*
   'Persetosa'
   White see *R. × alba*
   York and Lancaster see *R. × damascena*
   'Versicolor'
Rose Acacia see *Robinia hispida*

Rose of China see *Hibiscus rosa-sinensis*
Rose of Heaven see *Silene coeli-rosea*
Rose Pincushion see *Mammillaria
zeilmanniana*
Rose of Sharon see *Hypericum calycinum*
Roseroot see *Rhodiola rosea*

Rosmarinus rōs-ma-*reen*-us *Labiatae*. The L.
name from *ros* (dew) and *marinus* (of the
sea). Evergreen shrub.
   officinalis o-fi-ki-*nah*-lis. Sold as a herb.
   Rosemary. Mediterranean region.
   'Prostratus' pros-*trah*-tus. (= *R.
   lavendulaceus* hort.). Prostrate.

Rossioglossum ros-ee-ō-*glos*-um *Orchidaceae*.
After J. Ross, see *Lemboglossum rossii*.
Epiphytic orchids. Formerly included in
Odontoglossum. Mexico to Panama.
   grande grăn-dee (= *Odontoglossum grande*).
   Large. Tiger Orchid. Clown Orchid.
   Mexico, Guatamala.

Rough Bindweed see *Smilax aspera*

Rowan see *Sorbus aucuparia*
Royal Red Bugler see *Aeschynanthus pulcher*
Royal Paint Brush see *Scadoxus puniceus*
Royal Nodding Bells see *Streptocarpus
wendlandii*
Rubber Plant see *Ficus elastica*

Rubus *rub*-us *Rosaceae*. The L. name for the
blackberry. Deciduous and evergreen
shrubs.
   calycinoides see *R. pentalobus*
   cockburnianus kō-burn-ee-*ah*-nus. After
   Cockburn, see *Primula cockburniana*.
   China.
   deliciosus day-li-kee-*ō*-sus. Delightful (the
   flowers). W N America.
   idaeus ee-*die*-us. Of Mt. Ida. Raspberry.
   Europe, N Asia.
   laciniatus la-kin-ee-*ah*-tus. Deeply cut (the
   leaves). Cut-leaved Bramble. Cult.
   loganobaccus lō-ga-nō-*bah*-kus. After James
   Harvey Logan (1841–1928), an American
   judge who raised it, and L. *baccus* (a berry).
   Loganberry. Cult.
   nepalensis ne-pa-*len*-sis. Of Nepal.
   Himalaya.
   odoratus o-dō-*rah*-tus. Scented. E N
   America.
   pentalobus pen-ta-lo-bus (= *R. calycinoides*
   misapplied). With five lobes (the leaves).
   Taiwan.
   phoenicolasius foy-nee-ko-*lah*-see-us. With
   purple hairs (on the shoots). Wineberry. E
   Asia.

Rubus (continued)
spectabilis spek-*tah*-bi-lis. Spectacular.
Salmonberry. W N America.
thibetanus ti-bet-*ah*-nus. Of Tibet. W
China.
tricolor tri-ko-lor. Three-coloured. W
China.
ulmifolius ul-mi-*fo*-lee-us. *Ulmus*-leaved.
Europe.
'Bellidiflorus' be-li-di-*flō*-rus. *Bellis*-
flowered.

Rudbeckia rud-*bek*-ee-a *Compositae*. After
Olof Rudbeck the elder (1630–1702) and the
younger (1660–1740). Annual, biennial and
perennial herbs. Coneflower.
bicolor see *R. hirta*
fulgida ful-di-ga. Shining. United States.
deamii deem-ee-ee. After its discoverer,
Charles Clemon Deam (1865–1953).
hirta hir-ta. Hairy. Black-eyed Susan.
United States.
laciniata la-kin-ee-*ah*-ta. Deeply cut (the
leaves). N America.
maxima mahks-i-ma. Larger. United States.
purpurea see *Echinacea purpurea*
subtomentosa sub-tō-men-*tō*-sa. Somewhat
hairy. N America.
triloba tri-*lō*-ba. Three-lobed (the lower
leaves). Brown-eyed Susan. N America.

Rue see *Ruta graveolens*

Ruellia roo-*el*-ee-a *Acanthaceae*. After Jean
Ruel (1474–1537), French herbalist. Tender
herbs and shrubs.
ciliosa ki-lee-*ō*-sa. Fringed with hairs (the
calyx). SE United States.
devosiana da-vos-ee-*ah*-na. After Cornelius
de Vos (1806–95). Brazil.
macrantha ma-*krănth*-a. Large-flowered.
Christmas Pride. Brazil.
makoyana mă-koy-ah-na. After Jacob
Makoy, a Belgian nurseryman. Trailing
Velvet Plant. Brazil.
portellae por-*tel*-ie. After Francisco Portella
of Rio de Janeiro who sent plants to Kew
c 1880. Brazil.

Rugby Football Plant see *Peperomia argyreia*

Rumex *ru*-meks *Polygonaceae*. The L. name
for *R. acetosa*. Perennial herbs.
acetosa a-kay-*tō*-sa. Old name for plants
with acid leaves. Common Sorrel.
Europe, Asia, N America.
scutatus skoo-*tah*-tus. Shield-bearing.
French sorrel. C and S Europe.

Runner Bean see *Phaseolus vulgaris*
 Scarlet see *P. coccineus*
Rupturewort see *Herniaria glabra*

Ruscus *rus*-kus *Liliaceae* (*Asparagaceae*). The
L. name. Evergreen shrubs.
aculeatus a-kew-lee-*ah*-tus. Prickly.
Butcher's Broom. Europe, N Africa, W
Asia.
hypoglossum hi-pō-*glos*-um. Beneath the
tongue, the flowers are borne under a
tongue-like bract. S Europe, Turkey.
racemosus see *Danae racemosa*

Russelia rūs-*el*-ee-a *Scrophulariaceae*. After Dr
Alexander Russel (1715–68). Tender shrubs.
equisetiformis ek-wi-say-ti-*form*-is. (= *R.
juncea*). Like *Equisetum*. Coral Plant.
Mexico.
sarmentosa sar-men-*tō*-sa. Twiggy. C
America.

Ruta roo-ta *Rutaceae*. The L. name. Evergreen
sub shrub.
graveolens gra-*vee*-o-lenz. Strong-smelling.
Rue. SE Europe.

Rutabaga see *Brassica napus Napobrassica*

S

Saffron see *Crocus sativus*
Saffron Spike see *Aphelandra squarrosa*
Sage see *Salvia*
 Cardinal see *S. fulgens*
 Common see *S. officinalis*
 Gentian see *S. patens*
 Mealy-cup see *S. farinacea*
 Pineapple see *S. rutilans*
Sage Brush see *Seriphidium tridentatum*

Sagina sa-*geen*-a *Caryophyllaceae*. From L.
sagina (fodder), sheep were fed on a related
plant. Perennial herbs.
subulata soob-ew-*lah*-ta. (= *S. glabra*).
Awl-shaped (the leaves). Europe.
'Aurea' *ow*-ree-a. Golden

Sagittaria să-gi-*tah*-ree-a *Alismataceae*. From
L. sagitta (an arrow) referring to the shape
of the leaves. Aquatic perennial herbs.
Arrowhead.
latifolia lah-tee-*fo*-lee-a. Broad-leaved. N
America.
sagittifolia să-gi-ti-*fo*-lee-a. With arrow-
shaped leaves. Europe, Asia.

Sagittaria (continued)
'Flore Pleno' *flō*-ree-*play*-nō. (= *S. japonica* 'Flore Pleno'). With double flowers.
trifolia tri-*fo*-lee-a. Three-leaved, the leaves are three-lobed. N hemisphere.

Saguaro see *Carnegiea gigantea*

Saintpaulia saynt-*pawl*-ee-a *Gesneriaceae*. After Baron Walter von Saint Paul-Illaire (1860–1910), who discovered *S. ionantha*. Tender perennial herbs. African Violet. Tanzania.
confusa kon-*few*-sa. Confused (with another species).
grandifolia grănd-i-*fo*-lee-a. Large-leaved.
grotei grō-tee-ee. After Grote.
ionantha ee-on-*ănth*-a. With violet flowers.
shumensis shoom-*en*-sis. Of Shume, Tanzania.

Salal see *Gaultheria shallon*

Salix să-liks *Salicaceae*. The L. name. Deciduous trees and shrubs. Willow.
acutifolia a-kew-ti-*fo*-lee-a. With pointed leaves. Russia.
aegyptiaca ie-gip-tee-*ah*-ka. Egyptian. Musk Willow. W Asia.
alba ăl-ba. White (the leaves). White Willow. Europe, W Asia.
'Britzensis' brits-*en*-sis. Of Britz, near Berlin.
caerulea kie-*ru*-lee-a. Blue (the leaves). Cricket Bat Willow.
sericea ser-i-*kee*-a (= *S. a.* 'Argentea'). Silky. Silver Willow.
vitellina vi-te-*leen*-a. Egg-yolk yellow (the shoots). Golden Willow.
'Vitellina Pendula' see *S.* 'Chrysocoma'
× *boydii* boyd-ee-ee. After William Brack Boyd (1831–1918), Scottish naturalist who discovered it.
caprea kăp-ree-a. Of goats (the foliage was used for goat fodder). Goat Willow. Europe, NW Asia.
'Chrysocoma' kris-*o*-ko-ma. (= *S. alba* 'Vitellina Pendula'). Golden-haired, referring to the weeping, yellow shoots. Weeping Willow.
daphnoides dăf-*noi*-deez. Laurel-like. Violet Willow. Europe.
elaeagnos e-lee-*ăg*-nos. Gk. name of a willow, see *Elaeagnus*. Europe, W Asia.
exigua eks-*ig*-ew-a. Small. W N America.
fragilis *fră*-gi-lis. Fragile (the shoots). Crack Willow. Europe, Russia.

Salix (continued)
hastata hăs-*tah*-ta. Spear-shaped (the leaves). Europe, Asia.
'Wehrhahnii' vair-*hahn*-ee-ee. After H. R. Wehrhahn.
lanata lah-*nah*-ta. Woolly. Woolly Willow. N Europe, N Asia.
matsudana măt-soo-*dah*-na. After Matsuda. Peking Willow. N China.
'Pendula' *pen*-dew-la. Weeping.
'Tortuosa' tort-ew-*ō*-sa. Twisted (the shoots). Dragon's Claw Willow.
'Melanostachys' me-la-*no*-sta-kis. With black spikes.
pentandra pen-*tăn*-dra. With five stamens. Bay Willow. Europe, Asia.
purpurea pur-*pewr*-ree-a. Purple (the shoots). Purple Osier. Europe, N Asia.
repens ree-penz. Creeping. Creeping Willow. Europe, Asia.
reticulata ray-tik-ew-*lah*-ta. Net-veined (the leaves). Arctic and alpine N hemisphere.
sachalinensis sa-kah-lin-*en*-sis (= *S. udensis*). Of Sakhalin. NE Asia.
viminalis vee-min-*ah*-lis. With long, slender shoots. Common Osier. Europe, Asia.

Salmon Blood Lily see *Scadoxus multiflorus*
Salmonberry see *Rubus spectabilis*

Salpiglossus sal-pi-*glos*-us *Solanaceae*. From Gk. *salpinx* (a trumpet) and *glossa* (a tongue). Tender herb.
sinuata sin-ew-*ah*-ta. Wavy-edged (the leaves). Painted Tongue. Chile.

Salsify see *Tragopogon porrifolius*
Salt Tree see *Halimodendron halodendron*

Salvia săl-vee-a *Labiatae*. The L. name, from *salvus* (safe) referring to medicinal properties. Annual and perennial herbs and shrubs. Sage.
ambigens see *S. caerulea*
argentea ar-*gen*-tee-a. Silvery (the leaves). S Europe.
azurea a-*zew*-ree-a. Deep blue (the flowers). C United States.
grandiflora grănd-i-*flō*-ra. (= *S. pitcheri*). Large-flowered.
blepharophylla blef-a-rō-*fil*-la. With fringed leaves. Mexico.
caerulea see *S. guaranitica*.
farinacea fă-ree-*nah*-kee-a. Mealy. Mealy-cup Sage. SW United States.
fulgens *ful*-genz. Shining (the flowers). Cardinal Sage. Mexico.
greggii greg-ee-ee. After Dr John Gregg

Salvia (continued)
who discovered it in 1848. Texas,
Mexico.
guaranitica goo-ah-ran-it-ik-a (= *S.
caerulea*). Of Guara (Brazil). S America.
haematodes see *S. pratensis*
horminum see *S. viridis*
involucrata in-vo-loo-*krah*-ta. With bracts
around the flowers. Mexico.
　　'Bethellii' be-*thel*-ee-ee. After Mr
　　Bethell who raised it.
microphylla mik-rō-*fil*-la. Small-leaved.
Mexico.
nemorosa ne-mo-*rō*-sa. Of woods. Europe.
officinalis o-fi-ki-*nah*-lis. Sold as a herb.
Common Sage. S Europe.
　　'Icterina' ik-*te*-ri-na. Jaundice-yellow
　　(the leaves).
patens pǎ-tenz. Spreading (the flowers).
Gentian Sage. Mexico.
pratensis prah-*tayn*-sis. (= *S. haematodes*).
Of meadows. Europe.
rutilans roo-ti-lǎnz. Reddish (the flowers).
Pineapple Sage. Mexico.
sclarea sklah-ree-a. From L. *clarus* (clear)
the seeds and leaves are used in eye
lotions. Clary. S Europe.
splendens splen-denz. Splendid. Brazil.
uliginosa ew-li-gi-*nō*-sa. Of marshes. S
America.
viridis vi-ri-dis. (= *S. horminum*). Green
(the bracts of some forms). S Europe.

Salvinia sǎl-*veen*-ee-a *Salviniaceae*. After
Professor Antonio Maria Salvini
(1633–1729). Tender, floating ferns.
auriculata ow-rik-ew-*lah*-ta (= *S.
rotundifolia*). Auricled. Floating Fern. C
and S America.

Sambucus sǎm-*bew*-kus *Caprifoliaceae*. The L.
name. Deciduous shrubs. Elder.
canadensis kǎn-a-*den*-sis. Of E N America.
American Elder.
nigra nig-ra. Black (the fruits). Common
Elder. Europe, N Africa, W Asia.
racemosa rǎ-kay-*mō*-sa. In racemes, the
elongated inflorescence. Red-berried Elder.
Europe, W Asia, Siberia.

Sand Myrtle see *Leiophyllum buxifolium*

Sandersonia sahn-der-*son*-ee-a *Liliaceae*
(*Colchicaceae*). After John Sanderson
(c1820–1881), who discovered the following
in 1851. Tender perennial herb.
aurantiaca ow-rǎn-tee-*ah*-ka. Orange (the
leaves). Chinese Lanterns. S Africa.

Sandwort see *Arenaria*

Sanguinaria sang-gwi-*nah*-ree-a *Papaveraceae*.
From L. *sanguis* (blood) referring to the red
sap. Perennial herb.
canadensis kǎn-a-*den*-sis. Of EN America.
Bloodroot.

Sanguisorba sang-gwi-*sor*-ba *Rosaceae*. From
L. *sanguis* (blood) and *sorbeo* (to absorb), it
was said to stop bleeding. Perennial herbs.
Burnet.
canadensis kǎn-a-*den*-sis. Of Canada. N
America.
obtusa ob-*tew*-sa. Blunt (the leaflets). Japan.
tenuifolia ten-ew-i-*fo*-lee-a. With slender
leaflets. N Asia.

Sansevieria sǎn-sev-ee-*e*-ree-a *Agavaceae*.
After Raimond de Sangro, Prince of
Sanseviero, 18th-century patron of
horticulture. Tender, evergreen herbs.
trifasciata tri-fǎs-ee-*ah*-ta. In three bundles,
the flower clusters are in groups of 1 to
3. Mother-in-law's Tongue. E South
Africa.
　　'Hahnii' *hahn*-ee-ee. After Hahn.
　　'Laurentii' lo-*rent*-ee-ee. After Emile
　　Laurent (1861–1904), who discovered
　　it.

Santolina sǎn-to-*leen*-a *Compositae*. From
sanctum linum (holy flax) the L. name for *S.
rosmarinifolia*. Evergreen shrubs.
chamaecyparissus kǎ-mie-kew-pa-*ris*-us.
Dwarf cypress. Lavender Cotton.
Mediterranean region.
pinnata pin-*ah*-ta. Pinnate (the leaves).
Italy.
　　neopolitana nee-ah-po-li-*tah*-na. (= *S.
　　neapolitana*). Of Naples.
rosmarinifolia rōs-ma-reen-i-*fo*-lee-a. (= *S.
virens*). *Rosmarinus*-leaved. Holy Flax.
Mediterranean region.

Sanvitalia sǎn-vi-*tah*-lee-a *Compositae*. After
Frederico Sanvitali (1704–61). Annual herb.
procumbens prō-*kum*-benz. Prostrate.
Creeping Zinnia. C America.

Saponaria sǎ-pō-*nah*-ree-a *Caryophyllaceae*.
From L. *sapo (soap)*, a soap can be made
from *S. officinalis*. Annual and perennial
herbs.
caespitosa kie-spi-*tō*-sa. Tufted. Pyrenees.
calabrica ka-*lǎ*-bri-ka. Of Calabria, Italy. E
Mediterranean region.
ocymoides o-kim-*oi*-dees. Like *Ocimum*.
Rock Soapwort. S Europe.

Saponaria (continued)
officinalis o-fi-ki-*nah*-lis. Sold as a herb.
Bouncing Bet, Soapwort. Europe.
vaccaria see *Vaccaria hispanica*

Sapphire Flower see *Browallia speciosa*

Sarcococca sar-kō-*ko*-ka *Buxaceae*. From Gk.
sarcos (flesh) and *kokkos* (a berry) referring
to the fleshy fruits. Christmas Box, Sweet
Box.
confusa kon-*few*-sa. Confused (with other
species). China.
hookeriana hu-ka-ree-*ah*-na. After Sir
Joseph Hooker. Himalaya.
   digyna di-gi-na. With two styles. W
   China.
humilis hu-mi-lis. Low-growing. W
China.
ruscifolia rus-ki-*fo*-lee-a. *Ruscus*-leaved.
China.

Sarracenia să-ra-*sen*-ee-a *Sarraceniaceae*. After
Michael Sarrasin (1659–1734), French
botanist and physician. Carnivorous
perennial herbs. Pitcher Plant.
flava flah-va. Yellow (the flowers). Yellow
Pitcher Plant. SE United States.
leucophylla loo-kō-*fil*-la. (= *S. drummondii*).
White-leaved. Fiddler's Trumpets, Lace
Trumpets. SE United States.
purpurea pur-*pewr*-ee-a. Purple (the
pitchers). Northern Pitcher Plant. E N
America.

Sasa sah-sa *Gramineae*. The Japanese name.
Bamboo.
palmata pahl-*mah*-ta. Hand-like, the
spreading, clustered leaves. Japan.
tesellata see *Indocalamus tessellata*
veitchii veech-ee-ee. After Messrs Veitch for
whom Charles Maries introduced it in c
1880. Japan

Sasaella sas-ie-el-ă. *Graminae*. Diminutive
from *Sasa*. Bamboo. Japan.
ramosa ram-ō-sa (= *Arundinaria vagans*).
Branched. Azuma-Zasa.

Sassafras *săs*-a-frăs *Lauraceae*. Probably
adapted by French settlers from an
American Indian name. Deciduous tree.
albidum ăl-bi-dum. Whitish (the
undersides of the leaves). Sassafras. E N
America.

Satin Flower see *Clarkia amoena*
Satsuma see *Citrus reticulata*

Satureja săt-ew-*ray*-a *Labiatae*. The L. name.
Sub-shrub and annual herb.
hortensis hor-*ten*-sis. Of gardens. Summer
Savoury. Mediterranean region.
montana mon-*tah*-na. Of mountains.
Winter Savoury. S Europe.

Saucer Plant see *Aeonium undulatum*

Sauromatum sow-*ro*-ma-tum *Araceae*. From
Gk. *sauros* (a lizard) referring to the spotted
spathe and its long, tail-like appendage.
Tender perennial herb.
guttatum see *S. venosum*
venosum vay-*nō*-sum. Conspicuously
veined.

Savin see *Juniperus sabina*
Savoury, Summer see *Satureja hortensis*
   Winter see *S. montana*

Saxegothaea săks-ee-goth-*ee*-a *Podocarpaceae*.
After Prince Albert of Saxe-Coburg-Gotha.
Evergreen conifer.
conspicua con-*spik*-ew-a. Conspicuous,
distinguished. Prince Albert's Yew. Chile.

Saxifraga săks-*if*-ra-ga *Saxifragaceae*. From L.
saxum (a rock) and *frango* (to break), by
growing in rock crevices they appear to break
rocks. Perennial herbs. Saxifrage.
aizoides ie-zō-i-deez. Like *Aizoon*. Europe.
aizoon see *S. paniculata*
× *apiculata* a-pik-ew-*lah*-ta. *S. juniperifolia*
sancta × *S. marginata rocheliana*. With a
short, abrupt point (the leaves).
brunoniana see *S. brunonis*
brunonis broo-nō-nis. (= *S. brunoniana*)
After Robert Brown. Himalaya, S Tibet.
burseriana bur-sa-ree-*ah*-na.
After Joachim Burser (1583–1649),
German physician and botanist. E Alps.
callosa ka-*lō*-sa. Calloused, the lime-
encrusted leaves. W Mediterranean
region.
cochlearis kok-lee-*ah*-ris. Spoon-shaped
(the leaves). Alps.
cortusifolia kor-tew-si-*fo*-lee-a. *Cortusa*-
leaved. E Asia.
cotyledon ko-ti-*lay*-don. Like *Cotyledon*.
Alps, Pyrenees, NW Europe.
cuneifolia kew-nee-i-*fo*-lee-a. With the
leaves narrowed to the base. Pyrenees,
Alps, Carpathians.
diapensioides dee-a-pen-see-*oi*-deez. Like
Diapensia lapponica. SW Alps.
exarata ex-a-*ray*-ta. Furrowed, engraved
(the surface of the leaves).

Saxifraga (continued)
  moschata mos-*kah*-ta. Musky. C and S Europe.
  fortunei for-*tewn*-ee-ee. After Robert Fortune, see *Fortunella*. Japan
  frederici-augustii fred-er-*ee*-ki-aw-*gus*-tee-ee. For Frederic Augustus, of the (ex) Royal Family of Austria.
  grisebachii gree-za-*bahk*-ee-ee. After Professor August Heinrich Rudolph Grisebach (1814–79), German botanist. Balkan Peninsula.
  longifolia long-gi-*fo*-lee-a. Long-leaved. Pyrenees.
  oppositifolia o-po-si-ti-*fo*-lee-a. With opposite leaves. Europe.
  paniculata pa-nik-ew-*lah*-ta. (= *S. aizoon*). With flowers in panicles. C and S Europe, Norway.
  sarmentosa see *S. stolonifera*
  stolonifera sto-lō-*ni*-fe-ra. (= *S. sarmentosa*). Bearing stolons. Mother of Thousands. E Asia.
  umbrosa hort. see *S.* × *urbium*
  × *urbium* ur-bee-um. (= *S. umbrosa* hort.) *S. umbrosa* × *S. spathularis*. Of towns. London Pride.

Scabiosa skăb-ee-ō-sa *Dipsacaceae*. From L. *scabies* (itch, scurf), the rough leaves were said to cure it. Annual and perennial herbs. Scabious.
  alpina see *Cephalaria alpina*
  atropurpurea aht-rō-pur-*pewr*-ree-a. Deep purple. Sweet Scabious. S Europe.
  caucasica kaw-*kă*-si-ka. Of the Caucasus.
  columbaria ko-lum-*bah*-ree-a. Dove-like or of doves. Europe, N Africa, Asia.
  graminifolia grah-mi-ni-*fo*-lee-a. With grass-life leaves. S Europe.
  lucida loo-ki-da. Shining. C Europe.
  ochroleuca ok-rō-*loo*-ka. Yellowish-white. Europe, W Asia.

Scadoxus ska-*dox*-us *Amaryllidaceae*. Derivation uncertain. Bulbous or rhizomatous herbs, closely related to *Haemanthus* q.v. Tropical Arabia and Africa.
  multiflorus mul-tee-*flō*-rus. (= *Haemanthus multiflorus*). Many-flowered. Salmon Blood Lily. Tropical Africa.
    katharinae kath-a-*rin*-ie (= *Haemanthus katharinae*). After Mrs Katherine Saunders who collected in Natal. Blood flower. S Africa.
  puniceus pew-ni-kee-us (= *Haemanthus puniceus*). Reddish-purple (the flowers). Royal Paint Brush. E and S Africa.

Scarborough Lily see *Cyrtanthus elata*
Scarlet Leadwort see *Plumbago indica*
Scarlet Trompetilla see *Bouvardia longiflora*

Schefflera shef-le-ra *Araliaceae*. After J. C. Scheffler. Tender, evergreen shrubs and trees.
  actinophylla ăk-tin-ō-*fil*-la (= *Brassaia actinophylla*). With rayed leaves, referring to the radiating leaflets. Queensland Umbrella Tree. Queensland.
  arboricola ar-bo-*ri*-ko-la. (= *Heptapleurum arboricolum*). Growing on trees. Taiwan.
  digitata di-gi-*tah*-ta. Lobed like a hand (the leaves). Seven Fingers. New Zealand.
  elegantissima ay-le-găn-*tis*-i-ma. (= *Aralia elegantissima*). Most elegant. Finger Aralia, False Aralia. New Caledonia, Polynesia.

Schisandra skis-*ăn*-dra *Schisandraceae*. From Gk. *schizo* (to divide) and *aner* (a man) referring to the well-separated anther cells. Deciduous and evergreen climbers.
  glaucescens glow-*kes*-enz. Somewhat glaucous (the lower leaf surface). C China.
  grandiflora grănd-i-*flō*-ra. Large-flowered. Himalaya.
    rubiflora rub-ri-*flō*-ra. Red-flowered. China, NE India.
  propinqua prō-*ping*-kwa. Related to (another species). Himalaya.
  sinensis sin-en-sis. Of China.

Schizanthus skiz-*ănth*-us *Solanaceae*. From Gk. *schizo* (to divide) and *anthos* (a flower) referring to the deeply divided corolla. Annual herbs. Butterfly Flower. Poor Man's Orchid.
  pinnatus pi-*nah*-tus. Pinnate (the leaves). Chile.
  × *wisetonensis* wiez-ton-*en*-sis. Of Wiseton.

Schizocentron see *Heterocentron*

Schizophragma ski-zō-*frăg*-ma *Hydrangeaceae*. From Gk. *schizo* (to divide) and *phragma* (a wall), parts of the fruit fall away leaving it skeletonised. Deciduous climbers.
  hydrangeoides hi-drang-gee-*oi*-deez. Like *Hydrangea*. Japan.
  integrifolia in-teg-ri-*fo*-lee-a. Entire-leaved. C China.

Schizostylis ski-zō-*sti*-lis *Iridaceae*. From Gk. *schizo* (to divide) and *stylis* (a style), the style is deeply divided. Perennial herb.

Schizostylis (continued)
   coccinea kok–*kin*–ee–a. Scarlet (the flowers).
   Kaffir Lily. S Africa.

Schlumbergera shlum–*ber*–ga–ra *Cactaceae*.
After Frederick Schlumberger (1804–65),
Belgian horticulturist, explorer and plant
collector.
   bridgesii see × *buckleyi*
   × *buckleyi būk*–lee–ee. *S. russelliana* × *S.
   truncata.* After W. Buckley, a cactus
   grower. Garden origin. Christmas Cactus.
   gaertneri see *Hatiora gaertneri*
   truncata trung–*kah*–ta. (= *Zygocactus
   truncatus*). Abruptly cut off (the ends of
   the shoots). Claw Cactus, Crab Cactus.
   Brazil.

Schoenoplectus shoo–no–*plek*–tus *Cyperaceae*.
From Gk. *schoinos*, rush, and *pleko*, to plait,
an allusion to the mat-forming rhizomes.
Grass-like herbs. Cosmopolitan.
   lacustris lak–*us*–tris (= *Scirpus
   tabernaemontani*). Of, or pertaining to
   lakes (the habitat). Widespread.

Sciadopitys skee–a–*do*–pi–tis *Sciadopityaceae*.
From Gk. *skiados* (an umbel) and *pitys* (a fir
tree), the leaves appear in whorls like the ribs
of an umbrella.
   verticillata ver–ti–*lah*–ta. Whorled.
   Umbrella Pine. Japan.

Scilla skil–la *Liliaceae* (*Hyacinthaceae*). The Gk.
name for the sea squill, *Urginea maritima*.
Bulbous perennial herbs.
   bifolia bi–*fo*–lee–a. Two-leaved. S and C
   Europe, Caucasus, Turkey.
   hispanica see *Hyacinthoides hispanica*
   mischtschenkoana mi–cheng–kō–*ah*–na. (=
   S. tubergeniana). After Mischtschenko.
   Transcaucasus, NW Iran.
   monophyllos mo–nō–*fil*–los. One-leaved.
   SW Europe, N Africa.
   non-scripta see *Hyacinthoides non-scripta*
   ovalifolia see *Ledebouria ovalifolia*
   peruviana pe–roo–vee–*ah*–na. Originally
   thought to be from Peru. W
   Mediterranean region, Portugal.
   sibirica si–*bi*–ri–ka. Siberian. Siberian
   Squill. S Russia, Caucasus, W Asia.
   violacea see *Ledebouria socialis*

Scindapsus skin–*dăp*–sus *Araceae*. Gk. name
for an ivy-like plant. Tender, evergreen
climber.
   aureus see *Epipremnum aureum*
   pictus pik–tus. Painted (the leaves). Silver
   Vine. SE Asia.

Scindapsus (continued)
   'Argyraeus' ar–gi–*ree*–us. Silvery (the
   leaves).

Scirpus skir–pus *Cyperaceae*. The L. name.
Perennial herbs.
   tabernaemontani see *Schoenoplectus lacustris*

Scorpion senna see *Coronilla emerus*

Scorzonera skor–zo–*ne*–ra *Compositae*. From
French *scorzon* (a viper), the root was said
to cure snake bites. Biennial herb.
   hispanica his–*pah*–ni–ka. Spanish.
   Scorzonera. C and S Europe, S Russia.

Scotch Thistle see *Onopordum acanthium*
Screw Pine see *Pandanus*

Scrophularia skrō–few–*lah*–ree–a
Scrophulariaceae. From L. scrofulae (a swelling
of the neck glands) referring to medicinal
properties. Perennial herb.
   auriculata ow–rik–ew–*lah*–ta. (= *S.
   aquatica*). Auricled (the leaves). Water
   Figwort. W Europe, N Africa.

Scutellaria sku–te–*lah*–ree–a *Labiatae*. From L.
scutella (a small dish) referring to the
appearance of the calyx in fruit. Tender and
hardy perennial herbs.
   alpina ă–*peen*–a. Alpine. S and E Europe.
   baicalensis bie–ka–*len*–sis. Of Lake Baikal, E
   Siberia. E Asia.
   costaricana kos–ta–ree–*kah*–na. Of Costa
   Rica.
   indica in–di–ka. Indian. China, Japan.
   scordiifolia skor–dee–i–*fo*–lee–a. With leaves
   like *Scordium* (= *Teucrium*). Asia.

Sea Buckthorn see *Hippophae rhamnoides*
Sea Campion see *Silene uniflora*
Sea Daffodil see *Pancratium maritimum*
Sea Holly see *Eryngium maritimum*
Sea Kale see *Crambe maritima*
Sea Lily see *Pancratium maritimum*
Sea Squill see *Urginea maritima*
Sea Urchin see *Hakea laurina*
Sedge see *Carex*
   Great Pond see *C. riparia*

Sedum say–dum *Crassulaceae*. Classical name
for several succulent plants from L. *sedo* (to
sit). Tender and hardy succulents.
   acre ahk–ree. Sharp-tasting. Biting
   Stonecrop. Europe, N Africa, W and N
   Asia.
   adolphi a–*dolf*–ee. After Adolf Engler.
   Golden Stonecrop. Mexico.

Sedum (continued)
   album ăl-bum. White (the flowers).
   Europe, N Africa, W Asia.
   allantoides a-lăn-tō-i-deez. Sausage-like
   (the leaves). S Mexico.
   bellum bel-um. Beautiful. Mexico.
   brevifolium bre-vi-*fo*-lee-um. Short-leaved.
   S Europe, N Africa.
   cauticolum see *Hylotelephium cauticolum*
   dasyphyllum dăs-i-*fil*-lum. With hairy
   leaves. S Europe, N Africa.
   dendroideum den-*droi*-dee-um. Tree-like.
   Mexico.
   ewersii see *Hylotelephium ewersii*
   fastigiatum see *Rhodiola fastigiata*
   floriferum flō-*ri*-fe-rum. Floriferous. China.
   heterodontum see *Rhodiola heterodonta*
   kamtschaticum kămt-*shă*-ti-kum. Of
   Kamtchatka. NE Asia.
   lineare li-nee-*ah*-ree. Linear (the leaves). E
   Asia.
   lydium *li*-dee-um. Of Lydia, W Asia.
   middendorfianum mi-dan-dorf-ee-*ah*-num.
   After Middendorf see *Hemerocallis*
   middendorfiana. C Asia.
   morganianum mor-găn-ee-*ah*-num. After
   Dr Meredith Morgan who grew it soon
   after its discovery. Donkey's Tail. Mexico.
   oaxacanum wah-hah-*kah*-num. Of Oaxaca.
   Mexico.
   oreganum o-ree-*gah*-num. Of Oregon, it
   was discovered at the mouth of the
   Oregon River. W N America.
   pachyphyllum pă-ki-*fil*-lum. Thick-leaved.
   Jelly Beans. S Mexico.
   populifolium see *Hylotelephium populifolium*
   praealtum prie-*ăl*-tum. Very tall.
   reflexum rc-*fleks*-um. Reflexed (the leaves
   on flowering stems). Europe.
   rhodiola see *Rhodiola rosea*
   rosea see *Rhodiola rosea*
   rubrotinctum rub-rō-*tink*-tum. Red-tinged
   (the leaves). Christmas Cheer.
   rupestre see *R. reflexum*
   sieboldii see *Hylotelephium sieboldii*
   spathulifolium spăth-ew-li-*fo*-lee-um. With
   spatula-shaped leaves. W N America.
   spectabile see *Hylotelephium spectabile*
   spurium spew-ree-um. False. Caucasus.
   telephium see *Hylotelephium telephium*

Seersucker Plant see *Geogenanthus undatus*

Selaginella se-lah-gi-*nel*-a *Selaginellaceae*.
Diminutive of *selago* (*Lycopodium selago*).
Tender, moss-like plants.
   apoda a-*pod*-a. Stalkless (the strobili). E N
   America.
   involvens in-*vol*-venz. Rolled up. E Asia.

Selaginella (continued)
   kraussiana krows-ee-*ah*-na. After C.
   Ferdinand F. Krauss (1812–90), who
   introduced it. Spreading Clubmoss. S
   Africa.
   lepidophylla le-pi-dō-*fil*-a. With scale-like
   leaves. Resurrection Plant. SW United
   States, C America.
   pallescens pa-*les*-enz. (= *S. emmeliana*).
   Rather pale. Moss fern. N and S
   America.

Selenicereus se-lay-nee-*kay*-ree-us *Cactaceae*.
From Gk. *selene* (the moon) and *Cereus* q.v.,
they flower at night.
   coniflorus see *S. grandiflorus*
   grandiflorus grănd-i-*flō*-rus. Large-
   flowered. Queen of the Night. Jamaica,
   Cuba.
   megalanthus me-ga-*lănth*-us. Large-
   flowered. Peru.
   pteranthus te-*rănth*-us. With winged
   flowers. Princess of the Night. Mexico.
   setaceus say-*tah*-kee-us. Bristly. S America.
   wercklei vair-klee-ee. After Karl Werckle
   (1866–1924), who studied the flora of
   Costa Rica. Costa Rica.

Self Heal see *Prunella*

Selinum se-leen-um *Umbelliferae*. From Gk.
selinon (celery). Perennial herb.
   tenuifolium ten-ew-i-*fo*-lee-um. With
   finely divided leaves. Himalaya, Assam,
   Tibet.

Sempervivum sem-per-*veev*-um *Crassulaceae*.
From L. *semper* (always) and *vivus* (alive).
Succulent perennials. Houseleek.
   arachnoideum ă-răk-*noi*-dee-um. With hairs
   like a spiders-web. Cobweb Houseleek.
   Alps, Pyrenees.
   dolomiticum do-lo-*mi*-ti-kum. Of the
   Dolomites.
   marmoreum mar-*mo*-ree-um. Mottled.
   Balkans.
      rubrifolium ru-bri-*fo*-lee-um. Red-
      leaved.
   montanum mon-*tah*-num. Of mountains.
   Europe.
   soboliferum see *Jovibarba sobolifera*
   tectorum tek-*to*-rum. Growing on roofs.
   Common Houseleek, Hen and Chickens.
   Pyrenees to SE Alps.

Senecio se-ne-kee-ō *Compositae*. From L.
senex (an old man), referring to the fluffy,
white seed heads. Herbs, shrubs and tender
succulents.

Senecio (continued)
articulatus ar-tik-ew-*lah*-tus. (= *Kleinia articulata*). Jointed. Cradle Plant. S Africa.
cineraria ki-ne-*rah*-ee-a. (= *S. bicolor, S. maritimus. Cineraria maritima*). Ash-coloured, the leaves. W and C Mediterranean region.
clivorum see *Ligularia dentata*
compactus see *Brachyglottis compacta*
cruentus kroo-*en*-tus, see *Pericallis cruenta*
doronicum do-*ro*-ni-kum. From *Doronicum* q.v. Europe, N Africa.
fulgens see *Kleinia fulgens*
galpinii see *Kleinia galpinii*
greyi see *Brachyglottis greyi*
haworthii hay-*werth*-ee-ee. (= *Kleinia tomentosa*). After Haworth, see *Haworthia*. S Africa.
herreianus he-ree-*ah*-nus. After H. Herre of Stellenbosch. S Africa.
× *hybridus* see *Pericallis* × *hybrida*
kleinia see *Kleinia neriifolia*
laxifolius see *Brachyglottis laxifolius*
macroglossus măk-ro-*glos*-us. Large-tongued, referring to the long ray florets. Cape Ivy. S Africa.
mikanioides see *Delairea odorata*
monroi see *Brachyglottis monroi*
pendulus see *Kleinia pendula*
petasitis pe-ta-*see*-tis. Hat-like (the leaves). California Geranium. Mexico.
przewalskii see *Ligularia przewalskii*
rowleyanus rō-lee-*ah*-nus. After Gordon D. Rowley (born 1921), a succulent enthusiast. String of Beads. SW Africa.
scandens skăn-denz. Climbing. E Asia.
serpens ser-penz. (= *Kleinia repens*). Snake-like. S Africa.
tanguticus see *Sinacalia tangutica*

Senna sen-ā *Leguminosae* (*Caesalpinoideae*) From Arabic, *sanā*. Mostly tender trees, shrubs and herbs.
alata ah-*lah*-ta. Winged (the fruit). Tropics.
alexandrina al-ex-an-*dry*-na. Of Alexandria (Egypt). True senna, Alexandrian Senna. Mexico, Tropical Africa, India.
australis ow-*strah*-lis. Southern. Australia.
corymbosa ko-rim-*bō*-sa. With flowers in corymbs. S America.
marilandica mă-ri-*lănd*-i-ka. Of Maryland. American Senna. SE United States.

Sensitive Plant see *Mimosa pudica*

Sequoia se-*kwoy*-a *Cupressaceae*. After Sequoiah (1770–1843), son of a British merchant by a Cherokee Indian woman. Evergreen conifer.

Sequoia (continued)
sempervirens sem-per-*vi*-renz. Evergreen. Coast Redwood. California, Oregon.

Sequoiadendron se-kwoy-a-*den*-dron. *Cupressaceae*. From *Sequoia* q.v. and Gk. *dendron* (a tree). Evergreen conifer.
giganteum gi-*găn*-tee-um. Very large. Big Tree, Giant Redwood, Wellingtonia. California.

Seriphidium ser-i-fid-i-um *Compositae*. Gk. dimunitive of *seriphon*, a type of wormwood. Annuals, perennials and shrubs.
tridentatum tri-den-*tah*-tum (= *Artemisia tridentata*). Three-toothed (the leaves). Sage Brush. NW United States.

Service Tree see *Sorbus domestica*
    Wild see *S. torminalis*
Service Tree of Fontainbleau see *S. latifolia*

Serviceberry see *Amelanchier*

Setcreasia set-*krees*-ee-a *Commelinaceae*. Derivation obscure. Tender perennial herb.
pallida see *Tradescantia palida*.
striata hort. see *Callisia elegans*

Setiechinopsis see *Echinopsis*

Seven Fingers see *Schefflera digitata*
Shaddock see *Citrus maxima*
Shallon see *Gaultheria shallon*
Shallot see *Allium cepa* Aggregatum.
Shamrock Pea see *Parochetus communis*
Shasta Daisy see *Leucanthemum* × *superbum*
Sheep Laurel see *Kalmia angustifolia*
Sheep's Bit see *Jasione*
Shell Flower see *Moluccella laevis*

Shibataea shi-ba-*tie*-a *Gramineae*. After Keita Shibata (1877–1949), Japanese botanist. Bamboo.
kumasasa kew-ma-*sah*-sa. Japanese name of a bamboo. Japan.

Shooting Star see *Dodecatheon*

Shortia short-ee-a *Diapensiaceae*. After Charles W. Short (1794–1863), a Kentucky botanist. Evergreen, perennial herbs.
galacifolia ga-lăk-i-*fo*-lee-a. With leaves like *Galax*. Oconee Bells. E N America.
soldanelloides sol-da-nel-*oi*-deez. Like *Soldanella*. Japan.
uniflora ew-ni-*flō*-ra. With solitary flowers. Japan.

Shrimp Plant see *Justicea brandegeana*
Siberian Squill see *Scilla sibirica*

Sidalcea see–*dăl*–kee–a. *Malvaceae*. From *Sida*
and *Alcea*, related genera. Perennial herbs.
SW United States.
   candida kăn–di–da. White (the flowers).
   malviflora măl–vi–*flō*–ra. *Malva*–flowered.

Silene si–*lay*–nee *Caryophyllaceae*. Gk. name
for a related plant. Campion.
   acaulis a–*kaw*–lis. Stemless. Moss
   Campion. Europe.
   alpestris ăl–*pes*–tris. Of lower mountains. E
   Alps.
   armeria ar–*me*–ree–a. See *Armeria*. C, S and
   E Europe.
   coeli-rosea koy–lee–ro–see–a. (= *Lychnis coeli-*
   rosea). As the common name, Rose of
   Heaven. SW Europe.
   compacta com–*păk*–ta. Compact. SE
   Europe.
   pendula pen–dew–la. Pendulous (the
   flowers). Nodding Catchfly.
   Mediterranean region, W Asia.
   schafta shăf–ta. The native name. Caucasus.
   vulgaris vul–*gah*–ris. Common. Europe, N
   Africa, W Asia.
       maritima ma–*ri*–ti–ma see *S. uniflora*.
   uniflora yew–ni–*flor*–a. Single-flowered.

Silk Tree see *Albizia julibrissin*
Silky Oak see *Grevillea robusta*
Silver Bell see *Halesia*
Silver Berry see *Elaeagnus commutata*
Silver Crown see *Cotyledon undulata*
Silver Gum see *Eucalyptus cordata*
Silver Squill see *Ledebouria socialis*
Silver Torch see *Cleistocactus straussii*
Silver Tree see *Leucadendron argenteum*
Silver Vine see *Actinidia polygama*, *Scindapsus
pictus*

Silybum si–li–bum *Compositae*. From *silybon*
the Gk. name for a similar plant. Annual or
biennial herb.
   marianum mă–ree–*ah*–num. Of the Virgin
   Mary who is said to have caused the
   white mottling of the leaves by dropping
   milk on them. Our Lady's Milk Thistle.
   Mediterranean region.

Sinacalia sin–a–*kay*–lee–a *Compositae*.
Perennial herbs formerly included in *Senecio*.
China.
   tangutica tan–*gew*–tik–a (= *Senecio tanguticus*,
   Ligularia tangutica). Of Gansu (Kansu),
   China.

Sinarundinaria sin–a–run–din–ah–ree–a.
Graminae. From *Sino*, Chinese, and
Arundinaria. Bamboos. China, Himalayas.
   nitida ni–ti–da (= *Arundinaria nitida*).
   Shining (the leaves). Fountain Bamboo.

Sinningia si–*ning*–gee–a *Gesneriaceae*. After
William Sinning (1794–1874). Tender,
perennial herbs. Brazil.
   canescens cah–*nes*–enz (= *S. leucotricha*).
   With ash-grey hair. Brazil.
   cardinalis kar–di–*nah*–lis. (= *Rechsteineria
   cardinalis*). Scarlet. Cardinal Flower.
   eumorpha ew–*morf*–a. Of good shape.
   leucotricha see *S. canescens*
   macropoda ma–kro–po–da. (= *Rechsteineria
   cyclophylla*). With a large stalk.
   pusilla pu–*sil*–la. Dwarf.
   regina see *S. speciosa*
   speciosa spe–kee–*ō*–sa. Showy. Gloxinia.

Sisyrinchium si–si–*ring*–kee–um *Iridaceae*. The
Gk. name of a plant. Perennial herbs.
   angustifolium ang–gus–ti–*fo*–lee–um.
   Narrow-leaved. E N America.
   bellum be–lum. Beautiful. California.
   bermudiana ber–mew–dee–*ah*–na. An early
   name for the plant, from Bermuda.
   brachypus see *S. californicum*
   californicum kă–li–*forn*–i–kum. Of
   California. W N America.
   douglasii see *Olsynium douglasii*
   macounii ma–*koon*–ee–ee. After Professor
   John Macoun who first collected it.
   British Columbia.
   striatum stree–*ah*–tum. Striped (the
   flowers). Argentina, Chile.

Skimmia skim–ee–a *Rutaceae*. From the
Japanese name, *Miyami-Shikimi*. Evergreen
shrubs.
   anquetilia ang–kwe–*ti*–a. (= *S. laureola* hort.
   in part). An old generic name, a synonym
   of *Skimmia* after Anquetil-Duperron,
   friend of Decaisne who first described it.
   W Himalaya.
   × *foremanii* see *S. japonica*.
   japonica ja–*pon*–i–ka. Of Japan. Japan,
   Philippines.
       'Foremannii' After Foreman, a Scottish
   nurseryman.
       reevesiana reevz–ee–*ah*–na. After John
   Reeves (1774–1856), a tea inspector
   who introduced Chinese plants. China.
       'Rubella' ru–*bel*–la. Reddish (the flower
   buds).
   laureola low–ree–o–la. Like a small laurel.
   Himalaya, W China.

Skull-cap see *Scutellaria*
Skunk Cabbage see *Lysichiton americanum*, *Symplocarpus foetidus*
Slipper Flower see *Pedilanthes tithymaloides*
Slipperwort see *Calceolaria*
Sloe see *Prunus spinosa*
Small-leaved Gum see *Eucalyptus parvifolia*

Smilacina smee-la-*keen*-a *Liliaceae* (*Convallariaceae*). Diminutive of *Smilax* q.v. Perennial herbs. N America.
   racemosa ră-kay-*mō*-sa. With flowers in racemes. False Spikenard.
   stellata ste-*lah*-ta. Star-like (the flowers). Star-flowered Lily of the Valley.

Smilax smee-lăks *Liliaceae* (*Smilacaceae*). The Gk. name. Evergreen and deciduous climbers.
   aspera a-*spe*-ra. Rough (the stems). Rough Bindweed. Mediterranean region to W Asia.
   excelsa eks-*kel*-sa. Tall. E Europe, W Asia.
   rotundifolia ro-tun-di-*fo*-lee-a. Round-leaved. Horse Briar. E N America.

Smilax (of florists) see *Asparagus asparagoides*

Smithiantha smith-ee-*ănth*-a *Gesneriaceae*. After Matilda Smith (1854–1926), who drew for the Botanical Magazine. Tender perennial herbs. Temple Bells. Mexico.
   cinnabarina ki-na-ba-*reen*-a. Scarlet.
   fulgida *ful*-gi-da. Shining.
   multiflora mul-ti-*flō*-ra. Many-flowered.
   zebrina ze-*breen*-a. Striped (the leaves).

Smoke Tree see *Cotinus coggygria*
Snake Vine see *Hibbertia scandens*
Snake's-head Iris see *Hermodactylus tuberosus*
Snapdragon see *Antirrhinum majus*
Sneezeweed see *Helenium autumnale*
Sneezewort see *Achillea ptarmica*
Snow Gum see *Eucalyptus pauciflora niphophila*
   Tasmanian see *E. coccifera*
Snow-in-Summer see *Cerastium tomentosum*
Snow on the Mountain see *Euphorbia marginata*
Snowberry see *Symphoricarpos albus*
Snowdrop see *Galanthus*
   Common see *G. nivalis*
Snowdrop Tree see *Halesia*
   Mountain see *H. monticola*
Snowflake see *Leucojum*
   Spring see *L. vernum*
   Summer see *L. aestivum*

Snowy Mespilus see *Amelanchier ovalis*
Soapwort see *Saponaria officinalis*
   Rock see *S. ocymoides*

Soehrensia bruchii see *Lobivia bruchii*

Solanum so-*lah*-num *Solanaceae*. L. name of a plant. Annual and perennial herbs, shrubs and climbers.
   aviculare a-vik-ew-*lah*-ree. Of small birds. Kangaroo Apple. New Zealand, Australia.
   capsicastrum kăp-si-*kăs*-trum. Pepper-like (the fruit). False Jerusalem Cherry, Winter Cherry. Brazil.
   crispum kris-pum. Wavy-edged (the leaves). Chile, Peru.
   jasminoides yas-min-*oi*-dees. Jasmine-like. S America.
   melongena me-lon-*zhee*-na. From *melongene*, an old French name. Aubergine, Egg Plant. Africa, Asia.
   pseudocapsicum soo-dō-*kăp*-si-kum. False Capsicum. Jerusalem Cherry, Christmas Cherry. Old World tropics.
   tuberosum tew-be-*rō*-sum. Tuberous. Potato. Andes.

Soldanella sol-da-*nel*-la *Primulaceae*. Diminutive of Italian *soldo* (a small coin), referring to the rounded leaves. Perennial herbs.
   alpina ăl-*peen*-a. Alpine. Europe.
   minima *mi*-ni-ma. Smaller. E Alps.
   montana mon-*tah*-na. Of mountains. C and E Europe.
   pindicola pin-*di*-ko-la. Growing in the Pindus Mountains. NW Greece.
   pusilla pu-*sil*-la. Dwarf. Alps.
   villosa vi-*lō*-sa. Softly hairy. W Pyrenees.

Soleirolia so-lay-*rol*-ee-a *Urticaceae*. After Joseph Francois Soleirol (1796–1863), who collected in Corsica.
   soleirolii so-lay-*rol*-ee-ee. (= *Helxine soleirolii*). As above. Baby's Tears, Mind your own Business. W Mediterranean region.

Solenostemon sō-lee-nō-*stee*-mon. *Labiatae*. From Gk. *solen*, tube, and *stemon*, stamen; the stamens are joined at the base of the corolla tube. Shrubby or succulent herbs. Tropical Africa and Asia.
   scutellarioides skew-tel-*ah*-ree-oy-dees (= *Coleus scutellariodes, Coleus blumei* hort.) Resembling *Scutellaria* (Skullcap). Coleus. The parent of some 60 cultivars of Florists' Coleus, sometimes grouped

Solenostemon (continued)
under the name *Coleus blumei*. Malaysia,
SE Asia.

Solidago so-li-*dah*-gō *Compositae*. From L.
solido (to make whole or strengthen),
referring to medicinal properties. Perennial
herbs. Golden Rod.
canadensis dăn-a-*den*-sis. Of Canada. N
America.
virgaurea virg-*ow*-ree-a. (= *S.
brachystachys*). A golden rod. Europe, N
Africa, W Asia.

× **Solidaster** so-li-*dăs*-ter *Compositae*.
Intergeneric hybrid, from the names of the
parents. *Aster* × *Solidago*. Perennial herb.
luteus loo-tee-us. *Aster ptarmicoides* ×
Solidago sp. Yellow (the flowers).

Sollya so-lee-a *Pittosporaceae*. After Richard
Horsman Solly (1778–1858), English
botanist. Tender, evergreen climbers.
Australia.
heterophylla he-te-rō-*fil*-la. (= *S. fusiformis*).
With variable leaves. Australian Bluebell
Creeper.
parviflora par-vi-*flō*-ra. Small-flowered.

Solomon's Seal see *Polygonatum*

Sonerila so-*ne*-ri-la *Melastomataceae*. From
soneri-ila the Malabar name. Tender herbs.
margaritacea mar-ga-ri-*tah*-kee-a. Pearly,
the silvery-white spotted leaves. SE Asia.
orientalis o-ree-en-*tah*-lis. Eastern. Burma.

Sophora so-*fo*-ra *Leguminosae*. From the
Arabic name. Deciduous and evergreen trees
and shrubs.
davidii dă-*vid*-ee-ee. (= *S. viciifolia*). After
David, see *Davidia*. China.
japonica ja-*pon*-i-ka. Of Japan. Japanese
Pagoda Tree. China.
microphylla mik-rō-*fil*-la. Small-leaved.
New Zealand.
tetraptera tet-*răp*-te-ra. Four-winged (the
pod). Kowhai. New Zealand.

× **Sophrocattleya** sō-frō-*kăt*-lee-a *Orchidaceae*.
Intergeneric hybrids, from the names of the
parents. *Cattleya* × *Sophronitis*. Greenhouse
orchids.

× **Sophrolaelia** sō-frō-*lie*-lee-a *Orchidaceae*.
Intergeneric hybrids, from the names of the
parents. *Laelia* × *Sophronitis*. Greenhouse
orchids.

× **Sophrolaeliocattleya** sō-frō-lie-lee-ō-*kăt*-
lee-a *Orchidaceae*. Intergeneric hybrids, from
the names of the parents. *Cattleya* × *Laelia* ×
Sophronitis. Greenhouse orchids.

Sophronitis sō-*frō*-ni-tis. *Orchidaceae*. From
Gk. *sophron* (modest), the flowers are small.
Greenhouse orchids. Brazil.
cernua kern-ew-a. Nodding.
coccinea kok-*kin*-ee-a. (= *S. grandiflora*. *S.
rosea*). Scarlet.

Sorbaria sor-*bah*-ree-a *Rosaceae*. From *Sorbus*
q.v. a related genus. Deciduous shrubs.
aitchisonii see *tomentosa* var. *angustifolia*.
arborea ar-*bo*-ree-a. Tree-like. China.
sorbifolia sor-bi-*fo*-lee-a. N Asia.
tomentosa tō-men-*tō*-sa. Hairy. Himalaya.
angustifolia an-gŭs-ti-*fō*-lee-a (= *S.
aitchisonii*). With narrow leaves.
Afghanistan, Pakistan.

Sorbus sor-bus *Rosaceae*. L. name for the
service tree. (= *S. domestica*). Deciduous
trees and shrubs.
alnifolia ăl-ni-*fo*-lee-a. *Alnus*-leaved. E
Asia.
aria ah-ree-a. An old name for this tree.
Whitebeam. Europe.
'Chrysophylla' kris-ō-*fil*-la. Golden-
leaved.
'Lutescens' loo-*tes*-enz. Yellowish, the
creamy white young leaves.
'Majestica' mah-*yes*-ti-ka. (=
'Decaisneana'). Majestic.
aucuparia ow-kew *pah*-ree-a. From L. *avis*
(a bird) and *capere* (to catch), the fruits
attract birds. Mountain Ash, Rowan.
'Asplenifolia' a-splay-ni-*fo*-lee-a.
Asplenium-leaved.
'Edulis' ed-*ew*-lis. Edible (the fruit).
cashmiriana kash-mi-ree-*ah*-na. Of
Kashmir. W Himalaya.
commixta kom-*miks*-ta. Mixed together.
Japan, Korea.
cuspidata kus-pi-*dah*-ta. Abruptly sharp-
pointed (the leaves). Himalaya.
domestica do-*mes*-ti-ka. Cultivated. Service
Tree. S and E Europe, Caucasus, N
Africa.
esserteauiana e-ser-tō-ee-*ah*-na. After Dr
Esserteau, a bacteriologist who helped
Ernest Wilson in China. China.
huphensis hew-pee-*hen*-sis. Of Hubei
(Hupeh). China.
hybrida hib-ri-da. Originally thought to be
a hybrid. Scandinavia.
intermedia in-ter-*me*-dee-a. Intermediate
(between the wild service tree and a

Sorbus (continued)
mountain ash). Swedish Whitebeam.
Scandinavia.
× *kewensis* dew-*en*-sis. *S. aucuparia* × *S.
pohuashanenis.* Of Kew.
koehneana kur-nee-*ah*-na. After Bernard
Adalbert Emil Koehne (1848–1918). W
China.
latifolia lah-tee-*fo*-lee-a. Broad-leaved.
Service Tree of Fountainbleau. France.
poteriifolia po-te-ree-i-*fo*-lee-a. (= *S.
pygmaea*). With leaves like *Poterium.*
China, N Burma.
pygmaea see *S. poteriifolia*
reducta re-*duk*-ta. Dwarf. China.
sargentiana sar-jent-ee-*ah*-na. After
Sargent, see *Prunus sargentii*. China.
scalaris ska-*lah*-ris. Ladder-like (the leaves).
China.
× *thuringiaca* thu-ring-gee-*ah*-ka. *S. aria* ×
S. aucuparia. Of Thuringia, Germany.
Wild with the parents.
torminalis tor-mi-*nah*-lis. Effective against
colic. Wild Service Tree. Europe, N Africa,
SW Asia.
vilmorinii vil-mo-*rin*-ee-ee. After Maurice
Vilmorin, who received seeds from
Delavay in 1889. China.

Sorrel, Common see *Rumex acetosa*
    French see *R. scutatus*
Sorrel Tree see *Oxydendrum arboreum*
Southern Beech see *Nothofagus*
Southernwood see *Artemisia abrotanum*
Sowbread see *Cyclamen*
Spanish Bayonet see *Yucca aloifolia*
Spanish Shawl see *Heterocentron elegans*
Spanish Moss see *Tillandsia usneoides*

Sparaxis spa-*răks*-is Iridaceae. From Gk.
sparasso (to tear) referring to the lacerated
spathe bracts. Cormous perennial herbs.
Wandflower. S Africa.
tricolor tri-ko-lor. Three-coloured.
Harlequin Flower.

Sparmannia spar-*măn*-ee-a Tiliaceae. After Dr
Andreas Sparrman (1748–1820), Swedish
botanist. Tender shrub.
africana ăf-ri-*kah*-na. African. African
Hemp. S Africa.

Spartina spar-*teen*-a Gramineae. From Gk.
spartion (esparto grass). Perennial grass.
pectinata pek-ti-*nah*-ta. Comb-like (the one-
sided spikes). N. America.

Spartium *spar*-tee-um Leguminosae
(Papilionoideae). From Gk. *spartion* (esparto

Spartium (continued)
grass), they were both used for cordage.
Nearly leafless shrub.
junceum yung-kee-um. Rush-like. S
Europe, N Africa, W Asia.

Spartocytisus nubigenus see *Cytisus
supranubias*

Spathiphyllum spă-thi-*fil*-lum Araceae. From
Gk. *spathe* and *phyllon* (a leaf) referring to
the leaf-like spathe. Tender herb.
wallisii wo-*lis*-ee-ee. After Gustave Wallis
who introduced it from Colombia in
1824.

Spatterdock see *Nuphar advena*
Spearmint see *Mentha* × *spicata*

Specularia speculum-veneris see *Legousia
speculum-veneris*

Speedwell see *Veronica*
Speedy Jenny see *Tradescantia fluminensis*

Sphaeralcea sfie-*răl*-kee-a Malvaceae. From
Gk. *sphaira* (a globe) and *Alcea* a related
genus, referring to the spherical fruits. Sub-
shrubby perennials.
coccinea kok-*kin*-ee-a. (= *Malvastrum
coccineum*). Scarlet. W United States.
munroana mūn-rō-*ah*-na. After Munro. W
United States.
purpurata pur-pew-*rah*-ta. (= *Malvastrum
campanulatum*). Purplish (the flowers).
Chile.

Spice Bush see *Lindera benzoin*
Spider Flower see *Cleome hassleriana*
Spider Lily see *Hymenocallis*
    Golden see *Lycoris africana*
    Red see *Lycoris radiata*
Spider Plant see *Chlorophytum comosum*
Spike Heath see *Bruckenthalia spiculifolia*
Spinach see *Spinacia oleracea*
Spinach Beet see *Beta vulgaris*

Spinacia spee-*nah*-kee-a Chenopodiaceae. A
medieval L. name probably originally from
L. *spina* (a spine) referring to the spiny seeds.
Annual herb.
oleracea o-le-*rah*-kee-a. Vegetable-like.
Spinach. SW Asia.

Spindle Tree see *Euonymus europaeus*
Spinning Gum see *Eucalyptus perriniana*

Spiraea spee-*rie*-a Rosaceae. From Gk.

Spiraea (continued)
speiraira, a plant used in garlands. Deciduous shrubs.
   albiflora see *S. japonica* 'Albiflora'
   × *arguta* ar-*gew*-ta. Sharply toothed (the leaves). Bridal Wreath.
   × *billiardii* bi-lee-*ard*-ee-ee. After Billiard, the raiser.
   × *bumalda* see *S. japonica* 'Bumalda'
   japonica ja-*pon*-i-ka. Of Japan.
      'Albiflora' ăl-bi-*flō*-ra. (= *S. albiflora*). White-flowered.
      'Bullata' bu-*lah*-ta. With puckered leaves.
      'Bumalda' bew-*mahl*-da. (= *S.* × *bumalda*). After Bumaldus.
   nipponica ni-*pon*-i-ka. Of Japan.
   prunifolia proo-ni-*fo*-lee-a. *Prunus*-leaved. China.
   salicifolia să-li-ki-*fo*-lee-a. *Salix*-leaved. Bridewort. Europe to Japan.
   thunbergii thun-*berg*-ee-ee. After Thunberg, see *Thunbergia*. China.
   trilobata tri-lo-*bah*-ta. Three-lobed (the leaves). N Asia.
   × *vanhouttei* văn-*hoot*-ee-ee. *S. cantoniensis* × *S. trilobata*. After L. B. van Houtte (1810–76), Belgian nurseryman.

Spleenwort see *Asplenium*
   Black see *A. adiantum-nigrum*
   Ebony see *A. platyneuron*
   Green see *A. viride*
   Hanging see *A. flaccidum*
   Maidenhair see *A. trichomanes*
   Mother see *A. bulbiferum*
   Sea see *A. marinum*
Spotted Evergreen see *Aglaonema costatum*
Spreading Clubmoss see *Selaginella kraussiana*
Spring Meadow Saffron see *Bulbocodium vernum*
Spruce see *Picea*
   Black see *P. mariana*
   Brewer see *P. breweriana*
   Colorado see *P. pungens*
   Dragon see *P. asperata*
   Hondo see *P. jezoensis hondoensis*
   Norway see *P. abies*
   Serbian see *P. omorika*
   Sitka see *P. sitchensis*
   White see *P. glauca*
   Yezo see *P. jezoensis*
Spurge Laurel see *Daphne laureola*
Squirrel Tail Grass see *Hordeum jubatum*

Stachys stă-kis *Labiatae*. From Gk. *stachys* (a spike) referring to the inflorescence. Perennial herbs.

Stachys (continued)
   affinis a-*fee*-nis. Related to Chinese Artichoke. China.
   byzantina bi-zan-*teen*-a. (= *S. lanata. S. olympica*). Of Istanbul (Byzantium). Lamb's Tongue.
   corsica kor-si-ka. Of Corsica. Corsica, Sardinia.
   discolor dis-*kō*-lor (= *S. nivea*). Of two colours (the flowers are tinted pink or occasionally yellow). Caucasus.
   macrantha ma-*krănth*-a. Large-flowered. Bishop's Wort. Caucasus.
   monieri mo-nee-*e*-ree. (= *S. densiflora*). After Monier. Alps, Pyrenees.
   nivea see *S. discolor*

Stachyurus stă-kee-*ew*-rus *Stachyuraceae*. From Gk. *stachys* (a spike) and *oura* (a tail) referring to the slender racemes. Deciduous shrubs.
   chinensis chin-*en*-sis. Of China.
   praecox prie-koks. Early (flowering). Japan.

Standing Cypress see *Ipomopsis rubra*

Stanhopea stăn-*hō*-pee-a *Orchidaceae*. After Philip Henry, 4th Earl of Stanhope (1781–1855). Greenhouse orchids.
   costaricensis kos-ta-ree-*ken*-sis. Of Costa Rica.
   eburnea e-bur-*nee*-a (= *S. grandiflora*). Ivory-white (the flowers). N S America.
   grandiflora see *S. eburnea*.
   oculata ok-ew-*lah*-ta. With an eye. C America.
   tigrina ti-*green*-a. Striped like a tiger. Mexico.
   wardii word-ee-ee. After Mr Ward who sent it to England. C America.

Stapelia sta-*pel*-ee-a *Asclepiadaceae*. After Johannes Bodaeus von Stapel (died 1631). Tender succulents. S Africa.
   gigantea gi-*găn*-tee-a. Very large.
   grandiflora grănd-i-*flō*-ra. Large-flowered.
   hirsuta hir-*soo*-ta. Hairy (the corolla).
   pillansii pi-*lănz*-ee-ee. After Dr Neville Stewart Pillans (1884–1964) of Pretoria.
   variegata see *Orbea variegata*

Staphylea sta-*fi*-lee-a *Staphyleaceae*. From Gk. *staphyle* (a cluster) referring to the inflorescence. Deciduous shrubs. Bladder-nut.
   colchica kol-ki-ka. Of Colchis, W Asia. Caucasus.
   holocarpa ho-lo-*kar*-pa. With an unlobed fruit.
   pinnata pi-*nah*-ta. Pinnate (the leaves). SE Europe, W Asia.

Star-flowered Lily of the Valley see *Smilacina stellata*
Star of Bethlehem see *Ornithogalum umbellatum*
Star of the Veldt see *Dimorphotheca sinuata*
Starfish Plant see *Cryptanthus acaulis, Orbea variegata*
Stars of Persia see *Allium christophii*

Stauntonia stawn-*ton*-ee-a *Lardizabalaceae.* After Sir George Leonard Staunton (1737–1801). Evergreen climber.
hexaphylla heks-a-*fil*-la. With six leaves (leaflets). Japan, China.

Stenocactus sten-ō-*kăk*-tus *Cactaceae.* From Gk. *stenos* (narrow) and *Cactus* q.v. The following are often listed under *Echinofossulocactus.* Mexico.
coptonogonus kop-ton-ō-*gō*-nus. With notched ribs.
hastatus hăs-*tah*-tus. Spear-shaped (some of the spines).
lancifer lăn-ki-fer. Lance-bearing.
multicostatus mul-ti-kos-*tah*-tus. Many-ribbed.
violaciflorus vee-o-lah-ki-*flō*-rus. With violet flowers.
zacatecasensis ză-ka-te-ka-*sen*-sis. Of Zacatecas.

Stenocarpus sten-ō-*kar*-pus *Proteaceae.* From Gk. *stenos* (narrow) and *karpos* (a fruit). Tender shrub.
sinuatus sin-ew-*ah*-tus. Wavy-edged (the leaves). Firewheel Tree. E Australia.

Stenocereus steen-ō-*kay*-ree-us *Cactaceae.* From Gk. *stenos,* narrow, and *Cereus* q.v. Tree-like and shrubby cacti, formerly included in *Lemaireocereus.*
pruinosus proo-in-ō-sus (= *Lemaireocereus pruinosus*). Bloomed (the shoots). S Mexico.
thurberi thur-ba-ree (= *Lemaireocereus thurberi*). After George Thurber (1821–1890) who collected in the SW United States and Mexico. S Arizona.
treleasei tra-*lees*-ee-ee (= *Lemaireocereus treleasei*). After William Trelease (1857–1945), American botanist. S Mexico.

Stephanandra ste-fa-*năn*-dra *Rosaceae.* From Gk. *stephanos* (a crown) and *andros* (a man), the stamens form a wreath around the capsule. Deciduous shrubs.
incisa in-*kee*-sa. Deeply cut (the leaves). Japan, Korea.

Stephanandra (continued)
tanakae ta-*nah*-kie. After Tanaka, a Japanese botanist. Japan.

Stephanotis ste-fa-*nō*-tis *Asclepiadaceae.* Gk. name for myrtle which was used to make crowns, from *stephanos* (a crown) and *otos* (an ear) referring in this case to auricles in the staminal crown. Tender, evergreen climber.
floribunda flō-ri-*bun*-da. Profusely flowering. Wax Flower, Madagascar Jasmine. Madagascar.

Sternbergia stern-*berg*-ee-a *Amaryllidaceae.* After Count Kaspar von Sternberg (1761–1838), Austrian botanist. Bulbous perennial herbs.
candida kăn-di-da. White. SW Turkey.
clusiana klooz-ee-*ah*-na. After Clusius, see *Gentiana clusii.* W Asia.
colchiciflora kol-ki-ki-*flō*-ra. *Colchicum*-flowered. S Europe, W Asia.
fischeriana fi-sha-ree-*ah*-na. After Fischer who collected the type specimen. Caucasus, W Asia.
lutea loo-tee-a. Yellow. Mediterranean region to Iran, Caucasus.

Stetsonia stet-*son*-ee-a *Cactaceae.* After Francis Lynde Stetson of New York.
coryne ko-ri-nay. A club. Toothpick Cactus. Argentina.

Stewartia see *Stuartia*
Stinking Benjamin see *Trillium erectum*

Stipa stee-pa *Gramineae.* From Lat. *stipa* meaning tow or oakum, referring to the feathery inflorescence. *S. tenacissima* is esparto grass from which cordage is made. Perennial grasses.
barbata bar-*bah*-ta. Bearded. SW Europe.
calamagrostis kăl-a-ma-*gros*-tis. From Gk. *calama* (a reed) and *agrostis* (a grass). C and S Europe.
gigantea gi-*găn*-tee-a. Very large. Spain, Portugal, Morocco.
pennata pe-*nah*-ta. Feathery (the inflorescence). S and C Europe, Asia.

Stock see *Matthiola*
Brompton see *M. incana*
Night-scented see *M. longipetala bicornis*
Ten Weeks see *M. incana* Annua
Virginia see *Malcolmia maritima*

Stokesia stōks-ee-a *Compositae.* After Dr Jonathan Stokes (1755–1831), Edinburgh

Stokesia (continued)
physician and friend of Linnaeus the younger. Perennial herb.
  laevis lie-vis. Smooth. Stokes Aster. SE United States.

Stone Cress see *Aethionema*
Stonecrop see *Sedum*
  Biting see *S. acre*
  Golden see *S. adolphi*
Stranvaesia davidiana see *Photinia davidiana*
Strap Flower see *Anthurium crystallinum*

Stratiotes strǎ-tee-ō-teez *Hydrocharitaceae*. Gk. name for *Pistia stratiotes*. Aquatic herb.
  aloides a-lŏi-deez. Like *Aloe* (the leaves). Water Soldier. Europe.

Strawberry see *Fragaria*
  Alpine see *F. vesca*
  Garden see *F.* × *ananassa*
  Hautbois see *F. moschata*
  Mock see *Duchesnea indica*
Strawberry Tree see *Arbutus unedo*

Strelitzia stre-*lits*-ee-a *Strelitziaceae*. After Charlotte of Mecklenberg-Strelitz (1744–1818), Queen to George III. Tender herbaceous perennials. S Africa.
  alba ǎl-ba. (= *S. augusta*). White (the flowers).
  reginae ray-*geen*-ie. (= *S. parvifolia*). Of the Queen (see above). Bird of Paradise Flower, Crane Lily.

Streptocarpus strep-to-*kar*-pus *Gesneriaceae*. From Gk. *streptos* (twisted) and *karpos* (a fruit), the fruits are spirally twisted. Tender perennial herbs. Cape Primrose.
  dunnii dŭn-ee-ee. After its discoverer, Edward John Dunn (1844–1937). Red Nodding Bells. S Africa.
  holstii holst-ee-ee. After C. H. E. W. Holst (1865–94), a German gardener who travelled in E Africa. E Africa.
  × *hybridus* hib-ri-dus. Hybrid.
  polyanthus po-lee-*ǎnth*-us. Many-flowered. S Africa.
  rexii reks-ee-ee. After George Rex, son of George III, on whose property it was discovered. S Africa.
  saxorum sǎks-ō-rum. Growing on rocks. False African Violet. E Africa.
  wendlandii vend-*lǎnd*-ee-ee. After Hermann Wendland (1825–1903), a German botanist. Royal Nodding Bells. S Africa.

Streptosolen strep-to-*sō*-len *Solanaceae*. From

Streptosolen (continued)
Gk. *streptos* (twisted) and *solen* (a tube) the corolla tube is spirally twisted. Tender, evergreen shrub.
  jamesonii jaym-*son*-ee-ee. After Dr William Jameson, professor of botany at Quito. Marmalade Bush. N Andes.

String of Beads see *Senecio rowleyanus*

Strobilanthes stro-bi-*lǎnth*-eez *Acanthaceae*. From Gk. *strobilos* (a cone) and *anthos* (a flower) referring to the dense inflorescence. Tender shrub.
  dyerianus die-a-ree-*ah*-nus. After Sir William Thistleton-Dyer (1843–1928) who introduced it to Kew. Persian Shield. Burma.

Stromanthe strō-*mǎnth*-ee *Marantaceae*. From Gk. *stroma* (a bed) and *anthos* (a flower) referring to the inflorescence. Tender perennial herbs. Brazil.
  amabilis see *Ctenanthe amabilis*
  sanguinea sang-*gwin*-ee-a. Blood-red (the bracts).

Stuartia stew-*art*-ee-a *Theaceae*. (*Stewartia*). After John Stuart (1713–92), 3rd Earl of Bute. Deciduous shrubs and trees.
  malacodendron mǎ-la-kō-*den*-dron. Mallow tree. SE United States.
  ovata ō-*vah*-ta. Ovate (the leaves). SE United States.
  pseudocamellia soo-dō-ka-*mel*-ee-a. False Camellia. Japan.
    koreana ko-ree-*ah*-na. (= *S. koreana*). Of Korea.
  serrata se-*rah*-ta. Saw-toothed (the leaves). Japan.
  sinensis si-*nen*-sis. Of China.

Sturt's Desert Pea see *Clianthus formosus*

Stylophorum sti-*lo*-fo-rum *Papaveraceae*. From Gk. *stylos* (a style) and *phoros* (bearing) referring to the long style, a distinctive character of the genus. Perennial herb.
  diphyllum di-*fi*-lum. Two-leaved. Celandine Poppy. E United States.

Styrax sti-rǎks *Styracaceae*. The Gk. name for *S. officinalis*. Deciduous trees and shrubs.
  hemsleyanum hemz-lee-*ah*-num. After William Botting Hemsley (1843–1924). China.
  japonicum ja-*pon*-i-kum. Of Japan. Japan, China, Korea.

Styrax (continued)
obassia ō-*bă*-see-a. From the Japanese name. Japan, China, Korea.
wilsonii wil-*son*-ee-ee. After Wilson who introduced it in 1908, see *Magnolia wilsonii*. W Sichuan.

Sugar Beet see *Belta vulgaris*
Sugarberry see *Celtis laevigata*
Sumach see *Rhus*
  Smooth see *R. glabra*
  Stag's Horn see *R. typhina*
Summer Cypress see *Bassia scoparia*
Summer Hyacinth see *Galtonia candicans*
Summer Torch see *Billbergia pyramidalis*
Sun Plant see *Portulaca grandiflora*
Sun Rose see *Helianthemum*
Sundew see *Drosera*
Sundrops see *Oenothera fruticosa*
Sunflower see *Helianthus annuus*
Swamp Cypress see *Taxodium distichum*
Swan River Daisy see *Brachycome iberidifolia*
Swede see *Brassica napus* Napobrassica
Swedish Ivy see *Plectranthus australis*
Sweet Alyssum see *Lobularia maritima*
Sweet Bay see *Magnolia virginiana*
Sweet Bergamot see *Monarda didyma*
Sweet Box see *Sarcococca*
Sweet Briar see *Rosa rubiginosa*
Sweet Cicely see *Myrrhis odorata*
Sweet Fern see *Comptonia peregrina*
Sweet Flag see *Acorus calamus*
Sweet Four o'Clock Plant see *Mirabilis longiflora*
Sweet Gale see *Myrica gale*
Sweet Gum see *Liquidambar styraciflua*
Sweet Pea see *Lathyrus odoratus*
Sweet Pepper Bush see *Clethra alnifolia*
Sweet Potato Vine see *Ipomoea batatas*
Sweet Rocket see *Hesperis matronalis*
Sweet Sop see *Annona squamosa*
Sweet Sultan see *Amberboa moschata*
Sweet William see *Dianthus barbatus*
Sweet Woodruff see *Galium odoratum*
Swiss Chard see *Beta vulgaris*
Swiss Cheese Plant see *Monstera deliciosa*
Sycamore see *Acer pseudoplatanus*

Sycopsis si-*kop*-sis Hamamelidaceae. From Gk. *sykon* (a fig) and *-opsis* indicating resemblance. Evergreen shrub or tree.
sinensis si-*nen*-sis. Of China.

Symphoricarpos sim-fo-ree-*kar*-pos Caprifoliaceae. From Gk. *symphorein* (bear together) and *karpos* (a fruit) referring to the clustered fruits. Deciduous shrubs.
albus *ăl*-bus. White (the fruit). Snowberry. N America.

Symphoricarpos (continued)
laevigatus lie-vi-*gah*-tus. (= *S. rivularis*). Smooth (the leaves and shoots). W N America.
× *chenaultii* she-*nolt*-ee-ee. *S. orbiculatus* × *S. microphyllus*. After Chenault, French nurseryman who raised it.
orbiculatus or-bik-ew-*lah*-tus. Orbicular (the fruit). Indian Currant. United States, Mexico.

Symphytum sim-fi-tum Boraginaceae. The Gk. name from *symphysis* (growing together of bones) and *phyton* (a plant), it was reputed to heal broken bones. Perennial herbs. Comfrey.
caucasicum kaw-*kă*-si-kum. Of the Caucasus.
grandiflorum grănd-i-*flō*-rum. Large-flowered. Caucasus.
orientale o-ree-en-*tah*-lee. Eastern W Asia.
× *uplandicum* up-*lănd*-i-kum. *S. asperum* × *S. officinale*. Of Uppland, Sweden. Russian Comfrey. Caucasus.

Symplocarpus sim-plo-*kar*-pus Araceae. From Gk. *symploke* (a connection) and *karpos* (a fruit), the ovaries combine to make a single fruit. Perennial bog-garden herb.
foetidus foy-ti-dus. Foetid. Skunk Cabbage. N America, NE Asia, Japan

Symplocos sim-plo-kos Symplocaceae. From Gk. *symploke* (a connection), the stamens are united. Deciduous shrub.
paniculata pa-nik-ew-*lah*-ta. With flowers in panicles. Himalaya, China, Japan.

Syngonium sin-*gon*-ee-um Araceae. From Gk. *syn* (together) and *gone* (womb) referring to the fused ovaries. Tender climbers.
angustatum ang-gus-*tah*-tum. Narrowed. Arrowhead Vine. C America.
auritum ow-ree-tum. Eared (the outer leaf segments). Five Fingers. Caribbean.
podophyllum po-dō-*fil*-lum. With stoutly-stalked leaves. Arrowhead Vine. C America.
vellozianum see *S. podophyllum*

Synnema triflorum see *Hygrophila difformis*

Syringa si-*ring*-ga Oleaceae. From Gk. *syrinx* (a pipe) referring to the hollow stems. For the same reason, *Philadelphus* has long been known by this name. Deciduous shrubs. Lilac.
× *chinensis* chin-*en*-sis. *S. laciniata* × *S.*

Syringa (continued)
vulgaris. Originally thought to be a native of China. Rouen Lilac.
× *hyacinthiflora* hee-a-kinth-i-*flō*-ra. *S. oblata* × *S. vulgaris.* With hyacinth-coloured leaves.
josikaea jo-si-*kie*-a. After Baroness von Josika. E Europe.
julianae yoo-lee-*ah*-nie. Named by Schneider after his wife Juliana. W China.
laciniata la-kin-ee-*ah*-ta. Deeply cut (the leaves). W China.
meyeri may-a-ree. After F. N. Meyer who introduced it to America in 1908. N China (Cult.).
    'Palibin' *pă*-li-bin (= *S. palibiniana* hort. *S. velutina* hort.). From *S. palibiniana* which it was originally thought to be, after Ivan Vladimirovich Palibin (1872–1949), a Russian botanist.
palibiniana hort. see *S. meyeri* 'Palibin'.
× *persica* per-si-ka. *S. afghanica* × *S. laciniata.* Of Iran (Persia) where it has long been cultivated. Persian Lilac.
× *prestoniae* pres-*ton*-ee-ie. *S. reflexa* × *S. villosa.* After Dr Isabella Preston (1881–1965) of the Central Experimental Farm, Ottawa, who raised it in 1920.
reflexa re-*fleks*-a. Reflexed (the corolla lobes). C China.
velutina hort. see *S. meyeri* 'Palibin'.
vulgaris vul-*gah*-ris. Common. Common Lilac. C Europe.

T

Tacca *tă*-ka *Taccaceae.* From *taka* the Indonesian name. Tender perennial herbs.
chantrieri shon-tree-*e*-ree. After Chantrier Frères, French nurserymen. Cat's Whiskers, Devil Flower. SE Asia.
integrifolia in-teg-ri-*fo*-lee-a.(= *T. aspera*). With entire leaves. Bat Plant. SE Asia.
leontopetaloides lee-on-to-pe-ta-*loi*-deez. Like *Leontopetalon (Leontice)*. Old World tropics.

Tagetes ta-*gay*-teez *Compositae.* From *Tages* an Etruscan deity, the grandson of Jupiter, who sprang from the ploughed earth. Annual herbs. Marigold. Mexico, C America.
erecta e-*rek*-ta. Erect. African Marigold.
lucida loo-ki-da. Bright.
patula *păt*-ew-la. Spreading. French Marigold.
tenuifolia ten-ew-i-*fo*-lee-a. (= *T. signata*).

Tagetes (continued)
With finely divided leaves. Signet Marigold.

Tail Flower see *Anthurium crystallinum*
Tailor's Patch see *Crassula lactea*

Taiwania tie-*wahn*-ee-a *Cupressaceae.* From Taiwan. Semi-hardy, evergreen conifer.
cryptomerioides krip-to-me-ree-*oi*-deez. Like *Cryptomeria.* Taiwan.

Talinum ta-*leen*-um *Portulacaceae.* Derivation obscure, possibly an African name. Tender succulents. Fameflower.
caffrum *kăf*-rum. Of S Africa. Tropical and S Africa.
guadalupense gwah-da-loop-*en*-see. Of Guadalupe.
portulacifolium por-tew-lah-ki-*fo*-lee-um. *Portulaca*-leaved. India, Arabia, Africa.

Tamarix *tă*-ma-riks *Tamaricaceae.* The L. name. Deciduous shrubs. Tamarisk.
gallica *gă*-li-ka. Of France. W Europe, N Africa.
germanica see *Myricaria germanica*
parviflora par-vi-*flō*-ra. Small-flowered. SE Europe.
ramosissima rah-mō-*si*-si-ma. (= *T. pentandra*). Much branched. S Russia to China.
tetrandra tet-*răn*-dra. With four stamens. E Europe, W Asia.

Tanacetum tăn-a-*set*-um *Compositae.* From the medieval L. name of a herb. Perennial herbs.
argenteum ar-*gen*-tee-um. Silvery. Turkey.
balsamita băl-sa-*mee*-ta. (= *Balsamita major*). Bearing balsam. Alecost, Costmary. Caucasus, Iran.
coccinea kok-*kin*-ee-a (= *Chrysanthemum coccineum*). Scarlet (the flowers). Pyrethrum. W Asia.
corymbosum ko-rim-*bō*-sum. (= *Chrysanthemum corymbosum*). With flowers in corymbs. Europe.
densum den-sum. (= *Chrysanthemum densum*). Compact. W Asia.
haradjanii hă-ra-*dyahn*-ee-ee. (= *Chrysanthemum haradjanii*). After Haradjian. Turkey.
herderi her-da-ree. After Herder. Turkestan.
parthenium par-*then*-ee-um. (= *Chrysanthemum parthenium*). From *parthenion*, the Gk. name of a plant. Bachelor's Buttons. Balkan Peninsula.

Tanacetum (continued)
praeteritum prie-*te*-ri-tum. Overlooked.
Turkey.
vulgare vul-*gah*-ree. (= *Chrysanthemum
vulgare*). Common. Tansy. Europe, Asia.

Tanakaea tăn-a-*kie*-a *Saxifragaceae*. After
Yoshio Tanaka (1836–1916), Japanese
botanist. Perennial herb.
radicans rah-di-kănz. With rooting stems.
Japanese Foam Flower. China, Japan.

Tangerine see *Citrus reticulata*
Tansy see *Tanacetum vulgare*
Tarragon, French see *Artemisia dracunculus*
    Russian see *A. dracunculus inodora*
Tasmanian Blue Gum see *Eucalyptus
globulus*
Tassel Flower see *Emilia coccinea*
Tassel Hyacinth see *Muscari comosum*

Taxodium tăks-*o*-dee-um *Cupressaceae*. From
Taxus q.v. and Gk. *eidos* (resemblance).
Deciduous conifers. SE United States.
ascendens a-*sen*-denz. Ascending (the
ultimate shoots). Pond Cypress.
    Nutans *new*-tănz. Nodding (the
    ultimate shoots).
distichum dis-ti-kum. In two ranks (the
leaves). Swamp Cypress.

Taxus tăks-us *Taxaceae*. The L. name.
Evergreen trees. Yew.
baccata bah-*kah*-ta. Berry-bearing.
Common Yew. Europe, N Africa, W
Asia.
    'Dovastoniana' dūv-a-ston-ee-*ah*-na.
    After John Dovaston of West Felton
    where the original tree grows. West
    Felton Yew.
    'Fastigiata' fa-stig-ee-*ah*-ta. With
    upright branches. Irish Yew.
    'Standishii' stăn-*dish*-ee-ee. After the
    Standish nursery, Ascot who
    distributed it.
× *media* me-dee-a. *T. baccata* × *T. cuspidata*.
Intermediate (between the parents).
    'Hatfieldii' hăt-*feeld*-ee-ee. After T. D.
    Hatfield of the Hunnewell Arboretum,
    who raised it.
    'Hicksii' *hiks*-ee-ee. After Henry Hicks
    who introduced it to cultivation.

Tea Plant see *Camellia sinensis*
Tea Tree see *Leptospermum scoparium*
Teasel see *Dipsacus fullonum*

Tecoma te-*kŏ*-ma *Bignoniaceae*. From the
Mexican name. Tender shrub.

Tecoma (continued)
stans stănz. Erect. Yellow Elder. SE United
States, C and S America.

Tecomaria te-kŏ-*mah*-ree-a *Bignoniaceae*.
From *Tecoma* q.v. a related genus. Tender,
scrambling shrub.
capensis ka-*pen*-sis. Of the Cape of Good
Hope. Cape Honeysuckle. S Africa.

Tecophilaea te-ko-fi-*lie*-a *Tecophilaeaceae*.
After Tecophila Billoti, 18th-century Italian
botanical artist. Semi-hardy, cormous
perennial.
cyanocrocus see-ăn-ŏ-*krŏ*-kus. Blue Crocus.
Chilean Crocus, Chile.

Teddy Bear Plant see *Cyanotis kewensis*
Telegraph Plant see *Desmodium gyrans*

Telekia te-*le*-kee-a *Compositae*. After Samuel
Teleki di Szék, a Hungarian nobleman and
botanical patron. Perennial herbs.
speciosa spe-kee-ŏ-sa. (= *Buphthalmum
speciosum*). Showy. SE Europe, Caucasus,
W Asia.
speciosissima spe-kee-ŏ-*si*-si-ma. (=
Buphthalmum speciosissimum). Very showy.
N Italy.

Tellima te-li-ma *Saxifragaceae*. Anagram of
Mitella, a related genus. Perennial herb.
grandiflora grănd-i-*flŏ*-ra. Large-flowered.
W N America.

Telopea tay-*lŏ*-pee-a *Proteaceae*. From Gk.
telepos (seen from afar) referring to the
showy flowers. Evergreen, tender and semi-
hardy shrubs.
speciosissima spe-kee-ŏ-*si*-si-ma. Very
showy. Waratah. New South Wales.
truncata trung-*kah*-ta. Abruptly cut off (the
leaves). Tasmanian Waratah. Tasmania.

Temple Bells see *Smithiantha*
Tendergreen see *Brassica rapa* Perviridis

Tephrosia te-*fros*-ee-a *Leguminosae*
(*Papilionoideae*). From Gk. *tephros* (ash-
coloured) referring to the leaves. Tender
shrub.
grandiflora grănd-i-*flŏ*-ra. Large-flowered. S
Africa.

Tetracentron tet-ra-*ken*-tron *Tetracentraceae*.
From Gk. *tetra* (four) and *kentron* (a spur)
referring to the four-spurred fruit.
Deciduous tree.

Tetracentron (continued)
  sinense si-*nen*-see. Of China. China, Himalaya, N Burma.

Tetradium tet-ra-dee-um (= *Evodia, Euodia*) *Rutaceae*. From Gk. *tetra*, four; the fruit and flower parts occur in fours. Trees and shrubs. S and E Asia, W Malesia.
  daniellii dǎn-*yel*-ee-ee (= *Evodia daniellii, E. hupehensis*). After William Freeman Daniell (1818–65), a British army surgeon who found it in China. N China, Korea.
  hupehensis see *T. daniellii*

Tetrastigma tet-ra-*stig*-ma *Vitaceae*. From Gk. *tetra* (four) and *stigma*, the stigma is four-lobed. Tender, evergreen climber.
  voinierianum vwū-nee-e-ree-*ah*-num. After M. Voinier, chief veterinary surgeon with the French army in Hanoi. Chestnut Vine. Laos.

Teucrium toyk-ree-um *Labiatae*. From *teukrion*, the Gk. name. Herbs and shrubs. Germander.
  aroanum ǎ-rō-*ah*-num. Of Aroania. S Greece.
  chamaedrys ka-*mie*-dris. From *chamaidrys* (dwarf oak), the Gk. name. Wall Germander. Europe, SW Asia, N Africa.
  creticum kray-ti-kum. (= *T. rosmarinifolium*). Of Crete. E Mediterranean region.
  fruticans froo-ti-kǎnz. Shrubby. Shrubby Germander. W Mediterranean region.
  marum mah-rum. The Gk. name. Cat Thyme. W Mediterranean islands.
  polium po-lee-um. (= *T. aureum*). The Gk. name. S Europe.
  pyrenaicum pi-ray-*nah*-i-kum. Of the Pyrenees.
  subspinosum sub-spee-*nō*-sum. Somewhat spiny. Mallorca.

Thalictrum tha-*lik*-trum *Ranunculaceae*. Gk. name for a plant. Perennial herbs. Meadow Rue.
  alpinum ǎl-*peen*-um. Alpine. N hemisphere (arctic and alpine regions).
  aquilegiifolium ǎ-kwi-lee-gee-i-*fo*-lee-um. *Aquilegia*-leaved. Europe, Asia.
  delavayi de-la-*vay*-ee. After Delavay, see *Abies delavayi*. W China.
  dipterocarpum see *T. delavayi*
  flavum flah-vum. Yellow (the flowers). Europe, Asia.
    glaucum glow-kum. (= *T. speciosissimum*). Glaucous. Spain, Portugal.
  kiusianum kee-oo-see-*ah*-num. Of Kyushu. Japan.

Thamnocalamus tham-nō-*kal*-a-mus *Graminae*. From Gk. *thamnos*, bush, and *kalamos*, reed. Clump-forming bamboos. Himalaya, South Africa.
  spathaceus spǎth-a-*see*-us (= *Arundinaria murielae*). Spathe-like, referring to the shape of the leaves. Umbrella Bamboo. NW Himalaya.

Thelocactus thay-lō-*kǎk*-tus *Cactaceae*. From Gk. *thele* (a nipple) and *Cactus* q.v.
  bicolor bi-ko-lor. Two-coloured (the flowers). Glory of Texas. S Texas, N Mexico.
  lophothele lo-*fo*-thay-lay. With crested nipples. N Mexico.
  macdowellii mak-*dowl*-ee-ee (= *Neolloydia macdowellii*). After McDowell, a plant exporter of Mexico. N Mexico.
  setispinus say-tee-*speen*-us (= *Ferocactus setispinus*). With bristle-like spines. S Texas, N Mexico.

Thermopsis ther-*mop*-sis *Leguminosae*. From Gk. *thermos* (a lupin) and -*opsis* indicating resemblance. Perennial herbs.
  caroliniana see *T. villosa*
  lupinoides loo-peen-oi-deez. (= *T. lanceolata*). Like *Lupinus*. Siberia, Alaska.
  mollis mol-lis. Softly hairy. E United States.
  montana mon-*tah*-na. Of mountains. W United States.
  villosa vil-ō-sa (= *T. caroliniana*). With long, soft hair. Carolina Lupin. SE United States.

Thlaspi thlǎ-pee *Cruciferae*. Gk. name for a similar plant. Perennial, alpine herb.
  rotundifolium ro-tund-i-*fo*-lee-um. With round (basal) leaves. Alps.

Three Birds Flying see *Linaria triornithophora*
Thrift see *Armeria*
  Common see *A. maritima*
  Jersey see *A. alliacea*
Throatwort see *Campanula trachelium, Trachelium caeruleum*

Thuja thoo-ya *Cupressaceae*. Gk. name of a juniper. Evergreen conifers.
  koraiensis ko-rie-en-sis. Of Korea.
  occidentalis ok-ki-den-*tah*-lis. Western. Western White Cedar. E N America.
  orientalis see *Platycladus orientalis*
  plicata pli-*kah*-ta. Plaited, the appearance of the shoots. Western Red Cedar. W N America.
  standishii stǎn-*dish*-ee-ee. After John

Thuja (continued)
Standish (1814–75), a nurseryman for whom Robert Fortune introduced it. Japan.

Thujopsis thoo-*yop*-sis *Cupressaceae*. From *Thuja* q.v. and Gk. *-opsis* indicating resemblance. Evergreen conifer.
dolabrata do-lah-*brah*-ta. Hatchet-shaped (the leaves). Japan

Thunbergia thun-*berg*-ee-a *Acanthaceae*. After Carl Peter Thunberg (1743–1828), Dutch physician and botanist who introduced many Japanese plants. Tender climbers.
alata ah-*lah*-ta. Winged (the petioles). Black-eyed Susan. Tropical Africa.
coccinea kok-*kin*-ee-a. Scarlet. India.
fragrans frah-granz. Fragrant. India, Sri Lanka.
grandiflora grǎnd-i-*flō*-ra. Large-flowered. Clock Vine, Blue Trumpet Vine. India.
gregorii gre-*go*-ree-ee. After Dr J. W. Gregory who collected the type specimen. Tropical Africa.
mysorensis mie-sor-*ren*-sis. Of Mysore. Nilgiri Hills, India.
natalensis nǎ-ta-*len*-sis. Of Natal. S Africa.

Thunia thun-ee-a *Orchidaceae*. After Franz A. Graf von Thun (1786–1873). Greenhouse orchids.
alba ǎl-ba. White (the flowers). N India, Burma.
bensoniae ben-*son*-ee-ie. After the wife of Robert Benson (1822–94), who collected orchids in Burma. Burma.

Thymus *tiem*-us *Labiatae*. Gk. name. Perennial herbs and dwarf shrubs. Thyme.
azoricus see *caespitosus*
caespitosus kie-spi-*tō*-sus. (= *T. azoricus. T. micans*). Tufted. Portugal, Spain, Azores.
carnosus kar-*nō*-sus. (= *T. nitidus* hort.). Fleshy. Portugal.
cilicius si-*li*-kee-us. Of Cilicia (S Turkey).
× *citriodorus* kit-ree-o-*dō*-rus. *T. pulegioides* × *T. vulgaris*. Lemon-scented. Lemon Thyme.
doerfleri durf-la-ree. After J. D. Doerfler who discovered it in 1916. Albania.
drucei droos-ee-ee. After George Claridge Druce (1850–1932), English amateur botanist. Europe.
herba-barona her-ba-ba-*ron*-a. The Corsican name. Caraway Thyme. Corsica, Sardinia.
lanuginosus hort. see *T. pseudolanuginosus*

Thymus (continued)
membranaceus mem-bra-*nah*-kee-us. Membranaceous (the bracts). SE Spain.
micans see *T. caespitosus*
nitidus hort. see *T. carnosus*
praecox prie-koks. Early (flowering). SW and C Europe.
pseudolanuginosus soo-dō-lah-new-gi-*nō*-sus. False *T. lanuginosus* (woolly). Cult.
serpyllum ser-*pil*-lum. The L. name for thyme. Wild Thyme. Europe.
vulgaris vul-*gah*-ris. Common. Common Thyme, French Thyme, Garden Thyme. S Europe.

Tiarella tee-a-*rel*-la *Saxifragaceae*. Diminutive of Gk. *tiara* (a small crown) referring to the fruit. Perennial herbs.
cordifolia kor-di-*fo*-lee-a. With heart-shaped leaves. Foam Flower. E N America.
wherryi we-ree-ee. (= *T. collina*). After its discoverer, Edgar Theodore Wherry (born 1885), an American botanist. SE United States.

Tibouchina ti-boo-*chee*-na *Melastomataceae*. From the native name. Tender shrub.
urvilleana ur-vil-ee-*ah*-na. (= *T. semidecandra* hort.). After Jules Sébastian César Dumont d'Urville (1790–1844), a French naval officer. Glory Bush. Brazil.

Tickseed see *Coreopsis*
Tiger Lily see *Lilium lancifolium*
Tiger's Jaws see *Faucaria tigrina*

Tigridia ti-*gri*-dee-a *Iridaceae*. From L. *tigris* (a tiger), the flowers are spotted like the S American tiger (jaguar). Semi-hardy bulbous herbs.
pavonia pah-*vō*-nee-a. Peacock-like. Peacock Tiger Flower.
violacea vee-o-*lah*-kee-a. Violet. Mexico.

Tilia tee-lee-a *Tiliaceae*. The L. name. Lime, Linden. Deciduous tree.
cordata kor-*dah*-ta. Heart-shaped (the leaves). Small-leaved Lime. Europe.
'Euchlora' ew-*klō*-ra. From Gk. *eu* (good) and *chloros* (green) referring to the bright green leaves.
× *europaea* see *T. × vulgaris*
oliveri o-*li*-va-ree. After Oliver. C China.
'Petiolaris' pe-tee-o-*lah*-ris. With a petiole, the long petiole distinguishes it from *T. tomentosa*. Weeping Silver Lime.
platyphyllos plǎ-tee-*fil*-los. Broad-leaved. Broad-leaved Lime. Europe, SW Asia.

Tilia (continued)
'Rubra' *rub*–ra. Red (the young shoots). Red-twigged Lime.
tomentosa tō–men–tō–sa. Hairy (the undersides of the leaves). European White Lime. SE Europe.
vulgaris vul–*gah*–ris (= *T x europaea*) *T. cordata* × *T. platyphyllos* Common. Lime, Common Lime.

Tillandsia ti–*lǎndz*–ee–a *Bromeliaceae*. After Elis Til–Landz (died 1693), Swedish botanist. Tender epiphytic herbs.
cyanea see–*ǎn*–ee–a. Blue (the flowers). Pink Quill. Ecuador.
lindenii lin–den–ee–ee. After J. J. Linden, a Belgian nurseryman. Blue Flowered Torch. Peru.
tenuifolia ten–ew–i–*fo*–lee–a. (= *T. pulchella*). Slender-leaved. S America.
usneoides us–nee–*oi*–deez. Like *Usnea*, a lichen that hangs from trees in a similar way. Spanish Moss. SE United States, C and S America.

Tingiringi Gum see *Eucalyptus glaucescens*

Tithonia tee–*thō*–nee–a *Compositae*. After Tithonus, a beautiful youth and King of Troy. He was loved by Aurora who turned him into a grasshopper. Annual herb.
rotundifloria ro–tun–di–*fo*–lee–a. Round-leaved. Mexican Sunflower. Mexico, C America.

Toad Lily see *Tricyrtis hirta*
Toad Plant see *Orbea variegata*
Toadflax see *Linaria*
  Alpine see *L. alpina*
  Common see *L. vulgaris*
Toadshade see *Trillium sessile*
Tobacco Plant see *Nicotiana tabacum*

Tolmiea tol–*mee*–a *Saxifragaceae*. After Dr William Fraser Tolmie (1830–86), Scottish physician and botanist. Perennial herb.
menziesii men–*zeez*–ee–ee. After Menzies, see *Menziesia*. Piggy-back Plant. W N America.

Tomato see *Lycopersicum esculentum*
Toothache Tree see *Zanthoxylum americanum*
Torch Lily see *Kniphofia*

Torenia to–reen–ee–a *Scrophulariaceae*. After the Rev. Olof Toren (1718–53). Annual herbs.
baillonii see *T. flava*

Torenia (continued)
flava fla–va (= *T. baillonii*). Yellow (the flowers).
fournieri four–nee–*e*–ree. After Eugène Pièrre Nicolas Fournier (1834–84), French botanist. Wishbone Flower. Vietnam.

Torreya to–*ree*–a *Taxaceae*. After John Torrey (1796–1873), American botanist. Evergreen trees.
californica kǎ–li–*forn*–i–ka. Of California. California Nutmeg.
nucifera new–*ki*–fe–ra. Nut-bearing. Japan.

Tortoise Plant see *Dioscorea elephantipes*
Touch-me-not see *Impatiens noli-tangere*

Tovara tō–*vah*–ra *Polygonaceae*. After Simon a Tovar, 16th-century Spanish physician. Perennial herb.
virginiana vir–jin–ee–*ah*–na. Of Virginia. E N America.

Townsendia town–*zend*–ee–a *Compositae*. After David Townsend (1787–1858), American botanist. Perennial herbs.
exscapa eks–*skah*–pa. Without a scape. W N America.
formosa for–*mō*–sa. Beautiful. SW United States.
parryi pa–ree–ee. After Charles Christopher Parry (1823–90), American botanical explorer. W N America.

Trachelium tra–kay–lcc–um *Campanulaceae*. From Gk. *trachelos* (a neck) referring to supposed medicinal properties. Perennial herbs.
caeruleum kie–*ru*–lee–um. Deep blue (the flowers). Throatwort. Mediterranean region.

Trachelospermum tra–kay–lō–*sperm*–um *Apocynaceae*. From Gk. *trachelos* (a neck) and *sperma* (a seed) referring to the narrow seeds. Evergreen climbers.
asiaticum ah–see–*ah*–ti–kum. Asian. Japan, Korea.
jasminoides yas–min–*oi*–deez. Jasmine-like (the fragrant flowers). China, Japan.
'Japonicum' ja–*pon*–i–kum. (= *T. majus*. hort.). Of Japan.

Trachycarpus trǎ–kee–*kar*–pus *Palmae*. From Gk. *trachys* (rough) and *karpos* (a fruit). The hardiest palm.
fortunei for–*tewn*–ee–ee. (= *Chamaerops excelsa* hort.). After Robert Fortune who

Trachycarpus (continued)
sent plants to Kew. See *Fortunella*. Chusan
Palm, Fan Palm. C and S China.

Trachymene trǎ-kee-*may*-nee *Umbelliferae*.
From Gk. *trachys* (rough) and *meninx* (a
membrane) referring to the fruit. Annual
herb.
  caerulea kie-*ru*-lee-a. (= *Didiscus caeruleus*).
  Blue (the flowers). Blue Lace Flower.
  Australia.

Trachystemon trǎ-kee-*stay*-mon *Boraginaceae*.
From Gk. *trachys* (rough) and *stemon* (a
stamen) referring to the rough stamens of
one species. Perennial herb.
  orientalis o-ree-en-*tah*-lis. Eastern. SE
  Europe, W Asia.

Tradescantia trǎ-des-*kǎnt*-ee-a
Commelinaceae. After John Tradescant
(1608–62). Tender and hardy perennial
herbs.
  × *andersoniana* ǎn-der-son-ee-*ah*-na. (= *T.
  virginiana* hort.). After Anderson.
  blossfeldiana see *T. cerinthoides*
  cerinthoides ser-inth-*oy*-deez (= *T.
  blossfeldiana*). Resembling *Cerinthe*, or
  Honeywort. Flowering Inch plant. SE
  Brazil.
  fluminensis floo-min-*en*-sis. Of Rio de
  Janeiro. Speedy Jenny. Brazil.
    albiflora ǎl-bi-*flō*-ra. White-flowered.
    Inch Plant, Wandering Jew. S America.
  navicularis see *Callisia navicularis*
  pallida pǎ-li-da (= *Setcreasia pallida*). Pale
  (the flowers of some forms). Purple Heart.
  Mexico.
  sillamontana si-la-mon-*tah*-na. Of Cerro de
  la Silla, where it was found. White Velvet.
  Mexico.
  spathacea spa-tha-*kee*-a (= *Rhoeo spathacea*).
  Spathe-like (the large bracts). Boat Lily.
  W Indies, C America.
  zebrina ze-*breen*-a (= *Zebrina pendula*). The
  leaves are striped. Mexico.

Tragopogon trǎ-go-*pō*-gon *Compositae*. From
Gk. *tragos* (a goat) and *pogon* (a beard) referring
to the pappus. Biennial herb.
  porrifolius po-ri-*fo*-lee-us. With leaves like
  Allium porrum. Salsify. S Europe.

Trailing Arbutus see *Epigaea repens*
Trailing Watermelon Begonia see
Pellionia repens
Transvaal Daisy see *Gerbera jamesonii*

Trapa trǎ-pa *Trapaceae*. From L. *calcitrappa*

Trapa (continued)
(caltrop, a four-pointed weapon) referring
to the four-horned fruit of *T. natans*. Aquatic
perennials.
  bicornis bi-*korn*-is. Two-horned (the fruit).
  natans nǎ-tǎnz. Floating. Water Chestnut,
  Water Caltrop. Europe, Asia, Africa.

Traveller's Joy see *Clematis vitalba*
Treasure Flower see *Gazania rigens*
Tree of Heaven see *Ailanthus altissima*

Trichocereus see *Echinopsis*

Tricyrtis tri-*kur*-tis *Liliaceae* (*Tricyrtidaceae*).
From Gk. *tri* (three) and *kyrtos* (humped)
referring to the swollen bases of the three
outer petals. Perennial herbs.
  formosana for-mō-*sah*-na. (= *T. stolonifera*).
  Of Taiwan (Formosa).
  hirta hir-ta. Hairy. Toad Lily. Japan.
  macrantha ma-*krǎnth*-a. Large-flowered.
  Japan.
  macropoda ma-*kro*-po-da. With a large
  stalk. Japan.

Trifolium tri-*fo*-lee-um *Leguminosae*. The L.
name from *tri*- (three) and *folium* (a leaf),
the leaves have three leaflets. Perennial herb.
  repens ree-penz. Creeping. White Clover.
  Europe, Asia, N Africa.
    'Purpurascens Quadrifolium' pur-pew-
    rǎs-enz-*kwod*-ri-*fo*-lee-um. Purplish and
    four-leaved, the leaves have four
    bronze-purple leaflets.

Trillium tri-lee-um *Liliaceae* (*Trilliaceae*).
From L. *tri*- (three) they have three leaves
and the floral parts are in threes. Herbaceous
perennials.
  cernuum *ker*-new-um. Nodding (the
  flowers). E N America.
  erectum e-*rek*-tum. Erect (the flowers).
  Stinking Benjamin. E N America.
  grandiflorum grǎnd-i-*flō*-rum. Large-
  flowered. Wake Robin. E N America.
  ovatum ō-*vah*-tum. Ovate (the leaves). W
  N America.
  rivale ree-*vah*-lee. Growing by streams. W
  N America.
  sessile se-si-lee. Sessile (the flowers).
  Toadshade. N America.
  undulatum un-dew-*lah*-tum. Wavy-edged
  (the petals). Painted Wood-lily. E N
  America.

Triplet Lily see *Brodiaea coronaria*

Triteleia tri-te-*lay*-a *Liliaceae* (*Tricyrtidaceae*).

Triteleia (continued)
From Gk. *tri* (three) and *teleios* (perfect), the floral parts are in threes. Cormous perennial herbs.
 bridgesii bri-*jez*-ee-ee. After Thomas Bridges (1807–65), who collected in S America and California. W N America.
 crocea krō-kee-a. Saffron yellow. W N America.
 grandiflora grǎnd-i-*flō*-ra. Large-flowered. W N America.
 hyacinthina hee-a-kinth-*ee*-na. Hyacinth-coloured. W N America.
 ixioides iks-ee-*oi*-deez. Like *Ixia*. C California.
 laxa lǎks-a. (= *Brodiaea laxa*). Loose (the inflorescence). W N America.
 peduncularis pe-dunk-ew-*lah*-ris. With a flower stalk, referring to the long scape. California.
 x *tubergeniana* tew-ber-gen-ee-*ah*-na. *T. laxa* x *T. peduncularis*. After the van Tubergen nursery, where it was raised.
 uniflora see *Ipheion uniflorum*

Tritonia tri-*tō*-nee-a *Iridaceae*. From *Triton* (a weather cock), the stamens point in different directions. Semi-hardy or tender, perennial herbs. S Africa.
 crocata krō-*kah*-ta. Saffron yellow.
 x *crocosmiiflora* see *Crocosmia* x *crocosmiiflora*

Trochodendron tro-ko-*den*-dron *Trochodendraceae*. From Gk. *trochos* (a wheel) and *dendron* (a tree) referring to the arrangement of the stamens. Evergreen tree.
 aralioides a-rah-lee-*oi*-deez. Like *Aralia*. Japan, Taiwan, Korea.

Trollius tro-lee-us *Ranunculaceae*. From Swiss-German *Trollblume* (Globeflower). Perennial herbs. Globeflower.
 acaulis a-*kaw*-lis. Stemless. Himalaya.
 x *cultorum* kul-*to*-rum. Cultivated.
 europaeus oy-rō-*pie*-us. European. Europe, N America.
 pumilus pew-mi-lus. Dwarf. Himalaya, W China.

Tropaeolum tro-*pie*-o-lum *Tropaeolaceae*. From Gk. *tropaion* (a trophy), originally used for the trunk of a tree on which were fixed the shields and helmets of a defeated enemy. Linnaeus was reminded of this on seeing a plant growing on a post, the leaves representing the shields and the flowers the helmets. Hardy and semi-hardy, annual and perennial herbs.

Tropaeolum (continued)
 azureum a-*zew*-ree-um. Blue (the flowers). C Chile.
 majus mah-yus. Larger. Nasturtium. S America.
 peltophorum pel-*to*-fo-rum. Shield bearing, referring to the leaves. Andes.
 peregrinum pe-re-*green*-um. (= *T. canariense*). Foreign. Canary Creeper. Andes.
 polyphyllum po-lee-*fil*-um. With many leaves (leaflets). Chile.
 speciosum spe-kee-*ō*-sum. Showy. Flame Creeper. S Chile.
 tuberosum tew-be-*rō*-sum. Tuberous. N Andes.

Trout Lily see *Erythronium revolutum*

Trudelia trew-del-ee-a *Orchidaceae*. Derivation found. Epiphytic orchids formerly included in *Vanda*. India, Thailand.
 cristata kris-*tah*-ta (= *Vanda cristata*). Crested, the mid-lobe of the lip. Himalaya, Tibet, Assam.

Trumpet Vine see *Campsis radicans*

Tsuga tsoo-ga *Pinaceae*. From the Japanese name. Evergreen trees. Hemlock.
 canadensis kǎn-a-*den*-sis. Of E N America. Eastern Hemlock.
 heterophylla he-te-rō-*fil*-la. With variable leaves. Western Hemlock. W N America.
 mertensiana mer-tenz-ee-*ah*-na. After Karl Heinrich Mertens (1795–1830), a German botanist who discovered it in 1827. Mountain Hemlock. W N America.

Tuberose see *Polianthes tuberosa*
Tulip Tree see *Liriodendron tulipifera*
 Chinese see *L. chinense*

Tulipa tew-li-pa *Liliaceae*. From Turkish *tulband* (a turban), originally used in a descriptive sense but was thought to be a name. Bulbous perennial herbs. Tulip.
 acuminata a-kew-mi-*nah*-ta. Long-pointed (the petals). Horned Tulip. Cult.
 aucheriana ow-ka-ree-*ah*-na. After P. M. R. Aucher-Eloy (1792–1838). W Iran.
 batalinii bah-ta-*lin*-ee-ee. After A. F. Batalin (1847–96), A Russian botanist. C Asia.
 biflora bi-*flō*-ra. Two-flowered. SE Europe to C Asia and Siberia.
 clusiana klooz-ee-*ah*-na. After Clusius, see

Tulipa (continued)
  Gentiana clusii. Lady Tulip. Iran,
  Himalaya, Tibet.
    stellata ste-*lah*-ta (= *T. stellata*). Star-like
    (the flowers).
  eichleri see *T. undulatifolia*
  fosteriana fos-ta-ree-*ah*-na. After Foster. C
  Asia.
  greigii greeg-ee-ee. After General Greig,
  President of the Imperial Russian
  Horticultural Union. C Asia.
  humilis hu-mi-lis. (= *T. pulchella*). Low-
  growing. E Turkey, N Iraq, NW Iran.
    kaufmanniana kowf-măn-ee-ah-na. After
    General von Kaufmann, a governor-
    general of the region in which it was
    found. Water-lily Tulip. C Asia.
  linifolia leen-i-*fo*-lee-a. *Linum*-leaved. C
  Asia.
  orphanidea or-fa-*nid*-ee-a. After Dr
  Orphanides, professor of botany at
  Athens. SE Europe, W Turkey.
  praecox prie-koks. Early (flowering). ?C
  Asia.
  praestans prie-stahnz. Distinguished. C
  Asia.
  stellata see *T. clusiana stellata*
  sylvestris sil-*ves*-tris. Of woods. Italy, Sicily,
  Sardinia.
  tarda tar-da. Late. C Asia.
  tubergeniana tew-ber-gen-*ah*-na. After van
  Tubergen, the Dutch bulb nursery. C
  Asia.
  turkestanica tur-kes-*tahn*-i-ka. Of
  Turkestan. C Asia, NW China.
  undulatifolia un-dew-*lah*-ti-fō-lee-a (= *T.
  eichleri*). The leaf margins are undulate.
  Balkans, Greece, C Asia.
  urumiensis u-roo-mee-*en*-sis. Of Lake
  Urumia. NW Iran.
  wilsoniana wil-son-ee-*ah*-na. After Mr G.
  F. Wilson. C Asia.

Tunica saxifraga see *Petrorhagia saxifraga*
Tupelo see *Nyssa sylvatica*
Turnip see *Brassica rapa*
Turk's Cap Lily see *Lilium martagon*
Turtle Head see *Chelone*
Tutsan see *Hypericum androsaemum*

Tweedia twee-dee-a *Asclepiadaceae*. For J.
Tweedie (1775–1862). A single species of
tender herbaceous climber. S Brazil,
Uruguay.
  caerulea kie-*ru*-lee-a. (= *Oxypetalum
  caeruleum*). Deep blue (the flowers). Brazil,
  Uruguay.

Twin Flower see *Linnaea borealis*

Twinspur see *Diascia barberiae*

Tylecodon ty-le-*cō*-don *Crassulaceae*. An
anagram of *Cotyledon*. Succulent perennial
subshrubs. S Africa, Namibia.
  paniculatus pan-ik-yew-*lah*-tus (=
  Cotyledon paniculata). With flowers in
  panicles. S Africa.
  reticulatus ray-tik-yew-*lah*-tus (= *Cotyledon
  reticulatus*). Net-veined. S Africa.

Typha tee-fa *Typhaceae*. The Gk. name.
Aquatic or bog-garden perennial herbs.
Reed-mace.
  angustifolia ang-gus-ti-*fo*-lee-a. Narrow-
  leaved. Europe, Asia, N America.
  latifolia lah-tee-*fo*-lee-a. Broad-leaved.
  Europe, Asia, N Amercia.
  minima mi-ni-ma. Smaller. Europe.

U

Ugni ug-nee *Myrtaceae*. From the native
name. Shrubs formerly included in *Myrtus*.
Americas.
  molinae mo-lee-nee (= *Myrtus ugni*). After
  Giovanni Ignazio Molina (1740–1829),
  botanist. Chile, W Argentina.

Ulex ew-leks *Leguminosae (Papilionoideae)*.
The L. name. Spiny shrubs. Furze, Gorse,
Whin.
  europaeus oy-rō-*pie*-us. European.
  Common Gorse. W and C Europe.
  gallii gă-lee-ee. Of France. W Europe.
  minor mi-nor. Smaller. Dwarf Gorse. W
  Europe.

Ulmus ul-mus *Ulmaceae*. The L. name.
Deciduous trees. Elm.
  angustifolia ang-gus-ti-*fo*-lee-a. Narrow-
  leaved. Goodyer's Elm. S England, N
  France.
  carpinifolia kar-peen-i-*fo*-lee-a. *Carpinus*-
  leaved. Smooth Elm. Europe, N Africa,
  SW Asia.
  cornubiensis kor-new-bee-*en*-sis. Of
  Cornwall. Cornish Elm. SW England.
  glabra glăb-ra. Smooth (the bark). Wych
  Elm. Europe, W Asia.
    'Camperdownii' kăm-per-*down*-ee-ee.
    Of Camperdown House, Dundee,
    where it was raised. Camperdown Elm.
    'Exoniensis' eks-ō-nee-*en*-sis. Of
    Exeter where it was raised. Exeter Elm.

Ulmus (continued)
× *hollandica* ho-*lǎnd*-i-ka. Of Holland.
Dutch Elm.
  'Belgica' *bel*-gi-ka. Belgian. Belgian
  Elm.
procera prō-*kay*-ra. Tall. English Elm. S and
C England.
'Sarniensis' sar-nee-*en*-sis. Of Guernsey.
Jersey Elm, Wheatley Elm.

Umbellularia um-bel-ew-*lah*-ree-a *Lauraceae*.
From L. *umbella* (an umbel) referring to the
inflorescence. Evergreen tree.
  californica kǎ-li-*forn*-i-ka. Of California.
  California Laurel. W N America.

Umbrella Pine see *Sciadopitys verticillata*
Umbrella Plant see *Cyperus involucratus,
Darmera peltatum*
Umbrella Tree see *Magnolia tripetala*
  Ear-leaved see *M. fraseri*
Unicorn Plant see *Proboscidea louisianica*
Unicorn Root see *Aletris farinosa*

Urceolina ur-kee-o-li-na *Amaryllidaceae*.
From L. *urceolus* (a small pitcher) referring to
the shape of the flowers. Bulbous herb.
  latifolia lat-i-*fō*-lee-a. Broad-leaved.
  Peruvian Andes.

Urginea ur-*gin*-ee-a *Liliaceae (Hyacinthaceae)*.
From *Beni Urgin*, the name of an Arab tribe
in Algeria. Bulbous herb.
  maritima ma-*ri*-ti-ma. Growing near the
  sea. Sea Squill. Mediterranean region. W
  Asia.

Urn Gum see *Eucalyptus urnigera*
Urn Plant see *Aechmea fasciata*

Ursinia ur-*si*-nee-a *Compositae*. After
Johannes Heinrich Ursinus (1608–67), a
botanical author. Annual herbs. S Africa.
  anethoides a-nay-*thoi*-deez. Like *Anethum*.
  anthemoides ǎn-them-*oi*-deez. (= *U.
  pulchra*). Like *Anthemis*.
  cakilofolia ka-ki-li-*fo*-lee-a. With leaves like
  Cakile.

Utricularia ew-trik-ew-*lah*-ree-a
Lentibulariaceae. From L. *utriculus* (a small
bottle) referring to the bladders which trap
small, aquatic animals. Aquatic, carnivorous
herb. Bladderwort.
  vulgaris vul-*gah*-ris. Common. Europe,
  Asia, N America.

Uvularia oo-vew-*lah*-ree-a *Liliaceae
(Uvulariaceae)*. From *uvula* of anatomy, a

Uvularia (continued)
hanging structure in the throat, referring to
the hanging flowers. Perennial herbs.
Bellwort, Merrybells. N America.
  grandiflora grǎ-nd-i-*flō*-ra. Large-flowered.
  perfoliata per-fo-lee-*ah*-ta. With the base
  of the leaves pierced by the stem.
  sessilofolia se-si-li-*fo*-lee-a. With sessile
  leaves.

V

Vaccaria va-kah-*ree*-a *Caryophyllaceae*. From
L. *vacca* (a cow), it is said to have been used
for forage. Annual herb.
  hispanica his-*pan*-ik-a (= *V. pyramidata*) Of
  Spain. Cow Herb. S and C Europe.

Vaccinium va-*keen*-ee-um *Ericaceae*. A L.
name, variously stated to apply to either *V.
myrtillus* or a hyacinth. Deciduous and
evergreen shrubs.
  arctostaphylos ark-tō-*stǎ*-fi-los. Gk. name
  for a plant from *arktos* (a bear) and *staphyle*
  (a bunch of grapes) Caucasian
  Whortleberry. Caucasus, SE Europe.
  corymbosum ko-rim-*bō*-sum. With flowers
  in corymbs. Highbush Blueberry. E N
  America.
  delavayi de-la-*vay*-ee. After Delavay who
  discovered it, see *Abies delavayi*. SW
  China, Burma.
  deliciosum day-li-kee-*ō*-sum. Delicious (the
  fruits). NW North America.
  glaucoalbum glow-kō-*ǎl*-bum. Glaucous-
  white (the undersides of the leaves).
  Himalaya.
  macrocarpon mǎk-rō-*kar*-pon (= *Oxycoccus
  macrocarpus*). With large fruit. American
  Cranberry. E N America.
  myrtillus mur-ti-lus. A small myrtle.
  Bilberry, Whortleberry. Europe.
  nummularia num-ew-*lah*-ree-a. Coin-
  shaped (the leaves). Himalaya.
  ovatum ō-*vah*-tum. Ovate (the leaves). Box
  Blueberry. W N America.
  oxycoccus oks-ee-*kok*-us (= *Oxycoccus
  palustris*). With acid berries. Small
  cranberry. Europe, N Asia, N America.
  vitis-idaea vee-tis-ee-*die*-a. Grape of Mt Ida.
  Cowberry Europe, Asia, N America.

Valerian, Red see *Centranthus ruber*

Valeriana va-le-ree-*ah*-na *Valerianaceae*. The
medieval L. name, probably from L. *valere*

Valeriana (continued)
(to be healthy), referring to medicinal properties. Perennial herbs.
  arizonica ă-ri-*zo*-ni-ka. Of Arizona. SW United States, Mexico.
  montana mon-*tah*-na. Of mountains. Europe.
  officinalis o-fi-ki-*nah*-lis. Sold as a herb. Common Valerian. Europe, W Asia.
    sambucifolia săm-bew-ki-*fo*-lee-a. (= *V. sambucifolia*). *Sambucus*-leaved.
  phu foo. From Gk. *phou* (a kind of valerian). N Anatolia.
  saxatalis săks-*ah*-ti-lis. Growing among rocks. Alps, SE Europe.
  supina su-*peen*-a. Prostrate. Alps.

Valerianella va-le-ree-a-*nel*-la *Valerianaceae*. Diminutive of *Valeriana* q.v., a related genus. Annual herb.
  locusta lo-*kus*-ta. L. name for locust. Corn Salad, Lamb's Lettuce. Europe, N Africa, W Asia.

Vallisneria vă-lis-*ne*-ree-a *Hydrocharitaceae*. After Antonio Vallisnieri de Vallisnera (1661–1730). Aquarium plants.
  americana a-me-rik-*ah*-na (= *V. asiatica*, *V. gigantea*). Of America. E US, E Asia to Australia.
  asiatica see *V. americana*
  gigantea see *V. americana*
  spiralis spee-*rah*-lis. Spiralled (the female scape in fruit). S Europe, W Asia.

Vallota see *Cyrtanthus*

Vancouveria vang-koo-*ve*-ree-a *Berberidaceae*. After Captain George Vancouver (1758–98), British explorer. Perennial herbs. W United States.
  hexandra heks-*ăn*-dra. With six stamens.
  planipetala plah-ni-*pe-ta-la*. With flat petals.

Vanda văn-da *Orchidaceae*. From the Hindi name. Greenhouse orchids.
  amesiana see *Holcoglossum*
  coerulea kie-*ru*-lee-a. Deep blue. Himalaya, SE Asia.
  coerulescens kie-ru-*les*-enz. Bluish. Burma.
  crista see *Trudelia cristata*
  sanderiana see *Euanthe sanderiana*
  teres see *Papilionanthe teres*
  tricolor tri-ko-lor. Three-coloured. Java.

Veltheimia vel-*tie*-mee-a *Liliaceae* (*Hyacinthaceae*). After August Ferdinand von Veltheim of Brunswick (1741–1801). Tender, bulbous herbs. S Africa.

Veltheimia (continued)
  bracteata brak-tee-*ay*-ta (= *V. viridiflora*). With bracts. E Cape.
  capensis ka-*pen*-sis. Of the Cape of Good Hope.
  viridiflora see *V. bracteata*

Velvet Leaf see *Kalanchoe beharensis*
Velvet Plant see *Gynura aurantiaca*
  Trailing see *Ruellia makoyana*

× *Venidio–Arctotis hybrida*. A hybrid between *Arctotis fastuosa* and *A. venusta* q.v.

Venidium vay-ni-dee-um *Compositae*. From L. *vena* (a vein) referring to the ribbed fruits. Annual herbs. S Africa.
  decurrens day-*ku*-renz. (= *V. calendulaceum*). With the petioles running into the stem.
  fastuosum see *Arctotis fastuosa*

Venus's Fly Trap see *Dionaea muscipula*
Venus's Looking Glass see *Legousia speculum-veneris*

Veratrum vay-rah-trum *Liliaceae* (*Melianthaceae*). The L. name. Perennial herbs. False Hellebore.
  album ăl-bum. White. Europe, N Asia.
  nigrum nig-rum. Black. Europe, Asia.
  viride vi-ri-dee. Green. N America.

Verbascum ver-*băs*-kum *Scrophulariaceae*. The L. name. Biennial and perennial herbs. Mullein.
  arcturus ark-*tew*-rus. (= *Celsia arcturus*). From Gk. *arktos* (a bear) and *oura* (a tail). Cretan Bear's Tail. Crete.
  blattaria bla-*tah*-ree-a. The L. name from *blatta* (a moth). Moth Mullein. Europe, Asia.
  bombyciferum bom-bi-*ki*-fe-rum. Silky. W Asia.
  chaixii shay-zee-eel. After Dominique Chaix (1730–99), a French botanist. Nettle-leaved Mullein. S Europe.
  creticum kray-ti-kum. (= *Celsia cretica*). Of Crete. Cretan Mullein. Mediterranean region.
  densiflorum den-si-*flo*-rum. (= *V. thapsiforme*). Densely flowerd. Europe.
  dumulosum dew-mew-*lo*-sum. Like a small shrub. SW Turkey.
  longifolium long-gi-*fo*-lee-um. Long-leaved. Italy, SE Europe.
  nigrum nig-rum. Black. Dark Mullein. Europe.
  olympicum o-*lim*-pi-kum. Of Mt Olympus. Turkey.

Verbascum (continued)
phlomoides flo-*moi*-deez. Like *Phlomis*. C
and S Europe.
phoeniceum foy-*nee*-kee-um. Purple-red.
Purple Mullein. SE Europe, W Asia.
spinosum spee-*nō*-sum. Spiny. Crete.
thapsiforme see *V. densiflorum*
thapsus thǎp-sus. A town in Sicily, now
Magnisi. Aaron's Rod. Europe, Asia.
virgatum vir-*gah*-tum. Wand-like. W
Europe.

Verbena ver-*bay*-na *Verbenaceae*. L. name
for the foliage of ceremonial and
medicinal plants. Annual and perennial
herbs. Vervain.
bonariensis bo-nah-ree-*en*-sis. Of Buenos
Aires. S America.
× *hybrida* hi-bri-da. Hybrid. Rose Vervain.
officinalis o-fi-ki-*nah*-lis. Sold as a herb.
Common Vervain. Europe.
peruviana pe-roo-vee-*ah*-na. Of Peru.
Argentina, Brazil.
rigida ri-gi-da. Rigid. Lilac Vervain.
Brazil, Argentina.

Vernonia ver-*non*-ee-a *Compositae*. After
William Vernon (died c 1711), English
botanist who collected in Maryland.
Perennial herbs.
altissima see *V. gigantea*
arkansana ar-kan-*sah*-na (= *V. crinita*). Of
Arkansas. SC US.
crinita see *V. arkansana*
gigantea gi-*gǎn*-tee-a (= *V. altissima*). Very
large. SE US.
noveboracensis no-vee-be-ra-*ken*-sis. Of
New York. United States.

Veronica ve-ro-ni-ka *Scrophulariaceae*. After
St Veronica. Perennial herbs. Speedwell. See
also *Hebe*.
allionii ǎ-lee-ō-nee-ee. After Carlo Allioni
(1705–1804), Italian botanist. SW Alps.
armena ar-*me*-na. Of Armenia.
austriaca ow-stree-*ah*-ka. Austrian. C and
E Europe.
  teucrium toyk-ree-um. (= *V. teucrium*).
  From the resemblance to *Teucrium*.
  Europe.
bombycina bom-bi-*keen*-a. Silky. Lebanon.
cinerea ki-*ne*-ree-a. Grey. W Asia.
exaltata see *V. longifolia*
filiformis fee-li-*form*-is. Thread-like (the
pedicels). W Asia.
fruticans froo-ti-kǎnz. Shrubby. Europe.
gentianoides gen-tewe-a-*noi*-deez.
Gentiana-like. W Asia, Caucasus.
incana in-*kah*-na. Grey-hairy.

Veronica (continued)
longifolia long-gi-*fo*-lee-a. (= *V. exaltata*).
Long-leaved. Europe, Asia.
pectinata pek-ti-*nah*-ta. Comb-like (the
leaves). W Asia, SE Europe.
prostrata pros-*trah*-ta. Prostrate. Europe.
saturejoides sǎt-ew-ray-*oi*-deez. Like
Satureja. Balkan peninsula.
spicata spee-*kah*-ta. With flowers in spikes,
Europe, Asia.
teucrium see *V. austriaca teucrium*
virginica see *Veronicastrum virginicum*

Veronicastrum ve-ro-ni-*kǎs*-trum
Scrophulariaceae. From *Veronica* q.v. and L. -
aster (somewhat like or an inferior sort of).
Perennial herb.
virginicum vir-*jin*-i-kum. (= *Veronica
virginica*). Of Virginia. EN America.

Vervain see *Verbena*
  Common see *V. officinalis*
  Lilac see *V. rigida*
  Rose see *V.* × *hybrida*

Viburnum vee-*bur*-num *Caprifoliaceae*. The L.
name for *V. lantana*. Deciduous and
evergreen shrubs.
× *bodnantense* bod-nant-*en*-see. *V. farreri* ×
V. grandiflorum. Of Bodnant where the
most popular forms were raised.
× *burkwoodii* burk-*wud*-ee-ee. *V. carlesii* ×
V. utile. After Albert Burkwood of
Burkwood and Skipwith, the raisers.
carlesii karlz-ee-ee. After W. R. Carles who
collected the type specimen. Korea,
Japan.
cassinoides kǎ-si-*noi*-deez. Like *Ilex cassine*.
Withe-rod. E N America.
cinnamomifolium kin-a-mō-mi-*fo*-lee-um.
Cinnamomum-leaved. W Sichuan.
davidii dǎ-*vid*-ee-ee. After David, see
Davidia. China.
farreri fǎ-ra-ree. (= *V. fragrans*). After
Reginald John Farrer (1880–1920),
horticultural writer who collected in
China, Burma and the Alps. Japan,
China.
fragrans see *V. farreri*
× *globosum* glo-*bō*-sum. *V. calvum* × *V.
davidii*. Spherical (the habit).
grandiflorum grǎnd-i-*flō*-rum. Large-
flowered.
henryi hen-ree-ee. After Henry, who
discovered it in 1887, see *Illicium henryi*.
C China.
× *hillieri* hi-lee-a-ree. *V. erubescens* × *V.
henryi*. After Hillier's who raised it.

Viburnum (continued)
× *juddii jŭd*-ee-ee. *V. bitchiuense* × *V. carlessi.*
After the raiser William H. Judd.
lantana lan-*tah*-na. An old name for
Viburnum. Wayfaring Tree. Europe, N
Africa, W Asia, Caucasus.
macrocephalum măk-ro-*kef*-a-lum. Large-
headed. China.
opulus op-ew-lus. L. name for a kind of
maple. Guelder Rose. Europe, N Africa,
W Asia, Caucasus.
    'Xanthocarpum' zănth-ō-*kar*-pum.
    Yellow-fruited.
plicatum pli-*kah*-tum. Pleated (the leaves).
Japan, China.
    'Mariesii' ma-*reez*-ee-ee. After Maries
    who introduced it, see *Davallia mariesii.*
pragense prah-*gen*-see. *V. rhytidophyllum* ×
V. utile. Of Prague where it was raised.
× *rhytidophylloides* ri-ti-do-fi-*loi*-deez. *V.
lantana* × *V. rhytidophyllum*,. Like *V.
rhytidophyllum*
rhytidophyllum ri-ti-*do*-fi-lum. With
wrinkled leaves. China.
tinus teen-us. The L. name. Laurustinus.
Mediterranean region.

Vicai *vi*-kee-a *Leguminosae (Papilionoideae).* L.
name for vetch. Annual herb.
faba fă-ba. The L. name. Broad Bean. Cult.

Victoria vik-*tor*-ree-a *Nymphaeaceae.* After
Queen Victoria. Tender, aquatic perennial
herbs.
amazonica ă-ma-*zon*-i-ka. (= *V. regia*). Of
the Amazon. Royal Water Lily.
cruziana krooz-ee-*ah*-na. Of Santa Cruz.
Santa Cruz Water Lily. S America.

Vinca *ving*-ka *Apocynaceae.* From the L. name
vinca pervinca from *vincio* to bind) referring
to the shoots. Evergreen sub-shrubs.
Periwinkle.
major mah-yor. Larger. Greater
Periwinkle. Mediterranean region.
minor mi-nor. Smaller. Lesser Periwinkle.
Europe, W Asia, Caucasus.
rosea see *Catharanthus roseus*

Viola vee-o-la *Violaceae.* L. name for several
scented flowers. Annual and perennial herbs.
Pansy, Violet.
aetolica ie-*to*-lika. Of Aitolia, Greece.
Balkan peninsula.
bertolonii ber-to-*lōn*-ee-ee. After
Bertoloni, see *Bertolonia.* S Europe.
biflora bi-*flō*-ra. Two-flowered. Europe to
the Himalaya, China, Japan, W N
America.

Viola (continued)
cornuta kor-*new*-ta. Horned, referring to
the long spur. Horned Violet. Pyrenees.
cucullata see *V. obliqua*
gracilis gră-ki-lis. Graceful. Olympian
Violet. Balkan peninsula, W Asia.
hederacea he-de-rah-kee-a. Like *Hedera* (the
leaves). Australian Violet. Australia.
labradorica lăb-ra-*do*-ri-ka. Of Labrador. N
America.
obliqua ob-*li*-kwa (= *V. cucullata*).
Unequal-sided (the petals). Marsh Violet
N America.
odorata o-dō-*rah*-ta. Scented. Sweet Violet.
Europe, N Africa, Asia.
pedata pe-*dah*-ta. Like a bird's foot (the
leaves). Bird's-foot Violet. E United
States.
rupestris roo-*pes*-tris. Growing among
rocks. Europe.
saxatilis see *V. tricolor subalpina*
septentrionalis sep-ten-tree-ō-*nah*-lis.
Northern. E N America.
sororia so-*rō*-ree-a. Sisterly (i.e. resembling
another species). E N America.
tricolor tri-ci-lor. Three-coloured.
Heartsease. Europe.
    subalpina sub-ăl-*peen*-a. (= *V. saxatilis*).
    Sub-alpine. N Spain to the Crimea.
× *wittrockiana* vit-rok-ee-*ah*-na. After
Professor Veit Brecher Wittrock
(1839–1914), author of a history of the
cultivated pansy. Garden Pansy.

Violet see *Viola*
    Australian see *V. hederacea*
    Bird's-foot see *V. pedata*
    Horned see *V. cornuta*
    Marsh see *V. obliqua*
    Olympian see *V. gracilis*
    Sweet see *V. odorata*
Violet Cress se *Ionopsidium acaule*
Viper's Bugloss see *Echium vulgare*
Virginia Creeper see *Parthenocissus
quinquefolia*
Virginia Cowslip see *Mertensia maritima*
Viscaria alpina see *Lychnis alpina*

Viscum *vis*-kum *Viscaceae.* The L. name.
Evergreen semi-parasitic shrub.
album ăl-bum. White (the fruits).
Mistletoe. Europe, N Africa, temperate
Asia.

Vitaliana vi-tă-lee-*ah*-na *Primulaceae.* After
Vitaliano Donati (1717–62). Perennial herb.
primuliflora preem-ew-li-*flō*-ra. (=
Douglasia vitaliana). *Primula*-flowered. SE
Alps.

Vitex *vi*-teks *Verbenaceae*. The L. name for *V. agnus-castus*. Deciduous shrubs.
  agnus-castus ăn-yus-kăs-tus. An old name for this plant, from *agnos* the Gk. name and L. *castus*, both meaning chaste. Chaste Tree. Mediterranean region to C Asia.
  negundo ne-*gun*-do. The native name. S Asia.

Vitis *vee*-tis *Vitaceae*. L. name for the grape vine. Deciduous, woody climbers.
  amurensis am-ew-*ren*-sis. Of the Amur region. NE Asia.
  coignetiae kwūn-*yay*-tee-ie. After Mme Coignet, who collected seeds of it in its native Japan.
  labrusca lah-*brus*-ka. L. name for the wild grape vine. Northern Fox Grape. E N America.
  vinifera veen-*i*-fe-ra. Wine-bearing. Common Grape Vine. W Asia.
    'Apiifolia' ă-pee-i-*fo*-lee-a. *Apium*-leaved, *apium* being the L. for celery and parsley. Parsley Vine.

Vriesia *vreez*-ee-a *Bromeliaceae*. After Willem Hendrick de Vriese (1806–62), a Dutch professor of botany. Tender, evergreen herbs.
  carinata kă-ri-*nah*-ta. Keeled (the bracts). Painted Feather. Brazil.
  fenestralis fe-ne-*strah*-lis. Window-like (the net-veined leaves). Netted Vriesia. Brazil.
  hieroglyphica hee-e-rō-*gli*-fi-ka. Marked with hieroglyphs (the banded leaves). King of Bromeliads. Brazil.
  psittacina si-ta-*keen*-a. Parrot-like, the colour of the bracts and flowers. Dwarf Painted Feather. Brazil.
  regina ray-*geen*-a. Queen. Brazil.
  saundersii sawn-*derz*-ee-ee. After William Wilson Saunders (1809–79), a collector of rare plants. Brazil.
  splendens splen-denz. Splendid. Flaming Sword. French Guiana.

W

Wahlenbergia vah-lan-*berg*-ee-a *Campanulaceae*. After Georg Wahlenberg (1780–1851), Swedish professor of botany. Perennial herbs.
  albomarginata al-bō-mar-gi-*nah*-ta. White-margined. New Zealand.
  hederacea he-de-*rah*-kee-a. *Hedera*-like (the leaves). W Europe.
  matthewsii măth-*ewz*-ee-ee. After H. J.

Wahlenbergia (continued)
  Matthews who discovered it. New Zealand.
  saxicola see *W. albomarginata*

Wake Robin see *Trillium grandiflorum*

Waldsteinia văld-*stien*-ee-a *Rosaceae*. After Count Franz Adam Waldstein-Wartenburg (1759–1823), an Austrian botanist. Perennial herbs.
  fragarioides fra-gah-ree-*oi*-deez. Like *Fragaria*. E United States.
  ternata ter-*nah*-ta. In threes (the leaflets). Europe to Japan.

Wall Rue see *Asplenium ruta-muraria*
Wallflower see *Erysimum cheiri*
  Siberian see *E. allionii*
Walnut see *Juglans*
  Black see *J. nigra*
  Common see *J. regia*
  Japanese see *J. ailantifolia*
  Texan see *J. microcarpa*
Wandflower see *Dierama pulcherrimum, Sparaxis*
Wandering Jew see *Tradescantia fluminensis albiflora, Zebrina pendula*
Waratah see *Telopea speciosissima*
  Tasmanian see *T. truncata*
Wart Plant see *Haworthia venosa tesselata*

Washingtonia wo-shing-*ton*-ee-a *Palmae*. After George Washington (1732–99), first President of the United States. Tender Palms. Washington Palm.
  filifera fee-*li*-fe-ra. Thread-bearing, referring to the fibres that separate from the leaf margins. Desert Fan Palm. SW United States.
  robusta rō-*bus*-ta. Robust. Thread Palm. Mexico.

Water Avens see *Geum rivale*
Water Caltrop see *Trapa natans*
Water Chestnut see *Trapa natans*
  Chinese see *Eleocharis dulcis*
Water Crowfoot see *Ranunculus aquatilis*
Water Figwort see *Scrophularia auriculata*
Water Hyacinth see *Eichhornia crassipes*
Water Lettuce see *Pistia stratiotes*
Water Lily, Cape Blue see *Nymphaea capensis*
  Royal see *Victoria amazonica*
  Santa Cruz see *Victoria cruziana*
  White see *Nymphaea alba*
  Yellow see *Nuphar lutea*
Water Melon see *Citrullus lanatus*
Water Milfoil see *Myriophyllum*

Water Plantain see *Alisma plantago-aquatica*
Water Poppy see *Hydrocleys nymphoides*
Water Soldier see *Stratiotes aloides*
Water Sprite see *Ceratopteris thalictroides*
Water Starwort see *Callitriche*
Water Trumpet see *Cryptocoryne*
Water Violet see *Hottonia palustris*
Watercress see *Nasturtium officinale*

Watsonia wot-*son*-ee-a *Iridaceae*. After Sir William Watson (1715–87), London physician and botanist. Tender and semi-hardy perennial herbs. S Africa.
  fourcadei four-*kahd*-ee-ee. After H. G. Fourcade.
  meriana me-ree-*ah*-na. After Sybilla Merian, a Dutch naturalist.
  tabularis tăb-ew-*lah*-ris. Of Table Mountain.

Wattle see *Acacia*
  Cootamundra see *A. baileyana*
  Ovens see *A. pravissima*
  Queensland see *A. podalyriifolia*
  Rice's see *A. riceana*
  Silver see *A. dealbata*
  Sydney Golden see *A. longifolia*
Wax Flower see *Stephanotis floribunda*
Wax Myrtle see *Myrica cerifera*
Wax Plant see *Hoya carnosa*
  Miniature see *H. bella*
Wax Privet see *Peperomia glabella*
Wayfaring Tree see *Viburnum lantana*

Weigela vie-ge-la *Caprifoliaceae*. After Christian Ehrenfried von Weigel (1748–1831), a German botanist. Deciduous shrubs.
  florida flō-ri-da. Flowering. N China, Korea.
    'Foliis Purpureis' *fo*-lee-is pur-*pew*-ree-is. With purple leaves.
  praecox prie-koks. Early (flowering). NE Asia.

Wellingtonia see *Sequoiadendron giganteum*
Western Red Cedar see *Thuja plicata*
Western White Cedar see *Thuja occidentalis*
Whin see *Ulex*
White Paint-brush see *Haemanthus albiflos*
White Raintree see *Brunfelsia undulata*
White Snake-root see *Ageratina altissima*
White Velvet see *Tradescantia sillamontana*
Whitebeam see *Sorbus aria*
  Swedish see *S. intermedia*
Whorl Flower see *Morina longifolia*
Whortleberry see *Vaccinium myrtillus*
  Caucasian see *V. arctostaphylos*

Wilcoxia wil-*koks*-ee-a *Cactaceae*. After General Timothy E. Wilcox, American soldier and amateur botanist. Slender-stemmed cacti.
  poselgeri see *Echinocereus poselgeri*
  schmollii see *Echinocereus schmollii*

Willow see *Salix*
  Bay see *S. pentandra*
  Crack see *S. fragilis*
  Creeping see *S. repens*
  Cricket Bat see *S. alba caerulea*
  Dragon's Claw see *S. matsudana* 'Tortuosa'
  Goat see *S. caprea*
  Golden see *S. alba vitellina*
  Musk see *S. aegyptiaca*
  Peking see *S. matsudana*
  Silver see *S. alba sericea*
  Violet see *S. daphnoides*
  Weeping see *S.* 'Chrysocoma'
  White see *S. alba*
  Woolly see *S. lanata*
Willow Herb see *Epilobium*

× *Wilsonara* wil-son-*ah*-ra *Orchidaceae*. Integeneric hybrids, after Mr Alfred Gurney Wilson (c 1878–1957), Chairman of the RHS orchid committee. *Cochlioda* × *Odontoglossum* × *Oncidium*. Greenhouse Orchids.

Wineberry see *Rubus phoenicolasius*
Wingnut see *Pterocarya*
  Caucasian see *P. fraxinifolia*
Winter Aconite see *Eranthis hyemalis*
Winter Cherry see *Solanum capsicastrum*
Winter Heliotrope see *Petasites fragrans*
Winter Sweet see *Chimonanthus praecox*
Wintergreen see *Pyrola*
Winter's Bark see *Drimys winteri*
Wishbone Flower see *Torenia fournieri*

Wisteria wis-*te*-re-a *Leguminosae*. After Caspar Wistar (1761–1818), an American professor of anatomy. Deciduous climbers.
  floribunda flō-ri-*bun*-da. Profusely flowering. Japan.
  sinensis si-*nen*-sis. Of China.

Witch Grass see *Panicum capillare*
Witch Hazel see *Hamamelis*
  Chinese see *H. mollis*
  Japanese see *H. japonica*
Withe-rod see *Viburnum cassinoides*
Woad see *Isatis tinctoria*
Wolf's Bane see *Aconitum lycoctonum vulparia*
Wonga-wonga Vine see *Pandorea jasminoides*

Wood Sorrel see *Oxalis acetosella*
Woodbine see *Lonicera periclymenum*
Wormwood see *Artemisia absinthium*

Wulfenia wul-*fen*-ee-a *Scrophulariaceae*. After
Franz Xavier Freiherr von Wulfen
(1728–1805). Perennial herbs.
  baldaccii băl-*dăk*-ee-ee. After Antonio
Baldacci of the Bologna Botanic Garden.
Albania.
  carinthiaca ka-rinth-ee-*ah*-ka. Of Carinthia
Austria. E Alps, Balkan peninsula.

X

Xantheranthemum zănth-e-*rănth*-e-mum
Acanthaceae. From Gk. *xanthos* (yellow) and
Eranthemum, a related genus. Tender
perennial herb.
  igneum ig-nee-um. Fiery red. Peru.

Xanthoceras zănth-*o*-ke-ras *Sapindaceae*.
From Gk. *xanthos* (yellow) and *keras* (a horn)
referring to the yellow, horn-like appendages
between the petals. Deciduous shrub or
tree.
  sorbifolium sor-bi-*fo*-lee-um. *Sorbus*-leaved.
N China.

Xanthorhiza zănth-o-*reez*-a *Ranunculaceae*.
From Gk. *xanthos* (yellow) and *rhiza* (a root),
the roots are yellow. Deciduous shrub.
  simplicissima sim-pli-*ki*-si-ma. Most simple
i.e. unbranched. Yellowroot. E United
States.

Xeranthemum ze-*rănth*-e-mum *Compositae*.
From Gk. *xeros* (dry) and *anthos* (a flower)
referring to the dry, everlasting flowers.
Annual herb.
  annuum ăn-ew-um. Annual. Immortelle. S
Europe.

Y

Yam see *Dioscorea*
  Ornamental see *D. discolor*
Yarrow see *Achillea millefolium*
Yellow Adder's Tongue see *Erythronium americanum*
Yellow Archangel see *Lamium galeobdolon*
Yellow Elder sere *Tecoma stans*
Yellow Flag see *Iris pseudacorus*

Yellow Jessamine see *Gelsemium sempervirens*
Yellow Sage see *Lantana camara*
Yellow Wood see *Cladrastis sinensis*
Yellowroot see *Xanthorhiza simplicissima*
Yesterday Today and Tomorrow see *Brunfelsia calycina*
Yew see *Taxus*
  Common see *T. baccata*
  Irish see *T. baccata* 'Hibernica'
  West Felton see *T. baccata* 'Dovastoniana'
Youth and Old-age see *Aichryson* × *domesticum*

Yucca yoo-ka *Agavaceae*. The Caribbean name
for Cassava (*Manihot esculenta*) originally
thought to apply to *Y. gloriosa*. Evergreen,
hardy and tender perennial herbs and
shrubs.
  aloifolia ă-lō-i-*fo*-lee-a. *Aloe*-leaved.
Spanish Bayonet. S United States, Mexico,
W Indies.
  elephantipes e-le-*făn*-ti-pays. Like an
elephant's foot. Spineless Yucca. SW
United States, Mexico.
  filamentosa fee-lah-men-*tō*-sa. With
filaments (on the leaf margins). Adam's
Needle. SE United States.
  glauca glow-ka. Glaucous (the leaves). SE
United States.
  gloriosa glō-ree-*ō*-sa. Glorious. SE United
States.
  smalliana smawl-ee-*ah*-na. After John
Kunkel Small (1869–1938). SE United
States.
  whipplei wi-pal-ee. After Lieut. Amiel
Weeks Whipple (1818–63), who
explored N America. Our Lord's Candle.
California, N Mexico.

Yulan see *Magnolia denudata*

Yushania yoo-*shăn*-y-ă *Graminae*. Named for
the Yushi mountains, Korea. Bamboo.
Himalaya.
  anceps ăn-keps. Doubtful (the country of
origin). C and NW Himalaya.

Z

Zantedischia zăn-te-*dis*-kee-a *Araceae*. After
Francesco Zantedischi (born 1797), Italian
botanist. Tender and semi-hardy perennial
herbs.
  aethiopica ie-thee-*ō*-pi-ka. African. Arum
Lily. Transvaal.

Zantedischia (continued)
albomaculata ăl-bō-măk-ew-*lah*-ta. (= *Z. melanoleuca*). White-spotted (the leaves). Black-throated Arum, Spotted Arum. S Africa.
elliottiana e-lee-o-tee-*ah*-na. After Elliott. Golden Arum. S Africa.
pentlandii pent-*lănd*-ee-ee. Of Pentland House, home of Mr White who introduced it. Yellow Arum. S Africa.
rehmannii ray-*mahn*-ee-ee. After Rehmann. Pink Arum. S Africa.

Zanthoxylum zănth-*oks*-i-lum *Rutaceae*. From Gk. *xanthos* (yellow) and *xylon* (wood) referring to the yellow wood of some species. Deciduous trees.
ailanthoides ie-lănth-*oi*-deez. Like *Ailanthus*. Japan, China.
americanum a-me-ri-*kah*-num. American. Toothache Tree. E N America.
piperitum pi-pe-*ree*-tum. Pepper-like (the taste of the seeds). Japan Pepper. Japan. China.
planispinum plah-ni-*speen*-um. With flat spines. Japan, Korea, China.
schinifolium skeen-i-*fo*-lee-um. With leaves like *Schinus*. China, Korea, Japan.

Zauschneria see *Epilobium*

Zea zee-a *Gramineae*. From the Gk. name for a related plant. Annual grass.
mays mayz. From the Mexican name. Corn, Maize. Tropical America.

Zebra Basket Vine see *Aeschynanthus marmoratus*
Zebra Plant see *Aphelandra squarrosa*, *Calathea zebrina*, *Cryptanthus zonatus*

Zebrina ze-*breen*-a *Commelinaceae*. From *zebra* referring to the striped leaves. Tender, perennial herb.
pendula see *Tradescantia zebrina*

Zelkova zel-*kō*-va *Ulmaceae*. From the Caucasian name. Deciduous trees.
carpinifolia kar-peen-i-*fo*-lee-a. *Carpinus*-leaved. Transcaucasus, N Iran.
serrata se-*rah*-ta. Saw-toothed (the leaves). E Asia, Japan.

Zenobia zay-*no*-bee-a *Ericaceae*. After Zenobia, a Queen of Palmyra. Deciduous or semi-evergreen shrub.
pulverulenta pul-ve-ru-*len*-ta. Powdered,

Zenobia (continued)
referring to the glaucous-bloomed leaves. SE United States.

Zephyranthes ze-fi-*rănth*-eez *Amaryllidaceae*. From Gk. *zephyros* (the west wind) and *anthos* (a flower), they are natives of the W hemisphere. Tender and semi-hardy, bulbous herbs.
atamasco ă-ta-*măs*-kō. A native name. SE United States.
candida *kăn*-di-da. White (the flowers). Flower of the Western Wind. Argentina, Uruguay.
citrina ki-*tree*-na. Lemon-yellow. S America.
grandiflora grănd-i-*flō*-ra. Large-flowered. Mexico, Guatemala.
robusta see *Habranthus robustus*
rosea ro-see-a. Rose-coloured. W Indies, Guatemala.

Zigadenus zi-ga-*day*-nus *Liliaceae* (*Melianthaceae*). From Gk. *zygos* (a yoke) and *aden* (a gland) referring to the paired glands. Perennial herb.
elegans ay-le-gahnz. Elegant. N America.

Zingiber zing-gi-ber *Zingiberaceae*. Of ancient origin probably from pre-roman *srnga* (a horn) and *ver* (a root) giving L. *zingiber* and eventually English *ginger* which is obtained from the root. Tender herb.
officinale o-fi-ki-*nah*-lee. Sold as a herb. Ginger. SE Asia.

Zinnia zin-ee-a *Compositae*. After Johann Gottfried Zinn (1727–59). Annual herbs. Mexico.
angustifolia ang-gus-ti-*fo*-lee-a. Narrow-leaved.
elegans ay-le-gahnz. Elegant.

Zygocactus truncatus see *Schlumbergera truncata*

Zygopetalum zi-gō-pe-ta-lum *Orchidaceae*. From Gk. *zygos* (a yoke) and *petalon* (a petal), the callous at the base of the lip appears to be pulling the perianth segments together. Greenhouse orchids. Brazil.
crinitum kree-*nee*-tum. Long-haired.
intermedium in-ter-*me*-dee-um. Intermediate.
mackayi ma-*kay*-ee. After Mr Mackay of Dublin College Botanic Garden who imported it.

Acknowledgements

I am grateful to the Director, Royal Botanic Gardens, Kew for use of library facilities and to Nigel Taylor and Susyn Andrews of that institution for help on several, mainly nomenclatural, points. I would also like to thank Dr M. J. Harvey of the Department of Biology, Dalhousie University, Halifax, Canada for interesting and useful discussions on the subject of pronunciation.

The following publications (excluding taxonomic works) were freely consulted during the preparation of this work and would be interesting to anyone wishing to pursue the subject.

Desmond, R. 1977. *Dictionary of British and Irish Botanists and Horticulturists*. Taylor and Francis.

Glare, P. G. W. (ed.) 1982. *Oxford Latin Dictionary*. Oxford University Press.

Johnson, A. T. & Smith, H. A. 1972. *Plant Names Simplified*. Landsmans Bookshop Ltd.

Kidd, D. A. 1957. Collins Gem *Latin-English English-Latin Dictionary*. Collins.

Lewis, C. T. & Short, C. 1879. *A Latin Dictionary*. Oxford University Press.

Liddell, H. G. & Scott, R. 1843. *A Greek-English Lexicon*. Oxford University Press. New (9th) Edition revised by H. Stuart Jones & R. Mackenzie 1940.

Smith, A. W. & Stearn, W. T. 1972. *A Gardener's Dictionary of Plant Names*. Cassell.

Stearn, W. T. 1966. *Botanical Latin*. 2nd Edition 1973. David & Charles.

AMATEUR GARDENING POCKET REFERENCE SERIES:
The following titles are also available

The Hellyer Pocket Guide by Arthur Hellyer MBE VMH
Gardener's Fact Finder Edited by Graham Clarke

AN EXCITING NEW SERIES
Hamlyn, in association with *Amateur Gardening*, have also produced an
excellent series written by leading gardening authority **Stefan Buczacki**, a
well-known TV presenter, radio broadcaster, journalist and writer.
The easy-to-use pages are full of practical advice, inspirational ideas and
excellent full colour photographs.

Titles available:
Best Climbers
Best Foliage Shrubs
Best Shade Plants
Best Soft Fruit

To be published Spring 1995:
Best Water Gardens
Best Herbs

These books are available from all good bookshops or by mail order direct
from the Publisher. Payment can be made by credit card or cheque/postal
order in the following ways:

BY PHONE
Phone through your order on our special CREDIT CARD HOTLINE
on 0933 410511, quoting reference DPN. Speak to our customer service
team during office hours (9am to 5pm) or leave a message on the answer
machine, quoting your full credit card number, plus expiry date, and your full
name and address.

BY POST
Simply fill out the order form opposite and send it to:
REED BOOK SERVICES LTD, PO BOX 5, RUSHDEN, NORTHANTS,
NN10 6YX

GARDENING

I wish to order the following titles	Price	Quantity	Total
The Hellyer Pocket Guide ISBN 0 600 57698 1	£4.99		
Gardener's Fact Finder ISBN 0 600 57697 3	£4.99		
Best Climbers ISBN 0 600 57732 5	£4.99		
Best Foliage Shrubs ISBN 0 600 57735 X	£4.99		
Best Shade Plants ISBN 0 600 57734 1	£4.99		
Best Soft Fruit ISBN 0 600 57733 3	£4.99		
Postage and packing	£1.00		
Grand total			

Name _____ (block capitals)

Address _____

_____ Postcode _____

I enclose a cheque/postal order for £ _____
made payable to Reed Book Services Ltd
or please debit my Access ▢ Visa ▢ AmEx ▢ Diners ▢

Number ▢▢▢▢ ▢▢▢▢ ▢▢▢▢ ▢▢▢▢

by £ _____ Expiry date ▢▢ ▢▢

_____ Signature

- Offer available to UK only • While every effort is made to keep prices low, the publisher reserves the right to increase prices at short notice
- Your order will be dispatched within 28 days, subject to availability

Registered office Michelin House, 81 Fulham Road, London SW3 6RB. Registered in England No 1974080